MW01265422

Demographic Toxicity

Demographic Toxicity

Methods in Ecological Risk Assessment

EDITED BY

H. Reşit Akçakaya

John D. Stark

Todd S. Bridges

2008

Oxford University Press, Inc., publishes works that further
Oxford University's objective of excellence
in research, scholarship, and education.

Oxford New York
Auckland Cape Town Dar es Salaam Hong Kong Karachi
Kuala Lumpur Madrid Melbourne Mexico City Nairobi
New Delhi Shanghai Taipei Toronto

With offices in
Argentina Austria Brazil Chile Czech Republic France Greece
Guatemala Hungary Italy Japan Poland Portugal Singapore
South Korea Switzerland Thailand Turkey Ukraine Vietnam

Published by Oxford University Press, Inc.
198 Madison Avenue, New York, New York 10016

www.oup.com

Library of Congress Cataloging-in-Publication Data
Demographic toxicity : methods in ecological risk assessment (with CD-ROM)/
edited by H. Reşit Akçakaya, John D. Stark, and Todd S. Bridges.
p. cm.
Includes bibliographical references and index.
ISBN 978-0-19-533296-4
1. Pollution—Environmental aspects. 2. Pollution—Physiological effect.
3. Population biology. 4. Ecological risk assessment. I. Akçakaya, H. R. II. Stark,
John D. III. Bridges, Todd S.
QH545.A1D46 2008
577.27—dc22 2007023306

9 8 7 6 5 4 3 2 1

Printed in the United States of America
on acid-free paper

Contents

Contributing Authors

H. Reşit Akçakaya
Department of Ecology and Evolution
Stony Brook University
Stony Brook, New York, USA, and
Applied Biomathematics
Setauket, New York, USA

Mace G. Barron
U.S. Environmental Protection Agency
Office of Research and Development
National Health and Environmental Effects Research Laboratory
Gulf Ecology Division
Gulf Breeze, Florida, USA

Todd S. Bridges
Center for Contaminated Sediments
U.S. Army Corps of Engineers
Waterways Experiment Station
Vicksburg, Mississippi, USA

Barry Bunch
U.S. Army Corps of Engineers
Waterways Experiment Station
Vicksburg, Mississippi, USA

Thierry Caquet
INRA, Equipe Ecotoxicologie et Qualité des Milieux Aquatiques
UMR Ecologie et Santé des Ecosystèmes
Agrocampus, Rennes, France

Sandrine Charles
Université de Lyon, Université Lyon 1
Laboratoire de Biométrie et Biologie Evolutive (UMR 5558)
Centre National de la Recherche Scientifique
Villeurbanne, France

Arnaud Chaumot
Cemagref—Laboratoire d'Ecotoxicologie
Lyon, France

Jane Copeland
Computer Sciences Corporation
Narragansett, Rhode Island, USA

Marie-Agnès Coutellec
INRA, Equipe Ecotoxicologie et Qualité des Milieux Aquatiques
UMR Ecologie et Santé des Ecosystèmes
Agrocampus, Rennes, France

Scott Ferson
Applied Biomathematics
Setauket, New York, USA

Valery E. Forbes
Centre for Integrated Population Ecology
Department of Environmental, Social and Spatial Change
Roskilde University
Roskilde, Denmark

Douglas J. Fort
Fort Environmental Labs
Stillwater, Oklahoma, USA

Karen F. Gaines
Eastern Illinois University
Department of Biological Sciences
Charleston, Illinois, USA

T. Gleason
U.S. Environmental Protection Agency
Office of Research and Development
National Health and Environmental Effects Research Laboratory
Atlantic Ecology Division
Narragansett, Rhode Island, USA

Niklas Hanson
Department of Plant and Environmental Sciences
Göteborg University
Göteborg, Sweden

Lyndal L. Johnson
Northwest Fisheries Science Center
National Marine Fisheries Service
National Oceanic and Atmospheric Administration
Seattle, Washington, USA

Chris Klok
Centre for Ecosystem Studies
Alterra
Wageningen, The Netherlands

Anja B. Kristoffersen
National Veterinary Institute
Oslo, Norway

Anne Kuhn
U.S. Environmental Protection Agency
Office of Research and Development
National Health and Environmental Effects Research Laboratory
Atlantic Ecology Division
Narragansett, Rhode Island, USA

Laurent Lagadic
INRA, Equipe Ecotoxicologie et Qualité des Milieux Aquatiques
UMR Ecologie et Santé des Ecosystèmes
Agrocampus, Rennes, France

Wayne G. Landis
Institute of Environmental Toxicology
Huxley College of the Environment
Western Washington University
Bellingham, Washington, USA

J. D. Litzgus
Biology Department
Laurentian University
Sudbury, Ontario, Canada

John P. Lortie
Woodlot Alternatives, Inc.
Topsham, Maine, USA

Marina Mauri
Dipartimento di Biologia Animale
Università di Modena e Reggio Emilia
Modena, Italy

S. Jannicke Moe
Norwegian Institute for Water Research (NIVA)
Oslo Centre for Interdisciplinary Environmental and Social Research (CIENS)
Oslo, Norway

W. R. Munns, Jr.
U.S. Environmental Protection Agency
Office of Research and Development
National Health and Environmental Effects Research Laboratory
Atlantic Ecology Division
Narragansett, Rhode Island, USA

Diane E. Nacci
U.S. Environmental Protection Agency
Office of Research and Development
National Health and Environmental Effects Research Laboratory
Atlantic Ecology Division
Narragansett, Rhode Island, USA

Matthew C. Nicholson
U.S. Environmental Protection Agency, Region 3
Environmental Assessment and Innovation Division
Philadelphia, Pennsylvania, USA

James M. Novak
Eastern Illinois University
Department of Biological Sciences
Charleston, Illinois, USA

Annemette Palmqvist
Centre for Integrated Population Ecology
Department of Environmental, Social and Spatial Change
Roskilde University
Roskilde, Denmark

Daniela Prevedelli
Dipartimento di Biologia Animale
Università di Modena e Reggio Emilia
Modena, Italy

Sandy Raimondo
U.S. Environmental Protection Agency
Office of Research and Development
National Health and Environmental Effects Research Laboratory
Gulf Ecology Division
Gulf Breeze, Florida, USA

Steven A. Rego
U.S. Environmental Protection Agency
Office of Research and Development
National Health and Environmental Effects Research Laboratory
Atlantic Ecology Division
Narragansett, Rhode Island, USA

Roberto Simonini
Dipartimento di Biologia Animale
Università di Modena e Reggio Emilia
Modena, Italy

Julann A. Spromberg
Northwest Fisheries Science Center
National Marine Fisheries Service
National Oceanic and Atmospheric Administration
Seattle, Washington, USA

John D. Stark
Ecotoxicology Program
Department of Entomology
Washington State University
Puyallup, Washington, USA

Nils C. Stenseth
Centre for Ecological and Evolutionary Synthesis (CEES)
Department of Biology
University of Oslo
Oslo, Norway

Michael E. Thompson
Woodlot Alternatives, Inc.
Topsham, Maine, USA

W. Troy Tucker
Applied Biomathematics
Setauket, New York, USA

Steven Walters
University of Washington
Department of Urban Design and Planning
Seattle, Washington, USA

Demographic Toxicity

1

Demographic Toxicity

Assessing the Population-Level Impacts of Contaminants

H. REŞIT AKÇAKAYA
JOHN D. STARK

"Demographic toxicity" is the ecological impact of a pollutant or toxicant on the population(s) of a plant or animal species. Demographic toxicity is measured in terms of population-level endpoints, including decreased population size, decreased population growth rate, increased risk of decline or extinction, increased time to recovery, and decreased probability of recovery.

In recent years, ecological impacts of chemical toxicants have been increasingly assessed using population-level methods. This shift from studying effects on individuals to populations is gaining momentum with increasing emphasis on population-level assessments by regulatory agencies such as the U.S. Environmental Protection Agency. However, there are relatively few applications of population models in ecological risk assessments. One of the major difficulties assessors face is lack of case studies that they can use as a template or guide to aid in developing their own models. Our goal in organizing this book was to bring together a diverse array of applications of population models for assessing demographic toxicity. Each chapter uses one or several models to address ecotoxicological issues related to a particular species. These models are available on the CD-ROM included with the book.

The models are implemented in the population modeling software RAMAS GIS (Akçakaya 2005). The CD-ROM contains a demonstration version of the software. The appendix at the end of this book describes how to install the program and open, inspect, and run the models. This chapter includes an overview of the chapters, and two short introductions—to population modeling and to RAMAS GIS—and discusses limitations of, and alternatives to, these approaches.

Short Introduction to Population Modeling

Ecotoxicological risk assessments at the population level provide an evaluation of the impact of toxicants on plant and animal populations, using a variety of models. Modeling is a process of building simple, abstract representations (e.g., as mathematical equations) of complex systems (e.g., a biological population) to gain insights into how the system works, to predict how it will behave in the future, to guide further investigations, and to make decisions about how it can be managed. Assessing demographic toxicity involves using species-specific data and models to evaluate the impact of contaminants in terms of population-level endpoints such as population size, population growth rate, risk of decline or extinction, time to recovery, and probability of recovery. This book consists of examples of population models used in ecotoxicological assessments, but it is not an introductory textbook. This section provides only a brief introduction to modeling; several textbooks provide a full introduction to population modeling and risk assessment (e.g., Burgman et al. 1993; Akçakaya et al. 1999; Caswell 2001; Morris and Doak 2002) or review existing approaches (e.g., Pastorok et al. 2002).

Why population level?

The most important advantage of demographic toxicity assessments is ecological relevance. In most standard ecotoxicological tests, the endpoints are at the individual level and include growth, reproduction, and short-term survival. These variables are easy to measure, and these data are inexpensive to produce. As a result, they are widely used. However, the relevance of these short-term, individual-level measures to the integrity of ecosystems, and their interest to society are limited. Populations and communities represent the level of biological organization that is of greatest interest to the public and to the regulators (Ferson et al. 1996; Munns et al. 1997). In addition, any given individual-level measure does not necessarily correlate with population-level variables (Stark et al. 1997; Walthall and Stark 1997). In a review of studies in which toxicant effects were evaluated at both individual and population levels, Forbes and Calow (1999) showed that there was no consistent, predictable relationship between effects on individual-level and on population-level traits.

Other important advantages of population-level assessments of demographic toxicity include the relative ease with which cumulative effects and multiple stressors can be incorporated and overall impacts can be evaluated in the face of conflicting effects.

Cumulative effects and multiple stressors

Population models can easily incorporate cumulative effects on different "parts" of a population, for example, in different locations (subpopulations), on different life-history stages, or at different times of the year. These effects may be caused by the same pollutant or by different pollutants. Often, contaminants are not the only factors that affect populations. Exploitation (overfishing), habitat loss, habitat fragmentation, and isolation (decreased connectivity between populations) are some of the most common threats that face populations. Population models allow integration of all these threats to the species into one assessment framework and prevent threats being overlooked when separate assessments are made on single factors or on single pollutants, or single effects of a pollutant.

Conflicting impacts

In some cases, a single threatening factor (e.g., a contaminant) will have conflicting impacts. For example, fecundity may decrease and survival of some age classes may increase, or effects may change depending on the abundance (density) of the population. Single-factor assessments in these situations may give conflicting results or falsely imply that there is no discernable threat. Population models integrate all such impacts and report results in terms of the overall effect on the population of a range of stressors.

Despite the advantages of population-level assessments, there is also some disagreement about and resistance to their use. We believe the reasons for this include (1) difficulties in defining a population and (2) misunderstandings about the use of population-level methods. We discuss these two issues next.

Defining a population

Any population-level assessment requires an explicit definition of the population being assessed. In some cases, this assessment population is the same as a biological population, which can be defined as group of regularly interbreeding (i.e., panmictic) individuals. In other cases, the assessment population can be defined based on geographic or political boundaries, depending on the goals of the assessment. An assessment population can consist of a small part of a biological population, or it can include several biological populations (a metapopulation). The spatial extent (i.e., geographical range size and boundaries) of the assessment population is often more a social (thus, regulatory) issue than a scientific issue. If the assessment population is limited to the area affected by a particular threat (e.g., pollution from a certain source), the results will be more sensitive to the potential impacts but less relevant for the overall viability of the species. This may be preferred when the goal is to determine whether the threat in question is having a local effect on the species, or when a widespread or abundant species is used as an indicator of environmental stress. If, on the other hand, the goal is to determine whether the species in question is at risk of substantial decline or extinction due to the pollutant, then a larger spatial extent may be preferred. In this case, the affected area may be a small portion of the assessment population, which may include all or most of a species range. Thus, different scales and extents are needed for different purposes. In many cases, however, using more than one definition of assessment population would be a lot more informative. Thus, the assessment can be made with two or more models, with at least one model at a local scale, focusing on the area affected by the threat in question, and at least one model with a more regional focus to determine the population-level effects that may be relevant for the species as a whole.

Misunderstandings about population-level assessments

There are several arguments against the use and applicability of population-level methods in ecotoxicological assessments (e.g., Beyer and Heinz 2000; Beyer and Audet, n.d.). We briefly describe these arguments below and discuss why we believe they represent misunderstandings, and not fundamental reasons against the use of population-level methods.

1. Threatened species: One exception to the use of population-level methods that is frequently mentioned is threatened species. (Following the World Conservation Union [IUCN], we use "threatened" as a general term for different IUCN categories of endangerment at the species level: Vulnerable, Endangered, or Critically Endangered.) Regulations in some countries protect threatened species, especially those that are highly threatened, at the individual level. For example, in the United States, the National Oceanic and Atmospheric Administration and the Fish and Wildlife Service conduct consultations on species listed as threatened and endangered under the Endangered Species Act. Because all individuals of an endangered species are supposed to be protected from "take" under this law, individuals are important. However, importance of individuals (or even their protection under the law) does not mean that risk assessments for these species are done at the individual level. Actually, population-level methods are more developed and much more commonly applied in conservation biology, where the focus is on threatened species, than they are in ecotoxicology (Burgman et al. 1993). Population and metapopulation models are applied to threatened species to assess their level of endangerment, to evaluate relative efficacy of alternative management options, and to predict the impact of various human activities (Akçakaya et al. 2004a), none of which can be done as effectively at the individual level.

2. Tumors and deformations: Population-level effects are argued to be irrelevant when effects at the individual level (e.g., animals with tumors or malformations) suggest that the environment is seriously polluted. It is a mistake to infer that such situations call for individual-level assessment. This is because the concern of both the public and the regulators in such cases is not the health of the individuals that have these malformations, most of which would be dead (e.g., fished) by the time they are observed. Rather, the concern in most of these cases is the implications of these malformations for human health. It would indeed be a mistake to use population-level methods when the protection goal is not the wildlife population (or even the ecological community), but human health. For example, when the goal is to prevent an increased incidence of cancers in the human population, the number of cancer cases per million individuals is used as a measure.

3. The public values individuals: Lack of substantial impact at the population level is argued to be irrelevant when the public is upset to see individuals of highly visible species dead due to contamination, or when there is localized but high mortality such that there is perceptible decline in local abundance of a desired or valued species. For example, the American robin may be an abundant species in no risk of extinction, but hundreds of robins found dead in any region would be a major cause for concern. Again, such situations do not require that the assessments should be done at the individual level, because what is valued is not the individual organism, but either the quality of the environment for human health (see item 2 above) or the health of local populations. In the former case, neither population-level nor individual-level methods are appropriate, because the focus is not on the affected animal population, but on the (potentially) affected human population (see item 2). In the latter case, the issue is one of determining the correct assessment population. As discussed above (see "Defining a population"), the size and extent of the assessment population are not a scientific issue, but a social (thus, regulatory) one. If the citizens of a region, state, or county do not want an otherwise widespread and common species to become locally extinct from their area, a population-level assessment can (and should) be done at that spatial scale (perhaps also at other scales).

4. Not well developed: Population-level approaches are sometimes viewed as methods that "are still in the development phase and have not been shown to be reliable" (U.S. EPA 2004: 131). It is true that there are many new developments in population modeling, especially in terms of data analysis and parameter estimation (Holmes 2004; Dunham et al. 2006), detection and modeling of density dependence (Brook and Bradshaw 2006), incorporation of habitat and landscape dynamics (Akçakaya et al. 2004b, 2005; Wintle et al. 2005), and estimating natural variability by removing variance due to measurement error and sampling variability (Dennis et al. 2006). However, this shows only that population modeling is an active field; it does not show that such models are not reliable. Brook et al. (2000) validated population models in terms of their predictions of abundance and risks of decline. In this comprehensive and replicated evaluation, they estimated the parameters from the first half of each data set and used the second half to evaluate model performance. They found that predictions were accurate: The risk of population decline closely matched observed outcomes, there was no significant bias, and population size projections did not differ significantly from reality. Further, the predictions of five software packages they tested were highly concordant. Although validation of stochastic results may not be possible in every case, components of a model can be validated. In addition, some model results can be validated by comparing predicted values with those observed/measured in the field (e.g., McCarthy and Broome 2000; McCarthy et al. 2001).

5. Black-box: Another argument against the use of population models is that they are not well understood by most risk assessors, and there is no general consensus by risk assessors about the most appropriate models for use in ecological risk assessment. We agree that this is one of the main reasons for the rarity of population-level ecotoxicological assessments. However, lack of familiarity with population models does not mean that they are any less relevant or useful. We believe that the solution is training to increase the understanding and use of models.

6. It's the habitat, stupid! Habitat is regarded to be a "more basic" endpoint than populations. There are several problems with this viewpoint. First, habitat is a species-specific concept; it does not make sense outside the context of a particular species. Thus, any practical method for habitat-based assessment would also be at the population (or species) level. As a result, it is not clear what is more fundamental about habitat. Second, some threats do not involve the habitat (e.g., overfishing, introduced diseases), and thus protecting the habitat is necessary but not sufficient for protecting the populations and communities. Third, the habitat-population dichotomy is a false one; habitat is an integral component of population models, either indirectly (by determining population parameters, e.g., carrying capacity, survival, fecundity, dispersal) or directly (in explicitly habitat-based models; e.g., Akçakaya and Atwood 1997). Population models incorporate both habitat and demography and can assess threats that affect habitat, demography, or both.

The above discussion is not meant in any way to imply that populations are the only appropriate units of ecotoxicological assessment. As mentioned above, when the protection goal is not the population or the species (e.g., when the goal is human health and when other species are used only as indicators), then population-level approaches are not relevant. In addition, when practical and robust methods for higher level assessments (communities and ecosystems) are developed, assessments at these levels would be appropriate in many cases. Other reasons for not using population-level approaches include cases where a population-level assessment would be too expensive (compared to the cost of implementing the available protection or mitigation option) and cases where the presence of an impact is too obvious and does not need detailed quantification.

Overview of the Chapters

The chapters presented in this book cover a wide range of population models, species, and toxicants. In almost half of the chapters (7 of 16), the species studied are invertebrates. In the other chapters, there are five fish, two mammals, a bird, and an amphibian species. These species live in terrestrial, freshwater, and marine/estuarine systems. The toxicants evaluated include pesticides, polychlorinated biphenyls (PCBs), polycyclic aromatic hydrocarbons, metals, and pulp mill effluents. Hypothetical stressors are also evaluated in some of the chapters. Although all models use the same modeling platform, there is a variety of demographic and spatial population structures, types of variability incorporated (or not), and types of questions addressed.

In chapter 2, Moe, Kristoffersen, and Stenseth examine the effects of cadmium on stage-structured populations of the sheep blowfly *Lucilia sericata*. Stage-specific density, stochasticity levels, and toxicant exposure were varied in their models, producing different risk assessment outcomes. Results of this study indicate that stochasticity increased population fluctuations, which in turn increased both explosion and extinction risks. The type of density dependence also interacted with the levels of stochasticity. Toxicant exposure reduced mean abundances and explosions risks, but also reduced extinction risks, because the toxicant-exposed populations had lower fluctuations in abundances.

The effects of mercury stress on the endangered Florida panther, *Puma concolor coryi*, are explored in chapter 3 by Raimondo and Barron using a metapopulation model. The potential effects of mercury were examined by simulating reductions in either reproduction or reproduction and survival in panthers in the Everglades National Park. In the absence of mercury, the panther metapopulation was stable and had a high probability of persisting throughout their range. However, model simulations showed that low-level mercury stress increased the risk of extinction, but that extinction of the Everglades population would not affect the viability of the remaining metapopulation.

In chapter 4, Gaines and Novak developed a spatially explicit model for raccoon (*Procyon lotor*) populations residing in and around the U.S. Department of Energy's Savannah River site. Raccoons are hunted in this area and are possible contaminant vectors of the gamma-emitting radionuclide radiocesium (^{137}Cs) to the hunting public. Two different model simulations in RAMAS GIS were developed: One assumed that there was no recruitment from offsite populations after harvest, and the other assumed that "clean" raccoons from offsite would maintain these border populations at levels close to carrying capacity. Results of these simulations showed that the level of recruitment of clean individuals affects the population's estimated body burden of ^{137}Cs. If clean individuals are recruited and harvested before they reach equilibrium in the contaminated environment, then contaminant mobility will be minimized. However, if recruitment rate is low, then the survivors in each cohort would be original residents of the contaminated areas, having high burdens of ^{137}Cs regardless of harvest rate.

The effects of sublethal exposures of copper in soil on population viability of the earthworm (*Lumbricus rubellus*) in northwestern Europe is considered by Klok in chapter 5. Three scenarios were evaluated: a dynamic energy budget scenario, an additive scenario, and a single effect scenario. The level of population effects was found to be dependent to a large extent on the assumptions made on how sublethal effects on development and reproduction interact. The use of reproductive data only was found to strongly underestimate the population-level effect. Using reproduction and development data as independent and additive effects resulted in an overly optimistic estimate of consequences at the population level. The duration of the preadult stage was found to be a more informative indicator of the population-level consequences of copper in this study.

In chapter 6, Walters, Kuhn, Nicholson, Copeland, Rego, and Nacci illustrate how exposure to a spatially heterogeneous stressor can affect populations of the common loon (*Gavia immer*) in New Hampshire. Simulations included stressor-induced decreases in fecundity at distinct spatial locations. Lower metapopulation occupancy rates were observed in stressed populations, which also exhibited increased probabilities of decline relative to an unstressed population. Results of this study illustrate the importance of accounting for source–sink dynamics and the spatial heterogeneity in population and stressor distributions when assessing ecological risks to wildlife populations.

Chapter 7 describes population-level ecological risk models developed by Tucker, Litzgus, Ferson, Akçakaya, Thompson, Fort, and Lortie to explore the potential impact that PCB contamination might have on wood frog (*Rana sylvatica*) populations in Massachusetts. Using several scenarios of effect on the frog populations, the models predict that exposure to PCBs hastens population decline, reduces population numbers, and increases the likelihood of extinction.

Spromberg and Johnson use metapopulation models in chapter 8 to examine the potential effects of freshwater and estuarine contaminant exposure on lower Columbia River chinook salmon (*Oncorhynchus tshawytscha*) populations. Changes to demographic rates and heterogeneous contaminant distribution were used in their models. Fall chinook populations were connected by straying and were treated as metapopulations. Results of their models suggest that low-level straying between populations may be protecting some populations and depleting others while simultaneously masking the direct effects of contaminant exposure.

In chapter 9, Stark uses population recovery as a method to evaluate the effects of the pesticide spinosad on populations of the water flea *Daphnia pulex*. Exposure concentrations were based on a previous study and were equivalent to 48-hr acute mortality estimates of 4–12%. However, modeling results showed that the population growth rate (lambda, λ) values were reduced 34–97% compared to control population after exposure to these spinosad concentrations and that exposure resulted in recovery times of 1–11 weeks. The fact that *D. pulex* populations were so negatively affected after exposure to spinosad at concentrations that caused ≤12% mortality based on 48-hr mortality studies shows the importance of using a population modeling approach for the evaluation of pesticide effects.

In chapter 10, Coutellec, Caquet, and Lagadic describe 16 population models developed for the freshwater snail *Lymnaea stagnalis* to determine the simulated effect of a toxicant on population size where populations of the snail developed from various initial numbers of individuals and levels of genetic relatedness. Their goal was to estimate the consequences of demographic bottlenecks due to toxicants on further performances, in terms of population dynamics. Model predictions were similar among populations, and no relationship with initial abundances could be identified, except a tendency for initially large populations to show lower intrinsic growth rates. All populations showed positive and similar dynamics ($\lambda > 1$). No clear difference could be detected between initially large and small populations, or between the four types of initial effective population size, despite a slight (but nonsignificant) trend for initially larger populations to exhibit lower growth rates.

Chaumot and Charles, in chapter 11, explore the effects of cadmium on brown trout in southern France where chronic cadmium pollution occurs in one stretch of river. Both deterministic and stochastic models were developed for 15 populations of trout. Results of this study show that deterministic endpoints such as equilibrium abundances appear to seriously underestimate the effects of pollution to trout populations. However, extinction risk determined with stochastic models gave a more accurate assessment of cadmium effects. Furthermore, random variation in trout dispersal patterns adversely affected population dynamics, increasing extinction risk related to cadmium exposure.

Nacci, Walters, Gleason, and Munns use a series of stage-classified metapopulation models for the estuarine fish *Fundulus heteroclitus* (mummichog) to evaluate population effects associated with exposures to PCBs in chapter 12. Their results suggest that migrants from nearby stable or growing populations could partially support subpopulations in the highly contaminated sites. The results also indicate that only higher rates of dispersal (and/or high initial population sizes in source subpopulations) could successfully mitigate declines in contaminated subpopulations.

In chapter 13, Palmqvist and Forbes explore the potential effects of contaminant patchiness in sediments contaminated with the polycyclic aromatic hydrocarbon fluoranthene on the population dynamics of the sediment-dwelling marine polychaete *Capitella capitata* using metapopulation models. The relationships of population density, contaminant patchiness, and influence of life history were all considered in this study. Simulation models used in this study showed that patchiness may reduce the rate at which recolonization of contaminated sediments occurs. Furthermore, the degree to which patchiness slows down recolonization may depend on the dispersal abilities of resident species.

In chapter 14, Landis describes age-structured population models built for Cherry Point Pacific herring stock to examine the risk factors contributing to the decline of this important fish species. Results of the simulations suggest that the causative agent for the decline of the herring stock existed prior to 1982, with effects continuing until 2004. Results of Landis's analysis also indicate that the warming Pacific Decadal Oscillation is the best candidate for the main cause of the decline of the stock.

Male-biased populations of the eelpout (*Zoarces viviparus*) fish have been found on the Swedish Baltic coast, and this bias has been associated with pulp mill effluents. In chapter 15, Hanson describes population models showing that male-biased populations with sex ratios in the range of 30–50% females will result in substantial population reductions that ultimately increase the risk of extinction.

In chapter 16, Bridges, Akçakaya, and Bunch assess sediment quality and analyze the potential sediment toxicity to the estuarine amphipod *Leptocheirus plumulosus* to demonstrate the feasibility of using realistic, spatially structured models in ecotoxicological risk analysis at the population level. A stochastic metapopulation model based on the distribution and life-history traits of this species in the Gunpowder River in the upper Chesapeake Bay was developed in which dispersal rates were based on a hydrological model of the Gunpowder River. Thus, the stochastic metapopulation model was linked to a hydrological model.

In chapter 17, Simonini, Prevedelli, and Mauri investigate the effects of zinc and chromium on several key demographic parameters of the interstitial polychaete *Dinophilus gyrociliatus*. Sensitivity and elasticity analysis revealed that the net reproductive rate (R_0) was the demographic parameter most sensitive to metal exposure. Based on these results, the authors suggest that although population growth rate is the most often used index of population fitness, net reproductive rate should also be carefully considered in ecotoxicological analyses.

With this collection of chapters, the reader now has a series of case studies that include the actual models developed for each analysis. The accompanying models can be modified for other species/toxicants and can serve as a tool for teaching graduate-level courses in applied ecology, toxicology, and ecological risk assessment.

Short Introduction to Modeling with RAMAS GIS

The models presented in the book are implemented in RAMAS GIS software for developing either single-population or metapopulation models. A wide range of demographic and life-history characteristics are modeled in various chapters. As a result, different chapters use different sets of model features, options, and parameters. To provide a common introduction to these features, and to avoid repetition of model features in the following chapters, the basic components of the models are summarized in this section. Please consult the Help files on the CD-ROM for detailed information on any option or feature.

Demographic structure

The main characteristic of most of the models in this book is *demographic structure*, which refers to the way individuals in a population are grouped into a number of classes. Such models are also called *frequency based*, and they differ from individual-based models (which follow each organism) and occupancy models (which follow only which patches are occupied, but not the abundances of populations). Unstructured (scalar) models represent each population's abundance at a given time step with a single number (the total number of individuals in the population). Structured models represent the population's abundance with a set of numbers—one for each age class or stage. The parameters of the model (survival rates and fecundities) form a matrix. When it describes an age-structured population, this matrix is called a Leslie matrix; more generally it is referred to as a stage matrix or a stage transition matrix. RAMAS GIS allows age- or stage-structured models (i.e., Leslie or stage matrix) as well as unstructured (scalar) models, and models with sex structure (see below).

Structured population models are flexible and practical tools that can be used for making the connection between individual-level variables and population-level responses (Burgman et al. 1988; Ferson et al. 1989). Different population-level endpoints can be evaluated with structured models. Extinction risk, risk of decline, and chance of recovery are probabilistic (or stochastic) measures that are often used in the context of conservation biology (Burgman et al. 1993; Akçakaya et al. 1999) but are also relevant in ecotoxicological assessments (Ferson et al. 1996). A population-level deterministic measure is *population growth rate*, λ, which is the basis of life table response experiments (LTREs; Caswell 1996; Levin et al. 1996). Among other things, LTREs can be used to identify age-specific life-history traits (survival and fecundity) that have stronger effects on the population growth rate than do other parameters.

In addition, there may be separate stages for males and females. For such a model, the stages are defined in the Stages dialog (under the Model menu of the Metapopulation Model subprogram), which also includes stage-specific information such as the proportion of breeders in each age class or stage. The Sex Structure dialog is used to specify whether males, females, or both sexes are modeled; it also includes information about the number of female stages, the mating system (monogamous, polygynous, or polyandrous), and the degree of polygamy. For a monogamous mating system, fecundity is based on the minimum number of males and females in breeding classes. For polygynous and polyandrous mating systems, fecundity also depends on the degree of polygamy. This is represented with the average number of mates for males and females in polygynous and polyandrous mating systems, respectively. For example, if the number of females per male is specified as 2.0, then fecundity is based on the minimum of the number of females, or 2.0 times the number of males, at each time step.

Survival rates and fecundities, or the rate of transition among stages, are specified in the Stage Matrix dialog (under the Model menu). Each row (and column) of the stage matrix corresponds to one age class or stage. The element at the ith row and jth column of the matrix represents the rate of transition from stage j to stage i, including survival, growth, and reproduction. For more information about stage-based or matrix models, see Caswell (2001), Burgman et al. (1993), Akçakaya et al. (1999), and Akçakaya (2000a).

To use the stage matrix to make projections of population size, it is also necessary to specify the initial number of individuals in each age class. In RAMAS GIS, this is done in two steps for flexibility. First, the total initial abundance (of all stages or age classes in a patch) is entered in the Populations dialog. Second, the relative abundances of stages are entered in the Initial Abundances dialog. The stage matrix and initial abundances are sufficient to make a projection of the population's structure and abundance, but for many other calculations, it is necessary to specify which entries in the stage matrix are survivals and which are fecundities. The Constraints

Matrix (accessed with a button from the Stage Matrix dialog) is used to specify the proportion of each stage matrix element that is a survival rate (as opposed to a fecundity).

Variability

Fluctuations in environmental factors such as weather cause unpredictable changes in a population's parameters (e.g., survival rates and fecundities). This type of variation is called *environmental stochasticity*. In RAMAS GIS, environmental stochasticity is modeled by sampling the vital rates (fecundity and survival rates) and other model parameters (carrying capacity or dispersal rate) in each time step of the simulation from random distributions with given means and standard deviations. For example, the amount of random variability in stage matrix elements is specified in the Standard Deviation Matrix dialog. Variability in the carrying capacity is specified as the standard deviation of K parameters (under the "Density dep." tab in the Populations dialog), and variability in dispersal rates is specified in the Stochasticity dialog.

Extreme environmental events that adversely affect large proportions of a population are called *catastrophes*. Catastrophes may in some cases be considered to be a source of environmental variation that is independent of the normal year-to-year fluctuations in population parameters. In RAMAS GIS, there may be up to two catastrophes (which may be correlated; see the Stochasticity dialog).

Each catastrophe has the following parameters, which are specified either in the Catastrophes dialog or (if they are population specific) under the "Catastrophes" tab in the Populations dialog. Extent is either local (all populations affected independently) or regional (all populations affected at the same time). Probability can be a constant number or a time series representing a function of the number of years since the last catastrophe. Each catastrophe affects abundances, vital rates (i.e., stage matrix elements), carrying capacities, and/or dispersal rates. When a catastrophe occurs, its effect depends on stage-specific multipliers and local multipliers. Local multipliers can also be a function of the number of years since the last catastrophe (e.g., if the severity of fire depends on fuel accumulation). Catastrophes can spread among populations, either by dispersers (i.e., disease) or probabilistically (e.g., fire); the parameters for these modes of spread are specified in the Advanced Catastrophe Settings dialog (accessed with a button from the Catastrophes dialog).

When the number of individuals gets to be very small, there is another source of variation that becomes important even if the population growth rate remains constant. This variation is called *demographic stochasticity* and is modeled by sampling both the number of survivors or dispersers from binomial distributions and the number of offspring (recruits) from Poisson distributions (Akçakaya 1991), if the "Use demographic stochasticity" box (in the Stochasticity dialog) is checked.

Density dependence

As a population grows, increasingly scarce resources limit the growth potential of the population. Such dependence of population growth on "density" (or abundance) of the organisms can substantially affect the dynamics of the populations and its risks of decline and extinction (Ginzburg et al. 1990). In RAMAS GIS, density dependence

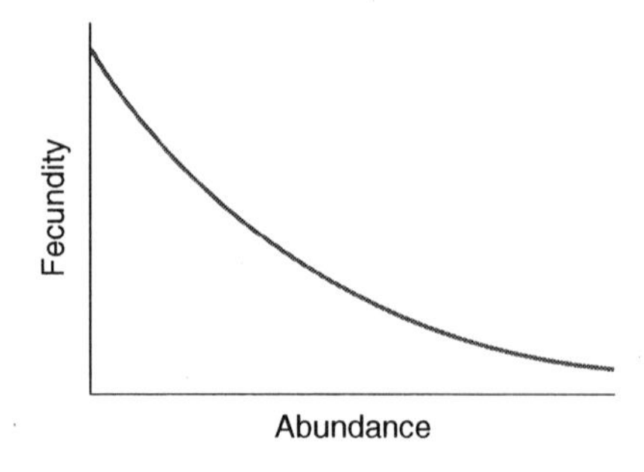

Figure 1.1. An example of density dependence: fecundity as a function of abundance.

is incorporated by allowing the matrix elements (vital rates) to decrease as population size increases. Thus, density dependence is a relationship that describes the growth of a population as a function of its density or abundance. Density dependence in RAMAS GIS is specified with the following parameters, specified either in the Density Dependence dialog, or (if they are population specific) under the "Density dep." tab in the Populations dialog):

1. The abundance on which the density dependence is based (i.e., the x-axis in figure 1.1)—the simplest choice is the total number of individuals in all stages. Density dependence can also be based on the abundance in a subset of stages. The stages that are the basis for density dependence are selected using the basis for "DD" parameter in the Stages dialog.
2. The vital rates affected (i.e., the y-axis in figure 1.1)—the choices are survivals, fecundities, or both.
3. The form of the density dependence function (i.e., the shape of the curve in figure 1.1)—the simplest type of density dependence involves truncating total abundance at a ceiling. Other types include Scramble (Ricker or logistic) and Contest (Beverton-Holt).
4. The parameters of the density dependence function—the basic parameters are the maximum rate of growth when density is low (R_{max}), and the carrying capacity (K). Other related parameters include the standard deviation of K, the temporal trend in K, and the effect of catastrophes on K.

In addition, RAMAS GIS accepts user-defined functions for density dependence. These are specified by the user as a DLL in the Density Dependence dialog.

Population growth may also be affected negatively as population size reaches very low levels. The factors that cause such a decline (e.g., difficulty in finding a mate or disruption of social functions) are collectively called *Allee effects*. Modeling Allee effects involves selecting one of the built-in functions that incorporate Allee effects and specifying an "Allee" parameter. A simpler way is to use a local extinction threshold greater than zero; see the parameters "Local threshold" in the Populations dialog, and "When abundance is below local threshold" in the Stochasticity dialog.

Spatial structure

A set of populations of the same species in the same general geographic area is called a *metapopulation*. Many species exist as metapopulations because of the patchy structure of their habitats, which may be natural or caused by human-induced habitat fragmentation. The dynamics of a metapopulation or a species depend not only on the factors discussed above (e.g., variability and density dependence) but also on the spatial variation in these factors and on other factors that characterize interactions among these populations. The additional factors that operate at the metapopulation or species level include the number and geographic configuration of habitat patches and the dispersal and spatial correlation among these patches.

In RAMAS GIS, spatial structure of a metapopulation is specified with three different sets of parameters: population-specific parameters, correlation, and dispersal.

Population-specific parameters

In the Populations dialog, the location of each population is entered with its x- and y-coordinates. Also in this dialog, many of the model parameters just described can be entered for each population, summarizing the spatial variability in these parameters. For example, each population may be assigned to a different stage matrix or to a different set of standard deviations. The probability and effect of catastrophes and of the parameters related to density dependence may also be specific to the population. Many of these parameters, including the location, carrying capacity, and size of the populations, may be based on the spatial distribution of the suitable habitat (see "Habitat relationships," below).

Correlation

Spatial correlation refers to the similarity (synchrony) of environmentally induced fluctuations in different populations. Independent (uncorrelated) fluctuations in survival and fecundity decrease the likelihood that all populations go extinct at the same time, compared to a case where the fluctuations were dependent—that is, synchronous (Akçakaya and Ginzburg 1991; Burgman et al. 1993; LaHaye et al. 1994). In most metapopulations, fluctuations in demographic rates are caused by factors such as rainfall and temperature that are often correlated even at relatively large distances.

In RAMAS GIS, spatial correlations among populations are specified as a matrix in the Correlations dialog. Each element of this matrix is the coefficient of correlation between the vital rates (stage matrix elements) of the two corresponding populations through time. Spatial correlation is often a function of the distance among the populations. Thus, the program offers an easy way to fill this matrix as a function of distances. The function is

$$C_{ij} = \boldsymbol{a}\exp(-D_{ij}^{\boldsymbol{c}}/\boldsymbol{b})$$

where C_{ij} is the coefficient of correlation between the *i*th and *j*th populations; $\boldsymbol{a}$, $\boldsymbol{b}$, and $\boldsymbol{c}$ are the function parameters; and D_{ij} is the distance between the two populations.

Dispersal

Dispersal among populations may lead to recolonization of empty patches (i.e., extinct populations) by immigration from extant populations. In RAMAS GIS, *dispersal* refers to movement of organisms from one population to another. It is equivalent to "migration" in other metapopulation models. It does not mean back-and-forth seasonal movement between two locations, and it does not refer to dispersal within a population (within a habitat patch). Dispersal rates are specified as the proportion (*not as the total number*) of dispersing individuals per time step from one population to another.

Dispersal rates depend on many factors, for example, species-specific characteristics (e.g., mode of seed dispersal), motility of individuals, and ability and propensity to disperse. Dispersal rate between any two populations may depend on a number of population-specific characteristics, including the distance between the populations, the type of habitat used during dispersal, and the density of the source population.

In RAMAS GIS, various input parameters are used to model dispersal:

1. In the Dispersal dialog, a dispersal matrix gives the proportion of individuals in each population that move (disperse) to other populations. Dispersal is often a function of the distance among the populations. Thus, the program offers an easy way to fill this matrix as a function of distances. The function is

$$M_{ij} = \boldsymbol{a}\exp(-D_{ij}^{c}/\boldsymbol{b}), \quad \text{if } D_{ij} \leq \boldsymbol{D}_{\max}$$
$$= 0, \quad \text{if } D_{ij} > \boldsymbol{D}_{\max}$$

   where M_{ij} is the proportion of individuals in the source population *j* that move to the target population *i*; $\boldsymbol{a}, \boldsymbol{b}, \boldsymbol{c}$, and $\boldsymbol{D}_{\max}$ are the function parameters; and D_{ij} is the distance between the two populations.
2. Rate of dispersal *out of* a population can be defined as a function of the (source) population's abundance, and rate of dispersal *into* a population can be defined as a function of the (target) population's carrying capacity (see density-dependent dispersal parameters under the "Density dep." tab in the Populations dialog).
3. Dispersal rates can be made specific to stage or age using the "Relative dispersal" parameter in the Stages dialog.
4. Dispersal rates can be specified to randomly fluctuate, using the "CV for dispersal" parameter, or the number of dispersers can fluctuate due to demographic stochasticity (see the Stochasticity dialog).
5. Human-mediated dispersal (translocation of individuals from high-density populations to empty or low-density patches, and introduction of individuals to patches or metapopulations that were previously inhabited by the same species) can be modeled in the Population Management dialog.

Habitat relationships

Modeling spatial structure with a metapopulation model requires identifying discrete populations and specifying their location, size, and other characteristics. RAMAS GIS provides functions for determining these aspects of the spatial structure of a metapopulation based on habitat maps (Akçakaya 2000b).

The habitat maps are entered in the Input Maps dialog under the Model menu of the Spatial Data subprogram. These maps are combined into a habitat suitability (HS) map using the HS function in the Habitat Relationships dialog.

The HS map (which can be viewed under the Results menu) is then used to calculate the spatial structure of the metapopulation, based on two parameters in the Habitat Relationships dialog. The "Habitat suitability threshold" is the minimum HS value below which the habitat is not suitable for reproduction and survival (although individuals may disperse or migrate through a habitat that has a lower HS than this threshold). "Neighborhood distance" is used to identify nearby cells that belong to the same patch. For an animal species, the neighborhood distance parameter may represent the foraging distance.

The program identifies clusters of suitable cells in the HS map. Suitable cells (as defined by the threshold parameter) that are separated by a distance less than or equal to the neighborhood distance are regarded to be in the same patch. Thus, neighborhood distance determines the spatial separation above which suitable cells (or clusters of suitable cells) are considered to belong to separate patches. A small neighborhood distance means that the species perceives the landscape as more patchy. Given the same HS map, either a higher threshold HS or a smaller neighborhood distance (or both) will result in a greater number of smaller patches and thus a more patchy landscape. The result of this process is that groups of cells are combined into patches that are based on the species-specific parameters.

In addition to the location and number of habitat patches, the HS map (and other maps) can be used to calculate parameters of the metapopulation model. For example, the carrying capacity of each patch can be based on the total HS in that patch (sum of the HS values of all cells in the patch), incorporating both the size of the patch and the quality of the habitat within the patch. Other parameters, such as fecundities and survival rates, may be based on the average HS (total HS divided by the area of the patch). These functions are specified in the Link to Metapopulation dialog. Parameters that are not based on habitat maps are specified in the Default Population dialog, and those that describe relationships among the populations are specified in the Dispersal and Correlation dialogs. After the program identifies habitat patches and calculates the patch-specific parameters, the results are saved as a metapopulation model file (see the File menu), which can be opened in the Metapopulation Model subprogram.

In many metapopulations, the landscape in which the species lives is not static but is changing due to natural processes (e.g., succession) and human activities (e.g., pollution and habitat loss). This results in a dynamic habitat in which patches may increase or decrease in size, split into or merge with other patches, and even disappear (due or habitat loss) or appear (due to habitat growth). In some cases, change in the habitat can be predicted based on long-term development or resource-use plans or on landscape models. These predictions can be summarized as a time series of maps—for example, one vegetation cover map for each year or decade.

If this kind of map series is available, RAMAS GIS allows the calculation of time series of carrying capacities and/or vital rates for each population. This can be used to model the effects of changes in habitat in time (e.g., habitat loss due to planned logging, or habitat increase due to forest growth). To do this, for each time step, the maps for that time step are used with the Spatial Data subprogram (as described under "Habitat relationships," above). The results are combined using the Habitat Dynamics subprogram.

References

Akçakaya, H. R. 1991. A method for simulating demographic stochasticity. *Ecological Modelling* 54:133–136.

Akçakaya, H. R. 2000a. Population viability analyses with demographically and spatially structured models. *Ecological Bulletins* 48:23–38.

Akçakaya, H. R. 2000b. Viability analyses with habitat-based metapopulation models. *Population Ecology* 42:45–53.

Akçakaya, H. R. 2005. *RAMAS GIS: Linking Spatial Data with Population Viability Analysis*, version 5.0. Applied Biomathematics, Setauket, New York.

Akçakaya, H. R., and J. L. Atwood. 1997. A habitat based metapopulation model of the California gnatcatcher. *Conservation Biology* 11:422–434.

Akçakaya, H. R., and L. R. Ginzburg. 1991. Ecological risk analysis for single and multiple populations. Pages 73–87 in: *Species Conservation: A Population-Biological Approach*. A. Seitz and V. Loeschcke, eds. Birkhauser Verlag, Basel.

Akçakaya, H. R., M. A. Burgman, and L. R. Ginzburg. 1999. *Applied Population Ecology: Principles and Computer Exercises using RAMAS EcoLab 2.0*, 2nd ed. Sinauer Associates, Sunderland, Massachusetts.

Akçakaya, H. R., M. A. Burgman, O. Kindvall, C. Wood, P. Sjögren-Gulve, J. Hatfield, and M. A. McCarthy (editors). 2004a. *Species Conservation and Management: Case Studies*. Oxford University Press, New York.

Akçakaya, H. R., V. C. Radeloff, D. J. Mladenoff, and H. S. He. 2004b. Integrating landscape and metapopulation modeling approaches: viability of the sharp-tailed grouse in a dynamic landscape. *Conservation Biology* 18:526–537.

Akçakaya, H. R., J. Franklin, A. D. Syphard, and J. R. Stephenson. 2005. Viability of Bell's sage sparrow (*Amphispiza belli* ssp. *belli*): altered fire regimes. *Ecological Applications* 15:521–531.

Beyer, W. N., and D. Audet. (n.d.). *Moving from the Individual to the Population Level in Environmental Risk Assessments*. Society for Environmental Toxicology and Chemistry. Available: www.setac.org/eraag/era_pop_discourse2.htm.

Beyer, W. N., and G. H. Heinz. 2000. Implications of regulating environmental contaminants on the basis of wildlife populations and communities. *Environmental Toxicology and Chemistry* 19:1703–1704.

Brook, B. W., and C. J. A. Bradshaw. 2006. Strength of evidence for density dependence in abundance time series of 1198 species. *Ecology* 87:1445–1451.

Brook, B. W., J. J. O'Grady, A. P. Chapman, M. A. Burgman, H. R. Akçakaya, and R. Frankham. 2000. Predictive accuracy of population viability analysis in conservation biology. *Nature* 404:385–387.

Burgman, M., H. R. Akçakaya, and S. S. Loew. 1988. The use of extinction models in species conservation. *Biological Conservation* 43:9–25.

Burgman, M. A., S. Ferson, and H. R. Akçakaya. 1993. *Risk Assessment in Conservation Biology*. Chapman and Hall, London.

Caswell, H. 2001. *Matrix Population Models: Construction, Analysis, and Interpretation*. Second Edition. Sinauer Associates, Sunderland, Massachusetts.

Caswell, H. 1996. Analysis of life table response experiments. 2. Alternative parameterizations for size- and stage-structured models. *Ecological Modelling* 88:73–82.

Dennis, B., J. M. Ponciano, S. R. Lele, M. L. Taper, and D. F. Staples. 2006. Estimating density dependence, process noise, and observation error. *Ecological Monographs* 76:323–341.

Dunham, A. E., H. R. Akçakaya, and T. S. Bridges. 2006. Using scalar models for precautionary assessments of threatened species. *Conservation Biology* 20:1499–1506.

Ferson, S., L. Ginzburg, and A. Silvers. 1989. Extreme event risk analysis for age-structured populations. *Ecological Modelling* 47:175–187.

Ferson, S., L. R. Ginzburg, and R. A. Goldstein. 1996. Inferring ecological risk from toxicity bioassays. *Water Air and Soil Pollution* 90:71–82.

Forbes, V. E., and P. Calow. 1999. Is the per capita rate of increase a good measure of population-level effects in ecotoxicology? *Environmental Toxicology and Chemistry* 18:1544–1556.

Ginzburg, L. R., S. Ferson, and H. R. Akçakaya. 1990. Reconstructibility of density dependence and the conservative assessment of extinction risks. *Conservation Biology* 4:63–70.

Holmes, E. E. 2004. Beyond theory to application and evaluation: diffusion approximations for population viability analysis. *Ecological Applications* 14:1272–1293.

LaHaye, W. S., R. J. Gutierrez, and H. R. Akçakaya. 1994. Spotted owl meta-population dynamics in Southern California. *Journal of Animal Ecology* 63:775–785.

Levin, L., H. Caswell, T. Bridges, C. DiBacco, D. Cabrera, and G. Plaia. 1996. Demographic responses of estuarine polychaetes to pollutants: life table response experiments. *Ecological Applications* 6:1295–1313.

McCarthy, M. A., and L. S. Broome. 2000. A method for validating stochastic models of population viability: a case study of the mountain pygmy-possum (*Burramys parvus*). *Journal of Animal Ecology* 69:599–607.

McCarthy, M. A., H. P. Possingham, J. R. Day, and A. J. Tyre. 2001. Testing the accuracy of population viability analysis. *Conservation Biology* 15:1030–1038.

Morris, W. F., and D. F. Doak. 2002. *Quantitative Conservation Biology: Theory and Practice of Population Viability Analysis*. Sinauer Associates, Sunderland, Massachusetts.

Munns, W. R., D. E. Black, T. R. Gleason, K. Salomon, D. Bengtson, and R. Gutjahr-Gobell. 1997. Evaluation of the effects of dioxin and PCBs on *Fundulus heteroclitus* populations using a modeling approach. *Environmental Toxicology and Chemistry* 16:1074–1081.

Pastorok, R. A., S. M. Bartell, S. Ferson, and L. R. Ginzburg (editors). 2002. *Ecological Modeling in Risk Assessment: Chemical Effects on Populations, Ecosystems, and Landscapes*. R. A. Lewis Publishers, Boca Raton, Florida.

Stark, J. D., L. Tanigoshi, M. Bounfour, and A. Antonelli. 1997. Reproductive potential: its influence on the susceptibility of a species to pesticides. *Ecotoxicology and Environmental Safety*. 37:273–279.

U.S. EPA. 2004. *An Examination of EPA Risk Assessment Principles and Practices*. Staff paper prepared for the U.S. Environmental Protection Agency by members of the Risk Assessment Task Force. U.S. Environmental Protection Agency, Office of the Science Advisor, Washington, DC. Available: www.epa.gov/osa/pdfs/ratf-final.pdf.

Walthall, W. K., and J. D. Stark. 1997. A comparison of acute mortality and population growth rate as endpoints of toxicological effects. *Ecotoxicology and Environmental Safety* 37:45–52.

Wintle, B. A., S. A. Bekessy, L. A. Venier, J. L. Pearce, and R. A. Chisholm. 2005. Utility of dynamic-landscape metapopulation models for sustainable forest management. *Conservation Biology* 19:1930–1943.

2

Lucilia sericata Laboratory Populations

Toxicant Effects Modified by Stage-Specific Density Dependence and Stochasticity

---

S. JANNICKE MOE
ANJA B. KRISTOFFERSEN
NILS C. STENSETH

The case study species for this chapter is the green-bottle blowfly *Lucilia sericata* (Meigen) (Diptera: Calliphoridae), a pest species that attacks sheep and causes significant economic damage worldwide. Pest organisms are typically opportunistic species with high reproductive capacity and rapid growth, which can lead to population outbreaks and large fluctuations in abundances. Here we use RAMAS (Akçakaya 2005) to explore which demographic factors promote population outbreaks for a laboratory system of *L. sericata*, and how the risks of population outbreaks are affected by toxicant exposure. Although extinction is not very likely for pest populations, analysis of extinction probabilities for different scenarios can nevertheless help identify the factors that suppress the population growth, which can clearly be an aim for control of pest species.

For management of pest species with distinct life-history stages (e.g., many insect species), it is important to understand the demographic processes of the different stages and how these can be affected differently by toxicants (e.g., pesticides), density dependence, and stochasticity. Moreover, whenever the economic damage is caused primarily by certain life-history stages, it is of interest to assess the probabilities of explosion and extinction separately for these stages.

Lucilia cuprina, a species closely related to our model species, was used in classical laboratory studies of density dependence and population dynamics by Nicholson (1950, 1954). The demography of *Lucilia* is therefore well known. Later experiments have explored how a toxicant (cadmium) affects long-term population dynamics of *L. sericata* (Daniels 1994) and how the toxicant exposure interacts with demographic processes (Moe et al. 2001, 2002a, 2002b). Cadmium was chosen as the toxicant because it is not metabolized in the organism (Simkiss et al. 1993), which made it possible to track the accumulation of the toxicant in individuals from different treatments (Moe et al. 2001). Although a pest species was used as a study organism for this laboratory system (Daniels 1994), the original purpose was not to study the effects of cadmium as a pesticide. Both the species and the toxicant were chosen simply because the characteristics mentioned above made them suitable for a long-term population-level ecotoxicological experiment. However, we assume that the conclusions reached in this study are applicable also for toxicants more generally, including pesticides.

For fluctuating populations like this blowfly model system, certain demographic processes are particularly important for generating the dynamics: overcompensating density dependence, delayed effects of density, and stochasticity. Stochastic forces may be of equal importance as deterministic forces in population dynamics (Bjørnstad and Grenfell 2001), but the role of stochasticity is often poorly understood. The importance of density dependence and stochasticity is also receiving increasing attention within population ecotoxicology and risk assessment (e.g., Grant 1998; Laskowski 2000; Forbes et al. 2001; Moe 2007). Although we have a good understanding of how toxicant exposure interacts with density dependence within one generation in the blowfly system (Moe et al. 2002a), it is more difficult to predict effects on long-term dynamics, and especially in combination with stochasticity. Here, we explore how toxicant exposure interacts with different types of delayed density dependence and with different levels of stochasticity in a population model, and analyze the roles of these three factors and their interactions for population-level endpoints that are relevant for risk assessment.

Most of the population models developed for the laboratory blowfly system have focused on adult dynamics (e.g., Maynard Smith 1974; Gurney et al. 1980; Readshaw and Cuff 1980; Kendall et al. 1999; Smith et al. 2000; Wood 2001). However, the dynamics of the larval stage may also be important when regarding the species as a pest organism, since this stage is often the main feeding stage. A model developed for field blowfly populations predicts that variations in adult mortality have a much greater effect on the population viability than do variations in juvenile mortality (Fenton and Wall 1997). Therefore, here we analyze population viability separately for the adult and larval stages.

Methods

Study species

Our case study is based on a combination of time-series and cohort experiments with populations of the blowfly *Lucilia sericata* (Meigen 1826) exposed to the toxicant cadmium. The time-series experiment (Daniels 1994) was designed to test the combined effects of toxicant exposure and density dependence on population dynamics and has been analyzed by Lingjærde et al. (2001). The cohort experiments (Moe et al. 2001, 2002a, 2002b) were designed to test predictions and assumptions of the semiparametric population model by Lingjærde et al. (2001), particularly regarding the density-dependent structures and the mechanisms underlying density-toxicant interactions.

The species is a primary infective agent for sheep myiasis in many countries and is therefore of epidemiological and economical importance (Rognes 1991). The species is also utilized in forensic entomology and in medicine (larvae may be used to heal chronically purulent wounds). In the field, the adult females oviposit in the fleece of sheep (Wall 1993). The larvae develop through three instars, feeding on epidermal tissues and skin secretions. During the last instar, the larvae migrate away from the sheep and burrow into the ground, where they undergo diapause before pupation. Adult flies may live for 20–40 days (Evans 1936). Females may lay up to 200 eggs per batch and may oviposit repeatedly (Wall 1993).

The ephemeral nature of the larval habitat has selected for opportunistic strategies in both the larval and the adult stages (Hanski 1987). The females have a very high reproductive potential, enabling them to exploit suitable oviposition sites quickly. The number of offspring is maximized (lifetime fecundity per female may be up to 2,000 offspring) while the offspring size is minimized; the larvae may therefore grow up in very high densities. Blowflies have the ability to accomplish the larval development in a short time, at the expense of size in the adult stage (Ullyett 1950), which may result from the high mortality risk in the larval stage (due to both short-lived resources and high competition). This particular life-history strategy makes blowflies a suitable study species for population-dynamic studies: The high fecundity (high intrinsic growth rate), strong competition in the larval stage (nonlinear density dependence) and delayed effect of competition in the adult stage (time lag between the density-dependent stimulus and its demographic responses) may all contribute to fluctuating dynamics in laboratory populations.

Experiments

The experimental design and protocol are based on an experiment by Nicholson (1954) and are described in more detail in Smith et al. (2000) and Lingjærde et al. (2001). The time-series data (Daniels 1994) consist of stage-specific counts of 12 laboratory populations of *Lucilia sericata*. The larvae were fed with a blood-based diet in limited amounts, whereas adults were fed with sugar and water ad libitum. The blood-based diet was also ingested by adult flies, which need proteins for reproduction but not for survival. The populations were divided into two treatment groups (six replicates per group), hereafter referred to as toxicant and control groups. In the toxicant group, cadmium acetate was added to the diet and ingested by both larvae and adults. The diet was replenished every two days. The following state variables were recorded every two days for 760 days: number of larvae, number of new pupae (defining a pupal cohort), mean individual weight per pupal cohort, and number of adults. In addition, the number of viable pupae (i.e., successfully emerged adults) per cohort was recorded until 20 days after pupation. The duration of the time-series experiment corresponds to approximately 20 generations.

The cohort experiments were designed similarly to the time-series experiments but had even-aged cohorts, in order to measure age-specific demographic rates. Three experiments were performed to examine density-dependent structures (Moe et al. 2002b), delayed versus direct effects of density on adult survival and reproduction

(Moe et al. 2001), and density–toxicant interactions, including stage-specific effects of both factors (Moe et al. 2002a).

Population Model

Simulation scenarios

The file Lucilia_8d.mp (in the CD-ROM accompanying this book) contains four model populations; these are referred to here as *model versions*.[1] The four model versions represent the two treatment groups (control and toxicant) combined with the two types of stage-specific density dependence (described below). For each of the four model versions, there are three levels of stochasticity (described below). The resulting 12 combinations of density dependence type, stochasticity, and toxicant treatment are referred to as *scenarios*. The purpose of comparing the scenarios is to explore the effects toxicant treatment on population viability under different combinations of stage-specific density dependence and stochasticity levels and, in particular, to investigate interactions among these three factors. In addition, we compare the responses of the larval and adult stages.

The initial densities are 120 adults (40 per age group; see next section). Each scenario is run in 100 replications. The duration of the simulations is 100 time steps (=800 days), and the initial 10 time steps are excluded from the results. The resulting simulated time series are of comparable length to the time-series experiment (~700 days).

Default model

Several dynamic models have been developed for describing this study system (summarized in Smith et al. 2000). An age-specific model with time step of two days, which is equal to the feeding and counting interval, seemed to give the best representation of density dependence and to reproduce the dynamics most successfully (Moe et al. 2005). However, the two-day model contains more than 30 two-day age groups, which makes it less suitable for the matrix formulation in RAMAS. Here we have used a simpler model with eight-day time steps, which corresponds approximately to the length both of the larval stage (eight to nine days) and of the pupal stage (seven to eight days) measured in the cohort experiments (Moe et al. 2001). The model has five stages: larvae (L), pupae (P), and three adult age groups ($A_1 + A_2 + A_3$) that are collectively referred to as adults (A).

The model parameter values (table 2.1) are based on a combination of previous population models (Smith et al. 2000; Lingjærde et al. 2001; Moe et al. 2005) and independent experiments (Moe et al. 2002a, 2002b). In the control model versions, the maximum reproductive rate (fA) is 20 offspring per capita per eight days, which corresponds to five offspring per female per day. This is well within reproductive capacity of *L. sericata* (at least 5–15 per day, depending on crowding during larval stage; Moe

[1] The populations in this model are not meant to be subpopulations in a metapopulation, so when running simulations for one of the populations, the other populations should be marked for exclusion.

Table 2.1. Parameter values for the four population models, as used in equations 2.1 and 2.2.

	Parameter									
Population	*fA*	*aA*	*bA*	*sL*	*aL*	*bL*	*sP*	*aP*	*bP*	*sA*
AF, control	20	0.012	3	0.9	0.003	1	0.9	0	1	0.7
AF, toxicant	10	0.012	3	0.8	0.003	1	0.8	0	1	0.67
PS, control	20	0	1	0.9	0.003	1	0.9	0.012	2	0.7
PS, toxicant	10	0	1	0.8	0.003	1	0.8	0.012	2	0.67

The time step is eight days. *f*, fecundity; *s*, survival; *a*, strength of density dependence; *b*, degree of compensation; *L*, larvae; *P*, pupae; *A*, adults. AF model version, larval density affects adult fecundity; PS model version, larval density affects pupal survival.

et al. 2002b). The maximum larval survival (*sL*) and pupal survival (*sP*) rates per eight days are both set to 0.9.

Density dependence

The basic formula used for density dependence is

$$N_{t+1} = N_t \cdot R \cdot \frac{1}{1+(aN_t)^b}, \qquad (2.1)$$

where N_t is the density at time t, R is the maximum reproductive rate, a is the "strength" of density dependence, and b is the degree of density-dependent compensation. In both model versions, larval density has a direct effect on larval survival. Larval survival is modeled by exactly compensating density dependence ($b = 1$; Contest competition).

In addition, the larval density has a delayed effect either on adult fecundity (model version AF; equation 2.2) or on pupal survival (model version PS; equation 2.3). For the PS version (equation 2.3), pupal survival is modeled as a function of larval density one time step ago. The density dependence is overcompensating ($b = 2$; Scramble competition). For the AF version (equation 2.2), adult fecundity for the three adult age groups is modeled as a function of larval density at two, three, and four time steps ago, respectively. The equations are implemented by a user-defined file (Lucilia_DD3.dll) for delayed density dependence.

Adult fecundity (AF) model version:

$$\begin{bmatrix} L_{t+1} \\ P_{t+1} \\ A^1_{t+1} \\ A^2_{t+1} \\ A^3_{t+1} \end{bmatrix} = \begin{bmatrix} 0 & 0 & fA \cdot \frac{1}{1+(aA \cdot L_{t-2})^{bA}} & fA \cdot \frac{1}{1+(aA \cdot L_{t-3})^{bA}} & fA \cdot \frac{1}{1+(aA \cdot L_{t-4})^{bA}} \\ sL \cdot \frac{1}{1+(aL \cdot L_t)^{bL}} & 0 & 0 & 0 & 0 \\ 0 & sP & 0 & 0 & 0 \\ 0 & 0 & sA & 0 & 0 \\ 0 & 0 & 0 & sA & 0 \end{bmatrix} \times \begin{bmatrix} L_t \\ P_t \\ A^1_t \\ A^2_t \\ A^3_t \end{bmatrix} \qquad (2.2)$$

Pupal survival (PS) model version:

$$\begin{bmatrix} L_{t+1} \\ P_{t+1} \\ A^1_{t+1} \\ A^2_{t+1} \\ A^3_{t+1} \end{bmatrix} = \begin{bmatrix} 0 & 0 & fA & fA & fA \\ sL \cdot \dfrac{1}{1+(aL \cdot L_t)^{bL}} & 0 & 0 & 0 & 0 \\ 0 & sP \cdot \dfrac{1}{1+(aP \cdot L_{t-1})^{bP}} & 0 & 0 & 0 \\ 0 & 0 & sA & 0 & 0 \\ 0 & 0 & 0 & sA & 0 \end{bmatrix} \times \begin{bmatrix} L_t \\ P_t \\ A^1_t \\ A^2_t \\ A^3_{t_t} \end{bmatrix} \tag{2.3}$$

Stochasticity

Three levels of stochasticity are used: level 0 has no stochasticity (standard deviation matrix StDevMat0); level 1 has stochasticity in pupal survival (StDevMat1); level 2 has stochasticity in both pupal survival and reproduction (StDevMat2). The standard deviations used are 0.09 for pupal survival (*sP*) and 1.5 for reproduction (*fA*), which correspond to approximately 10% of the respective parameter values (averaged for control and toxicant versions). These distributions represent environmental stochasticity and are added log-normally. In addition, demographic stochasticity is used in all simulations. "Stochasticity" in this chapter therefore refers to the environmental stochasticity in RAMAS.

Toxicant effects

The effects of toxicant exposure on demographic rates are based on the analyses by Moe et al. (2002a, 2005). Fecundity (*fA*) is reduced by 50%, larval and pupal survival rates (*sL* and *sP*) are reduced by approximately 10%, and adult survival rate (*sA*) is reduced by 4%.

Results

Abundances

As a validation of the model, the dynamics of the simulated series should be compared to the original data series. If the predicted dynamics are fundamentally different from the real, this indicates that the model has failed to properly represent the population structure, demographic processes (including density dependence), or stochasticity. Earlier models for our study system have been judged by their ability to reproduce (1) the correct mean densities, (2) the cyclic patterns with the right period in the adult densities, and (3) the effect of toxicant on both these properties (Smith et al. 2000; Moe et al. 2005).

The mean densities for real and predicted series are given in table 2.2. The total densities averaged across the different scenarios correspond well with the observed densities, for both control and toxicant populations. The real populations are affected by both types of density dependence (pupal survival and fecundity), but here we explore each type separately. The PS model versions (density-dependent pupal survival) generally

Table 2.2. Summary statistics for the simulated data from the 12 scenarios and for real data, for control and toxicant groups.

	Adult fecundity (AF)			Pupal survival (PS)			
Density dependence effects	0	1	2	0	1	2	Real data
Control							
Total abundance	708	712	616	616	469	471	614
A abundance	237	237	208	23	17	18	140
L abundance	350	353	302	447	346	348	368
P abundance	121	122	106	146	106	105	106
A variation (CV)	49%	50%	59%	80%	191%	204%	41%
L variation (CV)	89%	89%	109%	89%	191%	206%	66%
P variation (CV)	62%	63%	76%	44%	144%	155%	40%
A autocorrelation	0.71	0.69	0.26	0.49	0.20	0.19	0.18
A period (days)	63	64	69	61	64	63	72
Significant A cycles	100%	100%	86%	100%	73%	71%	83%
Toxicant							
Total abundance	437	437	399	368	359	362	375
A abundance	148	148	134	24	24	24	98
L abundance	201	201	185	240	238	241	174
P abundance	88	88	80	104	97	97	103
A variation (CV)	41%	41%	46%	56%	80%	74%	44%
L variation (CV)	68%	69%	92%	69%	81%	77%	75%
P variation (CV)	52%	52%	62%	33%	53%	49%	40%
A autocorrelation	0.69	0.63	0.20	0.25	0.16	0.17	0.12
A period (days)	59	60	67	61	62	63	102
Significant A cycles	100%	100%	80%	91%	76%	78%	50%

Stochasticity levels: 0, no stochasticity; 1, stochasticity in pupal survival (*sP*) only; 2, in both pupal survival and reproduction (*fA*). The rows contain mean total abundance, mean stage-specific abundances for adults (A), larvae (L), and pupae (P); CV (coefficient of variation = standard deviation/mean) of stage-specific abundances; and autocorrelation function, dominating period, and significance cycles of adult abundances. Each of the 12 scenarios was run in 100 replications.

predict somewhat too high densities, while the AF versions (density-dependent adult fecundity) predict too low densities. The PS versions (where recruitment to larval stage is not regulated) typically underestimate adult densities and overestimate larval and pupal densities, while the AF versions (where recruitment to adult stage is not regulated) overestimate adult densities. So the separation of these two density-dependent processes generates biased results in opposite directions, but this separation can still help us explore the effects of these two density-dependent processes.

Higher levels of stochasticity generally reduce the densities, particularly for control populations. The AF and PS populations respond differently to the different levels of stochasticity. The AF populations show little response to stochasticity level 1 (in pupal survival) but stronger response to stochasticity level 2 (stochasticity also in adult fecundity). For the PS control populations, the pattern is the opposite (strong response to stochasticity level 1, lower response to level 2). This means that the populations are more affected by stochasticity when it is added to a stage that is regulated by density dependence than when it is added to an unregulated stage. For the PS toxicant populations, mean abundances do not seem to be affected by stochasticity levels at all.

The toxicant versions of the models have generally lower abundances for all stages: Total abundances are reduced to 60–77% in the toxicant versions relative to control versions for the six pairs of scenarios (real populations, on average, reduced to 61%). Larval abundances are reduced to 54–79% (real populations, 47%), and adult abundances in the AF model versions are reduced to 62–64% (real populations, 70%). In the PS model versions, however, adult abundances are slightly increased. This indicates that in the toxicant model populations, where larval competition is reduced, the relief from density dependence in pupal survival completely compensates the negative toxicant impacts on the adult abundances. A similar compensating mechanism was probably operating in the real populations, although the compensation was not complete in the real case.

Population dynamics

In the real data, the control populations displayed more or less regular cycles in adult densities, as characterized by significant positive autocorrelation for periods around twice the generation time (60–70 days; table 2.2). The toxicant populations had more irregular and less significant fluctuations. Larval densities were highly variable in both treatment groups. This model also shows fluctuating behavior (figure 2.1), with clear cycles for the control populations and less significant cycles for the toxicant populations (figure 2.2, table 2.2). Analysis of autocorrelation of simulated densities shows that the average period is around 62–68 days (two generations) for both control and toxicant populations (figure 2.2, table 2.2). The toxicant populations have the same proportion of significantly cyclic populations, but the cycles are somewhat weaker (lower autocorrelation). The fact that the model succeeds in simulating the characteristics of the dynamics makes it plausible that the main demographic processes and toxicant effects are well represented in the model structure.

The amount of temporal variation in abundances is measured by the coefficient of variation (CV, standard deviation divided by mean). Adults have higher CVs in PS model versions than in AF versions, while the opposite is the case for larvae and pupae. This pattern is found in both control and toxicant versions. Stochasticity levels affect the AF and PS model versions differently. For AF versions, the CV increases little with stochasticity level 1, but more with level 2. This is consistent with the effect of stochasticity levels on mean abundances. For the PS control populations, effects of stochasticity on CVs are also consistent with the effects on mean abundances: strong effect of stochasticity level 1, little additional effect of level 2. For the PS toxicant populations, however, the effects of stochasticity levels on CVs are not consistent with the effects on mean abundances: CV first increases with stochasticity level 1, and then *decreases* somewhat with level 2. In the latter case, the increased stochasticity actually reduces the temporal variation of the populations. The scenarios with least variation have CVs in the same range as the real populations, while the other scenarios have higher CVs.

The simulated toxicant populations always have lower CVs than the corresponding control populations, in spite of the various differences among the scenarios. The reductions in vital rates thus result in lower population fluctuations, for all combinations of density dependence type and stochasticity level.

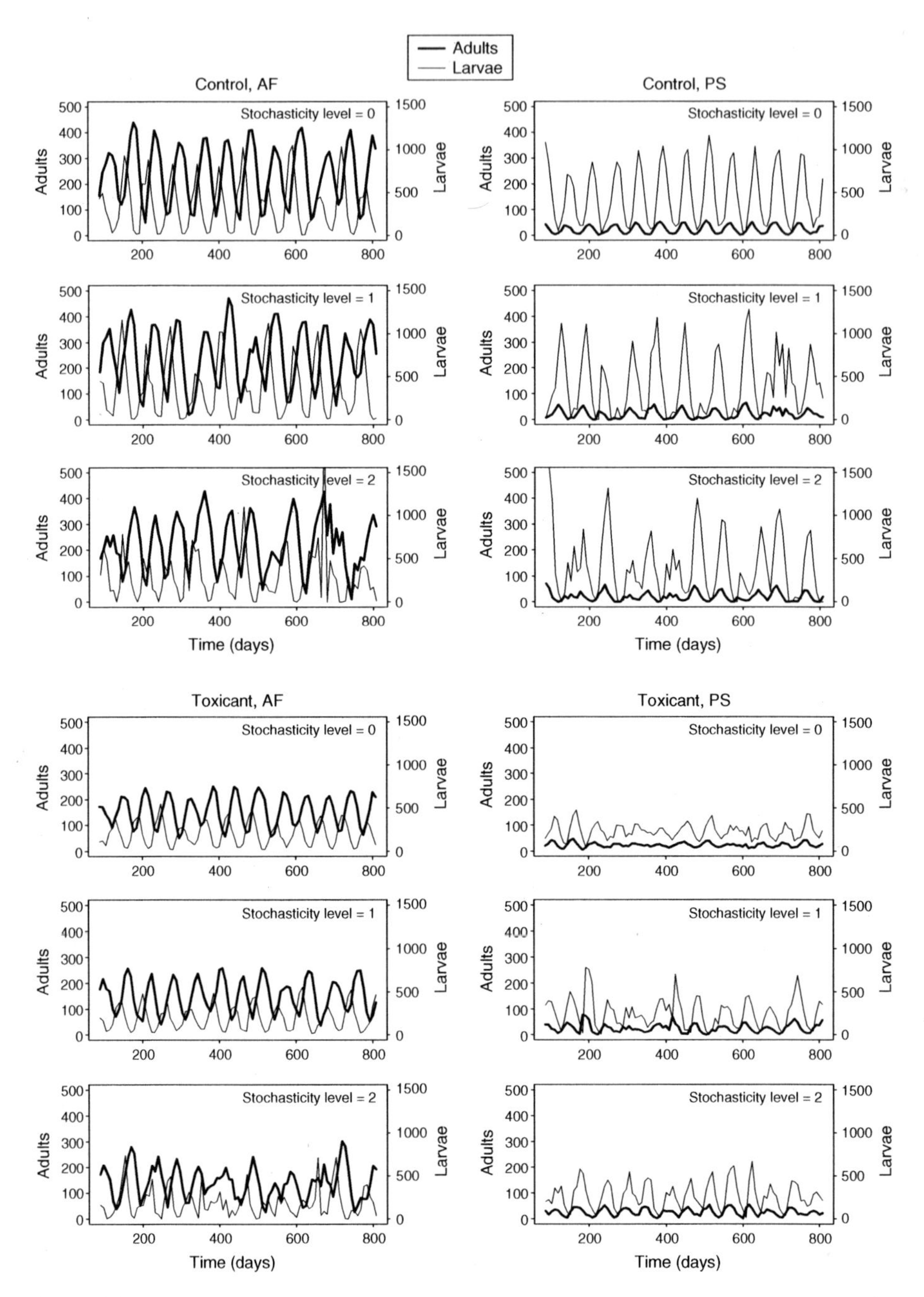

Figure 2.1. Simulated stage-specific abundance series for laboratory populations of the blowfly *Lucilia sericata*. The 12 model scenarios are represented by one random simulation from each scenario. In toxicant scenarios, the populations are exposed to cadmium. Model version AF has density-dependent adult fecundity (equation 2.2); model version PS has density-dependent pupal survival (equation 2.3). Stochasticity levels: 0, no stochasticity; 1, stochasticity in pupal survival only; 2, stochasticity in both pupal survival and reproduction. The thick black curves represent the adult stage (left y-axes), and the thin curves represent the larval stage (right y-axes). Note that the scales for adult and larval abundances differ.

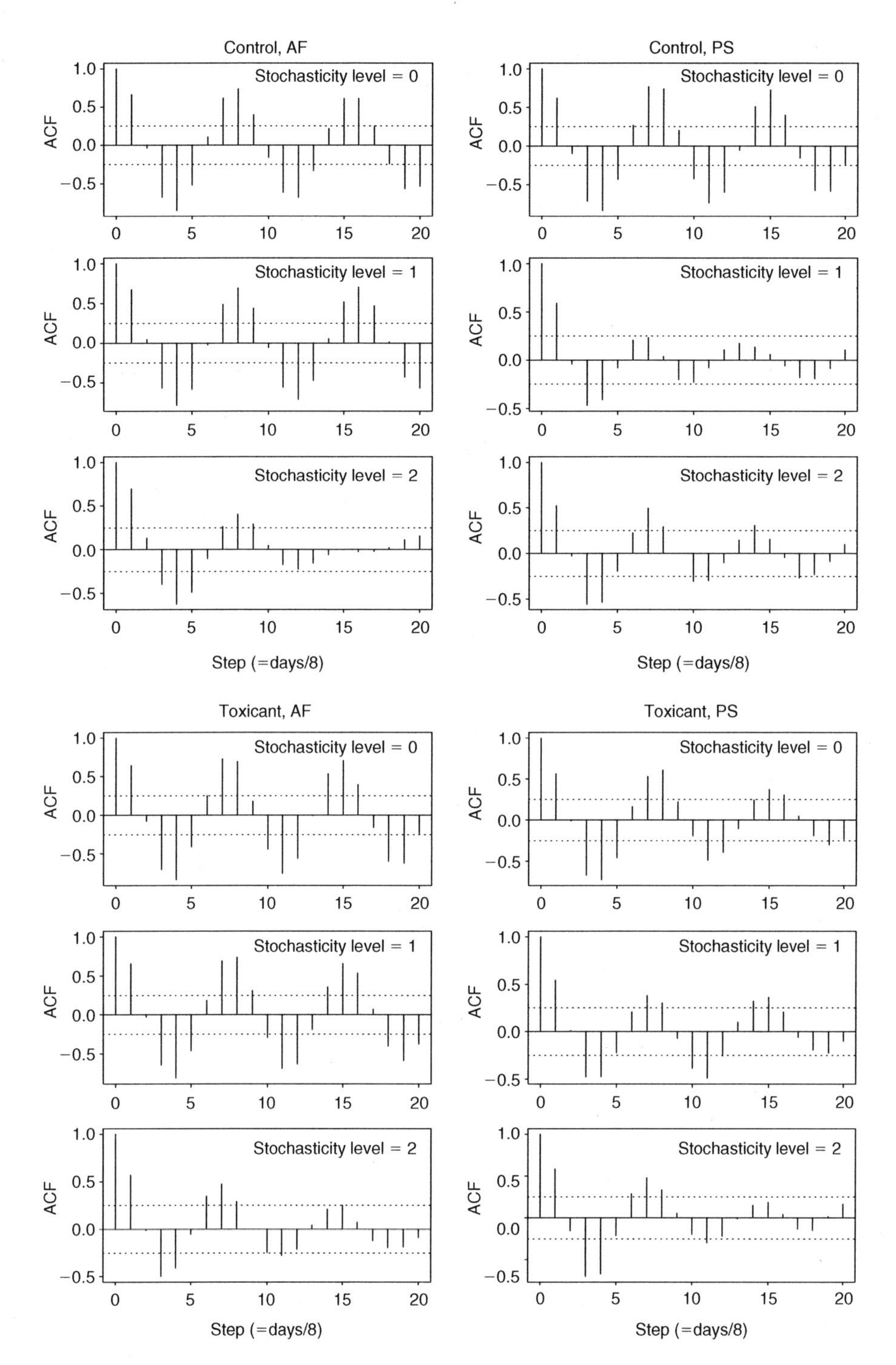

Figure 2.2. Autocorrelation function for simulated series of adult abundances (displayed in figure 2.1). Dotted lines represent 95% confidence intervals. The x-axes show the number of lags, in units of eight days. Stochasticity levels: 0, no stochasticity; 1, stochasticity in pupal survival (*sP*) only; 2, stochasticity in both pupal survival and reproduction (*fA*).

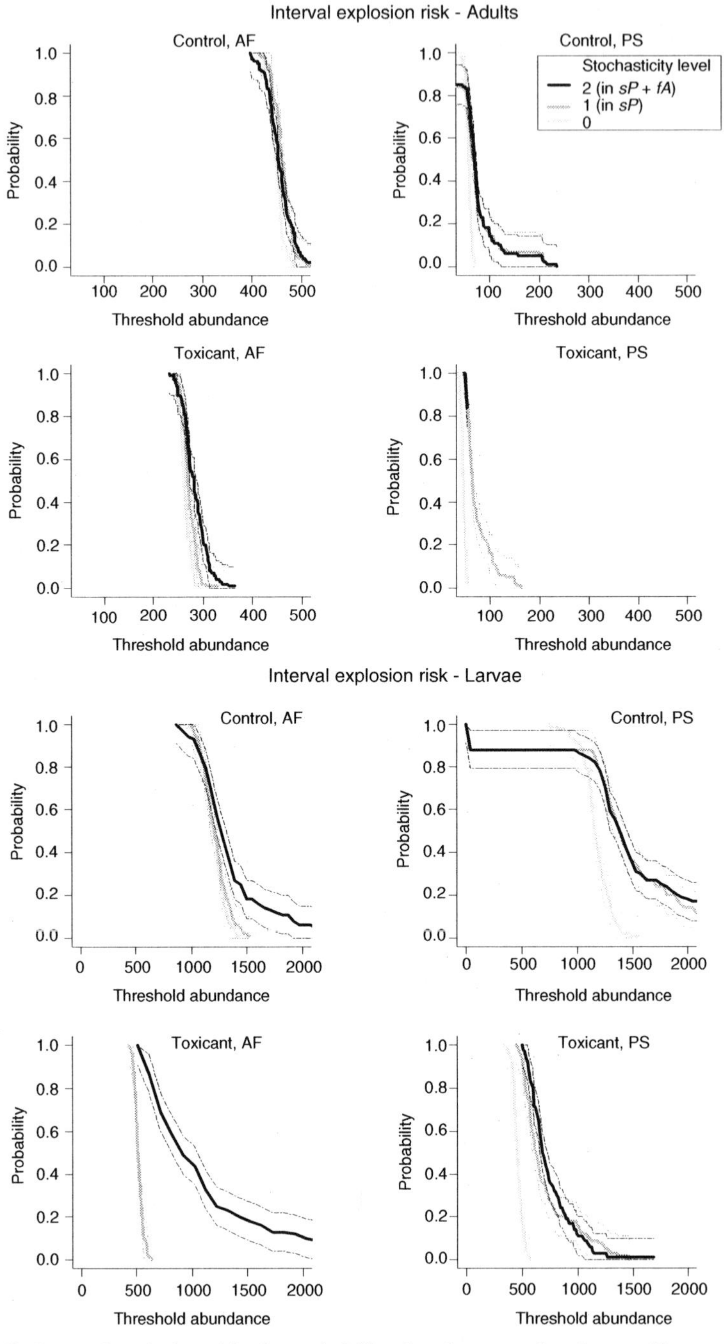

Figure 2.3. Interval explosion risk: the probability that the stage abundance will exceed a given range of abundances at least once during an interval corresponding to the simulation duration (800 days). Each x,y-point on the curve can be interpreted as "there is a *Y*% probability that the stage abundance will exceed *X* at least once during the simulation duration," where *Y*% is the probability from the y-axis and *X* is the threshold abundance from the x-axis. Dashed lines represent 95% confidence intervals. The shade of the curves indicates stochasticity level: light gray, no stochasticity; dark gray, stochasticity in pupal survival (*sP*) only; black, stochasticity in both pupal survival and reproduction (*fA*).

Table 2.3. Comparison of interval explosion risk for different scenarios: effects of density dependence type PS (affects pupal survival) versus AF (affects adult fecundity); effects of stochasticity (level 1 vs. 0: effect of stochasticity in pupal survival; level 2 vs. 1: effects of additional stochasticity in adult fecundity); and effects of toxicant versus control.

	Stochasticity level					
	0		1		2	
	D	*p*	*D*	*p*	*D*	*p*
Effects of density dependence type: PS vs. AF						
Adults control	–1.00	<0.001	–1.00	<0.001	–1.00	<0.001
Adults toxicant	–1.00	<0.001	–1.00	<0.001	–1.00	<0.001
Larvae control	–0.16	0.155	0.43	<0.001	0.24	0.006
Larvae toxicant	–0.57	<0.001	0.60	<0.001	–0.34	<0.001

	Level 1 vs. 0		Level 2 vs. 1	
	D	*p*	*D*	*p*
Effects of stochasticity				
Adults control AF	0.39	<0.001	0.20	0.037
Adults control PS	0.48	<0.001	0.09	0.813
Adults toxicant AF	0.29	<0.001	0.29	<0.001
Adults toxicant PS	0.88	<0.001	0.150	0.211
Larvae control AF	0.09	0.813	0.23	0.010
Larvae control PS	0.30	<0.001	0.06	0.994
Larvae toxicant AF	0.15	0.211	0.86	<0.001
Larvae toxicant PS	0.78	<0.001	0.23	0.010

	Stochasticity level					
	0		1		2	
	D	*p*	*D*	*p*	*D*	*p*
Effects of toxicant exposure vs. control						
Adults AF	–1	<0.001	–1	<0.001	–1	<0.001
Adults PS	–0.78	<0.001	–0.16	0.155	–1	<0.001
Larvae AF	–1	<0.001	–1	<0.001	–0.52	<0.001
Larvae PS	–1.00	<0.001	–0.79	<0.001	–0.8	<0.001

Test statistics: D, maximum difference in explosion risk between two scenarios (minus sign means reduced risk); *p*, *p*-value. Each scenario was run in 100 replications.

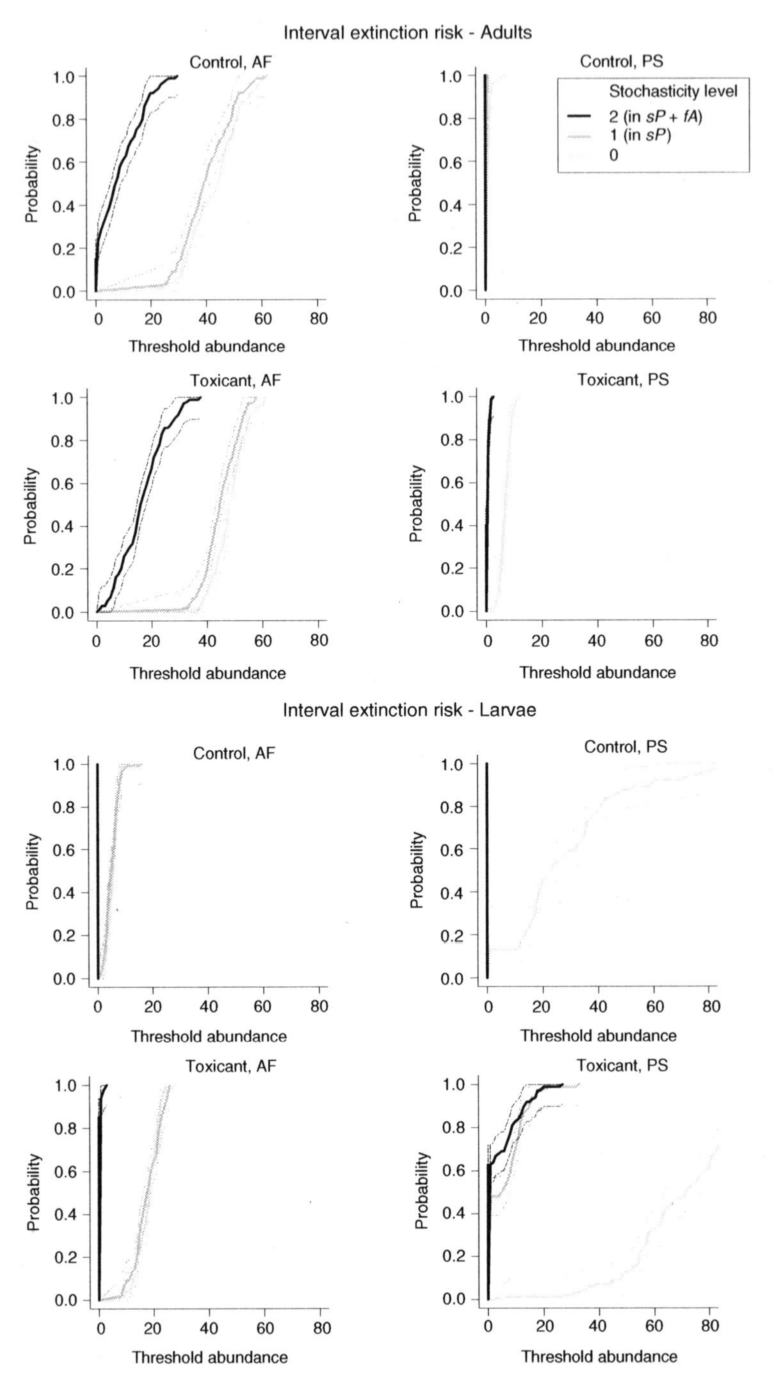

Figure 2.4. Interval extinction risk: the probability that the stage abundance will fall below a given range of abundances at least once during an interval corresponding to the simulation duration (800 days). Each x,y-point in the curve can be interpreted as "there is a *Y*% risk that the stage abundance will fall below *X* at least once during the simulation duration," where *Y*% is the probability from the y-axis and *X* is the threshold abundance from the x-axis. Dashed lines represent 95% confidence intervals. The shade of the curves indicates stochasticity level: light gray, no stochasticity; dark gray, stochasticity in pupal survival (*sP*) only; black, stochasticity in both pupal survival and reproduction (*fA*).

Table 2.4. Comparison of interval extinction risk for different scenarios (see table 2.3 for explanations).

	Stochasticity level					
	0		1		2	
	D	*p*	*D*	*p*	*D*	*p*
Effects of density dependence type: PS vs. AF						
Adults control	1.00	<0.001	1.00	<0.001	0.85	<0.001
Adults toxicant	1.00	<0.001	1.00	<0.001	0.97	<0.001
Larvae control	−0.86	<0.001	−1.00	<0.001	−0.00	1.000
Larvae toxicant	−0.98	<0.001	−0.72	<0.001	−0.34	<0.001

	Level 1 vs. 0		Level 2 vs. 1	
	D	*p*	*D*	*p*
Effects of stochasticity				
Adults control AF	0.19	0.054	0.95	<0.001
Adults control PS	0.81	<0.001	0.01	1.000
Adults toxicant AF	0.27	0.001	0.96	<0.001
Adults toxicant PS	0.99	<0.001	0.08	0.906
Larvae control AF	0.16	0.155	1	<0.001
Larvae control PS	0.87	<0.001	0.00	1.000
Larvae toxicant AF	0.12	0.468	1.00	<0.001
Larvae toxicant PS	0.98	<0.001	0.20	0.037

	Stochasticity level					
	0		1		2	
	D	*p*	*D*	*p*	*D*	*p*
Effects of toxicant exposure vs. control						
Adults AF	−0.32	<0.001	−0.34	<0.001	−0.47	<0.001
Adults PS	−0.95	<0.001	−0.51	<0.001	−0.97	<0.001
Larvae AF	−0.97	<0.001	−0.9	<0.001	−0.15	0.211
Larvae PS	−0.77	<0.001	−0.52	<0.001	−0.37	<0.001

Risk estimates

Since *Lucilia sericata* is regarded as a pest species, the probability of population explosion can be considered the most important risk, while population extinction can be considered a desired (but unlikely) outcome. The absolute values are not directly applicable to field populations, since the models are based on a laboratory system. Nevertheless, the difference between outcomes of various scenarios may be applicable also for field populations. The risk estimation for explosion is given as the probability of exceeding a threshold density during the simulation interval, and vice versa for extinction. The probabilities are calculated for the adult and larval stages separately. Since the population dynamics are nonstable, other risk calculations available in RAMAS such as the terminal explosion/extinction risk and the time to quasi explosion/extinction are less relevant.

The population explosion probabilities are displayed as a function of threshold densities in figure 2.3. Test results for risk comparisons are summarized in table 2.3. The two stages respond differently to the density dependence types: Adults reach higher explosion probability in the AF models than in the PS models (all $p < 0.001$), while the opposite is true for larvae ($p < 0.007$ in five of six cases).

Stochasticity generally increases explosion risk for both stages, although only slightly (maximum difference in probability averaged 0.34 for all scenarios). For adults, effects of stochasticity are more pronounced in AF models (all $p < 0.04$) than in PS models ($p > 0.2$ in two of four cases). For larvae, the opposite is the case: Effects of stochasticity are stronger in PS models ($p < 0.011$ in three cases) than in AF models ($p < 0.011$ in two cases). Explosion risks in the AF versus PS models are affected differently by the two levels of stochasticity (level 1, in pupal survival only, vs. level 2, in both pupal survival and reproduction). Stochasticity in pupal survival has a considerable impact for PS models (for both stages and both toxicant treatments; all $p < 0.001$) but less impact for AF models (adults, both $p < 0.001$; larvae, both $p > 0.2$). Adding stochasticity in reproduction, however, has little impact on PS models ($p > 0.2$ in three of four cases) but more impact on AF models (all $p < 0.04$). This pattern is similar for both the adult and larval stages.

The toxicant treatment consistently reduces the explosion risk for both adults and larvae, as could be expected from the reduction in vital rates. For adults, the effects of toxicant are more pronounced for the AF model (all $p < 0.001$) than for the PS model ($p < 0.001$ in two of three cases). The same is true for the larvae, but to a lesser degree (all $p < 0.001$).

The population extinction probabilities are likewise displayed in figure 2.4, and test results are summarized in table 2.4. Again, the two stages respond differently to the density-dependence types: Adults reach higher extinction probability in the PS models than in the AF models (all $p < 0.001$), while the opposite is true for larvae ($p < 0.001$ in five of six cases).

Stochasticity generally increases extinction risk for both stages, as could be expected. For adults, effects of stochasticity are more pronounced in AF models (all $p < 0.04$) than in PS models ($p > 0.2$ in two of four cases). For larvae, the opposite is the case: Effects of stochasticity are stronger in PS models ($p < 0.011$ in three cases) than in AF models ($p < 0.011$ in two cases). Extinction risks in the AF versus PS models are affected differently by the two levels of stochasticity, in a similar way as described above for explosion risks. Stochasticity in pupal survival has a considerable impact for PS models (for both stages and both toxicant treatments; all $p < 0.001$) but less impact for AF models ($p > 0.05$ in three of four cases). Adding stochasticity to fecundity has little impact on PS models ($p > 0.9$ in three of four cases) but more impact on AF models (all $p < 0.001$). Again, this pattern is similar for both the adult and larval stages.

The toxicant treatment consistently *reduces* the extinction risk for both adults and larvae ($p < 0.001$ in 11 of 12 cases). This may seem counterintuitive since the simulated toxicant populations have lower abundances, and one could therefore expect higher extinction risk. However, this result must be seen in connection with the reduction

Table 2.5. Summary of results: effects of the different factors (density dependence type, stochasticity level, and toxicant) on different population-level endpoints related to risk assessment.

	Abundances		Variation		Explosion		Extinction	
Treatment	Adults	Larvae	Adults	Larvae	Adults	Larvae	Adults	Larvae
Density dependence in PS (vs. AF)	–	+	+	+	–	–/+[a]	+	–
Stochasticity level 1 (sP) vs. 0	0/–	0/–	0/+	0/+	+	0/+	+	0/+
Stochasticity level 2 ($sP + fA$) vs. 1	–/0	–/0	+/0	+/0	+/0	+	+/0	+
Toxicant (vs. control)	–/+	–	–	–	–	–	–	–

The symbols indicate positive (+), negative (–), or no effect (0). Two symbols per cell indicate an interaction with density dependence type (results for AF/PS model versions, respectively).
[a]Both positive and negative effects, but no consistent interaction with other factors.

in population fluctuations (as measured by CV). For the toxicant populations in this model, the reduction in fluctuations must have contributed positively to the population viability, in such a way that it outweighed the negative contribution from reduced vital rates and reduced average abundances.

Table 2.5 gives a qualitative summary of the impacts of all factors (density dependence, stochasticity, and toxicant) on the population-scale endpoints used in this analysis (abundances, variation, explosion risk, and extinction risk). Table 2.5 also gives an overview of consistent interactions between two factors (mainly between density dependence and stochasticity) and shows where the larval and adult stages are affected differently. Certain patterns can be noted to summarize the main results: (1) For some of the endpoints, larval versus adult stages are affected differently by the different types of density dependence, as well as by the different levels of stochasticity. (2) Higher stochasticity results in both higher explosion risk and higher extinction risk. (3) Toxicant exposure reduces extinction risk as well as explosion risk.

Discussion

In this study, we investigated how toxicant exposure in combination with density dependence and stochasticity affects the population viability of a pest species with fluctuating dynamics. Since the dynamics of this species is known to be driven by stage-specific demographic processes, we have simulated scenarios with different types of stage-specific density dependence and stochasticity and analyzed the dynamics of the larval and adult stages separately.

Our assumptions about stage-specific delayed effects of larval density (on pupal survival or adult fecundity, respectively) affected the risk assessment differently for the two main stages. A model that includes both effects of density dependence might be the most appropriate for this system (see Moe et al. 2005), but for our study here, separating the two types of delayed responses made it possible to analyze the effects of either type on population viability. The adult populations generally had better viability in the adult fecundity (AF) model versions, where reproduction is density dependent but recruitment from pupal to adult stage is unregulated, than in the pupal survival (PS) versions. The viability was better in terms of higher mean densities, higher explosion rates, and lower extinction rates. Conversely, the larval populations generally had better viability in the PS versions, where recruitment to the larval stage is unregulated. Yet, the average total population abundances were similar for both the AF and PS versions. This implies that a proper formulation of stage-specific density dependence is particularly important when one is interested in risk assessment for a specific stage rather than for the whole population. This should be relevant, for example, for management of insect pest species where one of the stages causes most of the damage.

Stochasticity generally increased population fluctuations in our model, and thus both explosion risks and extinction risks. There was an important interaction between stochasticity and density dependence: Stochasticity had much stronger impacts on population fluctuations when added to a demographic process that had been regulated by density dependence, than when added to an unregulated process.[2] As an implication, one should consider how interactions between stochasticity and density dependence in a population model affect the risk assessment. For example, if environmental stochasticity in reality has most importance for a density-dependent stage, but is modeled in an unregulated stage, then the real effects of stochasticity on extinction risk may be underestimated.

The toxicant exposure in our model generally reduced mean abundances and the explosion risks, as could be expected. However, toxicant exposure also reduced the extinction risks, which can be explained by reduced fluctuations (as measured by the CVs). Populations that reach extremely high abundances may subsequently fall down to very low abundance, if the density dependence is strong enough. The toxicant exposure prevents the extreme peaks and may therefore also prevent some of the lowest troughs. This has implications for management of pests with this kind of dynamics. Although pesticide exposure may prevent population explosions above some threshold abundance, it may also make reductions below a lower threshold less likely.

The importance of density dependence in ecotoxicology and risk assessment has been demonstrated by several experimental studies (reviewed in Forbes et al. 2001; Moe 2007), whereas the role of environmental stochasticity is more challenging to test experimentally. One possible approach is temporal variation of the food supply in combination with a stressor. A relevant study was performed with the soil mite *Sancassania berlesei* (Cameron and Benton 2004; Benton et al. 2004): Populations were reared in either constant or varying environments and were also subject to perturbations (stage-specific "harvesting"). The population-level responses depended on whether there was environmental variability and on which stage was manipulated, and the authors concluded that predicting such responses will require the incorporation of density dependence and environmental stochasticity in population models. Similar experiments could be performed with toxicant exposure instead of harvesting.

Our analysis of population viability for the sheep pest *Lucilia sericata* should generally have relevance for stage-structured species where density dependence in one

[2] In RAMAS, the stochasticity is added after density-dependent regulation. In models where stochasticity is added first, its effects might be more damped by the following density dependence.

stage negatively affects vital rates in later stages. The conclusions will be most relevant for opportunistic species with high fecundity, which is usually the case for pest species, and for species where adult fecundity is determined by growth in the juvenile feeding stage, which is the case for many insect species. The implications of these stage-specific analyses are particularly important for pest species where only one of the stages is being monitored (e.g., free-living adults) while another stage causes more of the damage (e.g., larvae living inside of host plants or animals).

In our case study, the toxicant was applied to both the larval and the adult stages. Experiments revealed that exposure of larvae had both direct effects during the larval stage and delayed effects in the pupal and adult stages (Moe et al. 2001). For simplicity, we have modeled only the direct toxicant effects here. A more realistic model could include also stage-specific toxicant exposure (which may be the case for a pesticide) and delayed toxicant effects. Such a model could give a better understanding of how stage-specific toxicant effects interact with stage-specific density dependence and stochasticity. Such knowledge may aid the design of efficient management programs for stage-structured pest species.

For natural blowfly populations, temperature is also an important predictor of developmental rates (Wall et al. 1993), which should be considered in field population models. In our current model, the stage durations are fixed, so that all individuals develop to the next stage or die at the end of the stage. In a field version of the model, the stage structure could be more flexible, and effects of temperature variations on developmental rates could be incorporated by allowing a proportion of the individuals in a stage to remain in this stage for one more time step (i.e., positive diagonal values in the stage matrix).

Since our case study is based on a laboratory system, the GIS components of RAMAS were not relevant for this analysis. Blowfly populations in nature, however, have a clearly patchy population structure: The larvae are confined to discrete "patches" (sheep) and have low ability for migration. Although adults can move freely among the "patches," females nevertheless tend to aggregate on certain patches for oviposition (Fenton et al. 1999; Cruickshank and Wall 2002). This behavior optimizes the survival of young larvae but may also increase the competition among older larvae (Hanski 1987; Moe et al. 2002b). A metapopulation approach should also be relevant for the purpose of pest management. Model simulations show that as the degree of larval aggregation increases, the number of sheep that are struck decreases. This suggests that higher heterogeneity in susceptibility within a sheep flock (or heterogeneity in habitat suitability, from the flies' point of view) restricts strikes to relatively few sheep (Fenton et al. 1999). Natural sheep blowfly populations should therefore be very suitable for metapopulation modeling in RAMAS, including, for example, stage-specific differences in migration rates and density dependence. To our knowledge, spatially explicit models are not yet developed for risk assessment for this group of pest species.

References

Akçakaya, H. R. 2005. *RAMAS GIS: Linking Spatial Data with Population Viability Analysis* (version 5.0). Applied Biomathematics, Setauket, New York.

Benton, T. G., Cameron, T. C., and Grant, A. 2004. Population responses to perturbations: Predictions and responses from laboratory mite populations. Journal of Animal Ecology 73(5): 983–995.

Bjørnstad, O. N., and Grenfell, B. T. 2001. Noisy clockwork: Time series analysis of population fluctuations in animals. Science 293: 638–643.

Cameron, T. C., and Benton, T. G. 2004. Stage-structured harvesting and its effects: An empirical investigation using soil mites. Journal of Animal Ecology 73(5): 996–1006.

Cruickshank, I., and Wall, R. 2002. Aggregation and habitat use by *Lucilia* blowflies (Diptera: Calliphoridae) in pasture. Bulletin of Entomological Research 92: 153–158.

Daniels, S. 1994. Effects of Cadmium Toxicity on Population Dynamics of the Blowfly *Lucilia sericata*. Ph.D. thesis, University of Reading, UK.

Evans, A. C. 1936. The physiology of the sheep blow-fly *Lucilia sericata* Meig. (Diptera). Transactions of the Royal Entomological Society of London 85:363–377.

Fenton, A., and Wall, R. 1997. Sensitivity analysis of a stochastic model for the sheep blow-fly *Lucilia sericata*. Journal of Applied Ecology 34: 1023–1031.

Fenton, A. Wall, R., and French, N. P. 1999. The effects of oviposition aggregation on the incidence of sheep blowfly strike. Veterinary Parasitology 83: 137–150.

Forbes, V. E., Sibly, R. M., and Calow, P. 2001. Toxicant impacts on density-limited populations: A critical review of theory, practice, and results. Ecological Applications 11: 1249–1257.

Grant, A. 1998. Population consequences of chronic toxicity: Incorporating density dependence into the analysis of life table response experiments. Ecological Modelling 105: 325–335.

Gurney, W. S. C., Blythe, S. P., and Nisbet, R. M. 1980. Nicholson's blowflies revisited. Nature 287: 17–21.

Hanski, I. 1987. Nutritional ecology of dung- and carrion-feeding insects. In *Nutritional Ecology of Insects, Mites, Spiders, and Related Invertebrates* (ed. F. Slansky and J. G. Rodriguez), pp. 837–884. Wiley, New York.

Kendall, B. E., Briggs, C. J., Murdoch, W. W., Turchin, P., Ellner, S. P., McCauley, E., Nisbet, R. M., and Wood, S. N. 1999. Why do populations cycle? A synthesis of statistical and mechanistic modeling approaches. Ecology 80: 1789–1805.

Laskowski R. 2000. Stochastic and density-dependent models in ecotoxicology. In *Demography in Ecotoxicology* (ed. J. Kammenga and R. Laskowski). Wiley, Chichester.

Lingjærde, O. C., Stenseth, N. C., Kristoffersen, A. B., Smith, R. H., Moe, S. J., Read, J. M., Daniels, S., and Simkiss, K. 2001. Exploring the density-dependent structure of blowfly populations by nonparametric additive modeling. Ecology 82(9): 2645–2658.

Maynard Smith, J. 1974. *Models in Ecology*. Cambridge University Press, London.

Moe, S. J. 2007. Density dependence in ecological risk assessment. In *Population-Level Ecological Risk Assessment* (ed. L. W. Barnthouse, W. R. Munns Jr. and M. T. Sorensen). CRC Press, Boca Raton, FL.

Moe, S. J., Stenseth, N. C., and Smith, R. H. 2001. Effects of a toxicant on population growth rates: Sublethal and delayed responses in blowfly populations. Functional Ecology 15: 712–721.

Moe, S. J., Stenseth, N. C., and Smith, R. H. 2002a. Density-dependent compensations in blowfly populations give indirectly positive effects of a toxicant. Ecology 83(6): 1597–1603.

Moe, S. J., Stenseth, N. C., and Smith, R. H. 2002b. Stage-specific density dependence in blowfly populations: Experimental assessment of estimates from non-parametric time series modelling. Oikos 98: 523–533.

Moe, S. J., Kristoffersen, A. B., Smith, R. H., and Stenseth, N. C. 2005. From patterns to processes and back: Analysing density-dependent responses to an abiotic stressor by statistical and mechanistic modelling. Proceedings of the Royal Society Series B 272(1577): 2133–2142.

Nicholson, A. J. 1950. Population oscillations caused by competition for food. Nature 165: 476–477.

Nicholson, A. J. 1954. An outline of the dynamics of animal populations. Australian Journal of Zoology 2: 9–65.

Readshaw, J. L., and Cuff, W. R. 1980. A model of Nicholson's blowflies cycles and its relevance to predation theory. Journal of Animal Ecology 49: 1005–1010.

Rognes, K. 1991. Blowflies (Diptera, Calliphoridae) of Fennoscandia and Denmark. In *Fauna entomologica Scandinavica*. Brill, Leiden.

Simkiss, K., Daniels, S., and Smith, R. H. 1993. Effects of population density and cadmium toxicity on growth and survival of blowflies. Environmental Pollution 81: 41–45.

Smith, R. H., Daniels, S., Simkiss, K., Bell, E. D., Ellner, S. P., and Forrest, M. B. 2000. Blowflies as a case study in non-linear population dynamics. In *Chaos in Real Data: The Analysis of Non-linear Dynamics in Short Ecological Time Series* (ed. J. N. Perry, R. H. Smith, I. P. Woiwod, and D. Morse), pp. 137–172. Kluwer Academic Publishers, Dordrecht.

Ullyett, G. C. 1950. Competition for food and allied phenomena in sheep-blowfly populations. Philosophical Transactions of the Royal Society of London, Series B 234: 77–174.

Wall, R. 1993. The reproductive output of the blowfly *Lucilia sericata*. Journal of Insect Physiology 39: 743–750.

Wall, R., French, N. P., and Morgan, K. L. 1993. Predicting the abundance of the blowfly *Lucilia sericata* (Diptera, Calliphoridae). Bulletin of Entomological Research 83: 431–436.

Wood, S. N. 2001. Partially specified ecological models. Ecological Monographs 71: 1–25.

3

Population-Level Modeling of Mercury Stress in the Florida Panther (*Puma concolor coryi*) Metapopulation

SANDY RAIMONDO
MACE G. BARRON

The Florida panther (*Puma concolor coryi*) is an endangered species that has been the focus of a national conservation effort since 1967, when it was listed as a federally endangered species (U.S. Fish and Wildlife Service 1999). The Florida panther's geographic distribution, which once extended throughout the southeastern United States, is now limited to a small area in southern Florida. A census taken in 2003 indicated that the entire Florida panther population may consist of fewer than 90 individuals (McBride 2003), indicating increased need for effective conservation and restoration efforts. The primary concerns of panther conservation have been decreasing habitat and genetic depression associated with low population abundance. Conservation efforts have focused on preserving and expanding suitable habitat and increasing genetic diversity within the population (U.S. Fish and Wildlife Service 1999; Beier et al. 2003). Florida panthers have been radio tracked since 1981, and several metapopulation models have been developed for conservation and land management plans (Root 2004; Maehr et al. 2002a; Kautz et al. 2006). However, an additional threat to the vitality of panther populations is the increased risk of anthropogenic stressors, such as contaminant exposure, which may be exacerbated by the geographic isolation of the populations in southern Florida.

Population-level assessments using spatially structured models have had relatively limited application in ecological risk assessment and have generally not been used to assess contaminant risks of special status species (e.g., Beissinger and Westphal 1998). In the Everglades of southern Florida, high mercury (Hg) contamination has been documented during the last two to three decades (Frederick et al. 2004; Porcella et al. 2004), and elevated methyl mercury (MeHg) levels have been documented in panthers (Newman et al. 2004; Land et al. 2004). Screening-level probabilistic risk assessments of Hg in Everglades food webs concluded that historical Hg exposures posed significant risks to piscivorous wildlife and to panthers, but current risks of Hg were low (Duvall and Barron 2000; Barron et al. 2004). The spatial distribution of Hg risks in southern Florida likely reflects the heterogeneous distribution of environmental Hg, as well as different feeding ecologies of panthers in the Everglades, Big Cyprus National Preserve (BCNP), and northern areas (Maehr et al. 1990; Roelke et al. 1991). For example, Roelke (1991) noted a "health cline" that separated panthers in the northern part of their range from the Everglades population. The decrease in general fitness of Everglades panthers was thought to be associated with a shift in their diet from primarily deer to increased reliance on raccoons and armadillos, which increased panther exposure to MeHg via the food chain (Roelke 1991).

The objective of this research was to assess population-level impacts and metapopulation extinction probability under various scenarios of Hg exposure in Florida panthers. A female-only, spatially explicit, stochastic metapopulation model was developed in RAMAS GIS using the most recently estimated demographic parameters obtained from more than 20 years of tracking data. The metapopulation was composed of four distinct populations that represent the current panther distribution in southern Florida. Several Hg exposure scenarios were developed and linked to reductions in either reproduction or reproduction and survival in the Everglades and BCNP populations. Exposure simulations projected the dynamics of both individual populations, as well as the metapopulation, and were used to develop a probabilistic population-level assessment of Hg stress on Florida panthers.

Methods

Study species and area

The Florida panther is a medium-sized subspecies of the mountain lion, *Puma concolor.* Male panthers may weigh as much as 68 kg, whereas females average 34 kg. Panthers are solitary animals with large home ranges (male, 435–650 km^2; females, 193–396 km^2) (Beier et al. 2003; Comiskey et al. 2002) that overlap to some degree. The home range of resident males typically encompasses the home ranges of several breeding females (Comiskey et al. 2002). The current distribution of the Florida panther is restricted to southern Florida, particularly the Big Cypress National Preserve (BCNP) and Everglades National Park (ENP) areas. These areas encompass a mosaic of habitat types, including cypress swamps, hardwood hammocks, pine flatwoods, seasonally flooded prairies, and freshwater marshes. Panthers inhabit virtually all available natural habitats with the exception of coastal mangrove forests (Comiskey et al. 2002) and are the top predator in these ecosystems, preying on deer, wild hogs, marsh rabbits, and raccoons (Beier et al. 2003).

In the 1980s, the Florida panther population was believed to be as low as 30–50 individuals (Comiskey et al. 2002). Low population abundance was attributed to reduced habitat area and habitat quality, and low genetic diversity associated with the small population size. Florida panthers possessed a high frequency of crooked tails, cowlicks, and cryptorchidism, traits associated with inbreeding depression that are not observed in other subspecies (for review, see Beier et al. 2003). In 1995, a genetic restoration project was initiated to address the inbreeding depression by introducing eight Texas panthers (*P. c. stanleyana*) into southern Florida. By 2003, the verified population consisted of 87 panthers, approximately tripling the pre-introgression population (McBride 2003). Descendants of Texas panthers possessed higher survival rates and lower incidences of genetic anomalies than did offspring of pure Florida panther descent (Land et al. 2004).

Florida panthers have been tracked using radio telemetry collars since 1981. As of 2004, 132 panthers had been radiocollared since the inception of the study, and data have been collected on more than 55,000 locations and 169 neonate kittens in den sites (Beier et al. 2003; Land et al. 2004). Telemetry data have been used to estimate survival rates, fecundity, dispersal, and home ranges of panthers (Land et al. 2004) and develop population viability models (Maehr et al. 2002a; Root 2004) and land management models (Comiskey et al. 2004; Meegan and Maehr 2002; Kautz et al. 2006) for use in panther conservation.

Metapopulation model

Historically, vegetation maps have been overlaid with panther tracking data to determine optimal panther habitat and dispersal patterns (Maehr and Cox 1995). The conclusion of Maehr and Cox (1995) that panthers prefer and depend on forested habitats has been relied on heavily to develop previous spatially explicit models applied in panther conservation (Maehr and Deason 2002; Maehr et al. 2002a; Meegan and Maehr 2002; Root 2004). The reliability of these habitat models has become a matter of scientific debate (Shrader-Frechette 2004; Beier et al. 2003; Comiskey et al. 2002, 2004), and more recent field reports have emphasized the importance to panthers of the mosaic of habitats within expansive tracts of undisturbed land large enough to support the panthers' large home ranges (McBride 2000, 2001, 2002). Comiskey et al. (2002) identified four areas of undeveloped land that supported panther populations in southern Florida: (1) east of State Road 29, (2) north of Interstate 75, (3) BCNP south of I-75, and (4) ENP. These populations are separated by natural and man-made barriers that impede panther dispersal to varying degrees.

To develop a metapopulation model of the Florida panther based on the four centers of activity identified in Comiskey et al. (2002), we used three habitat layers to delineate optimal panther habitat. The first layer consisted of protected and undeveloped lands, including national and state parks, wildlife preserves, and the Seminole Indian reservation. Distribution of historical panther den sites shows federal, state, and tribal lands containing 95% of den locations (figure 3.1). Most private lands were excluded from this layer because development in southern Florida continues to expand (Florida Legislature Office of Economic and Demographic Research, http://edr.state.fl.us/population.htm, accessed March 2006), resulting in questionable reliability of these areas as long-term panther habitat. The exception to this was a small tract of private land connecting BCNP and Okaloacoochee Slough State Forest. This tract contained a number of historic den locations and is considered by the Florida panther sub-team of the U.S. Fish and Wildlife Service Multi-species Ecosystem Recovery Implementation Team (MERIT) as part of the population located north of I-75 (Jane Comiskey, personal communication). The second layer consisted of major highways and roads that fragment the protected lands with a 100-m buffer around each road. Female panthers

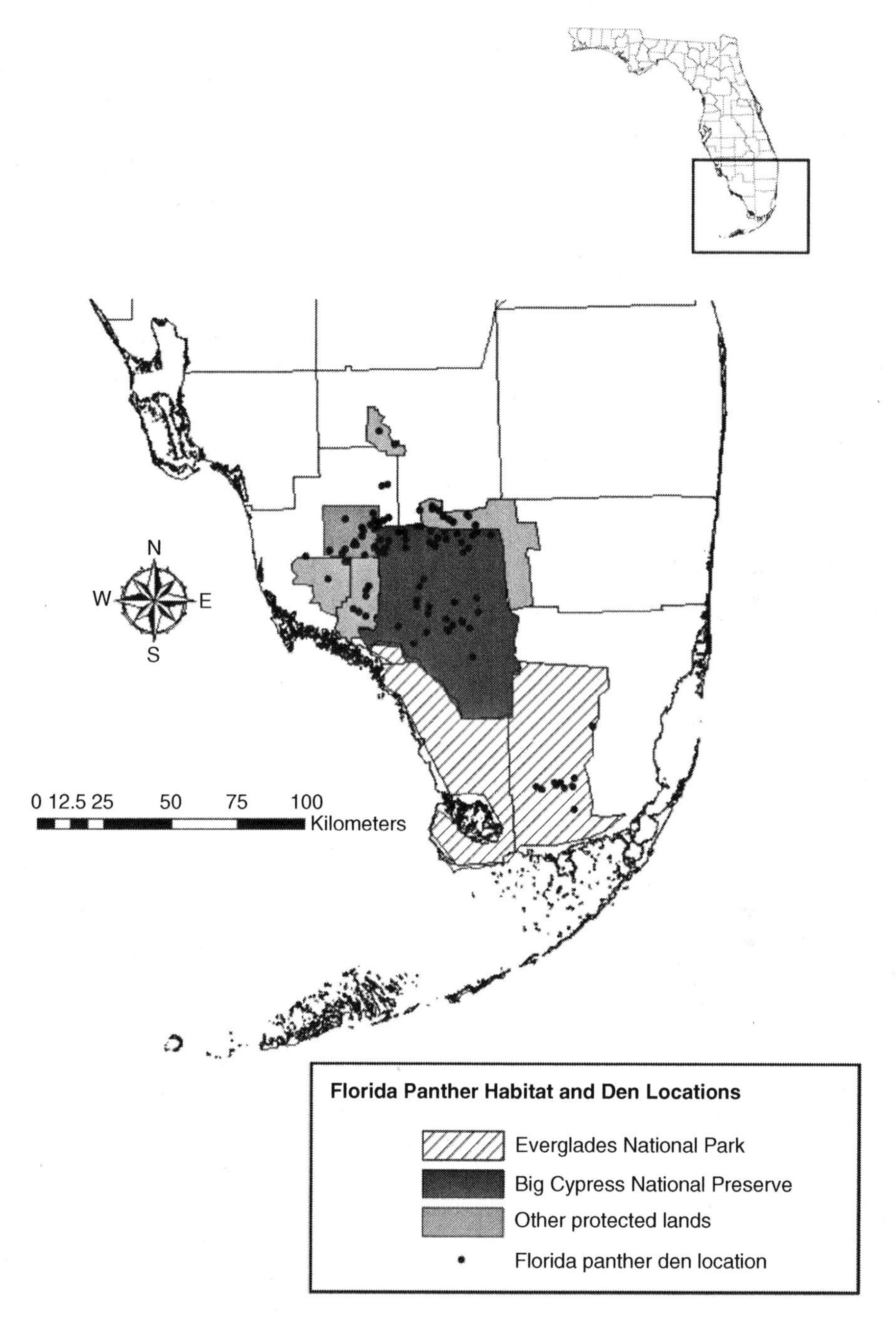

Figure 3.1. Historical den locations and protected lands representing Florida panther (*Puma concolor coryi*) distribution in southern Florida.

are reluctant to cross major roads, which act as barriers to movement and dispersal, and female offspring generally establish home ranges that overlap the mother's range (Comiskey et al. 2002). Only 2% of historical den locations are within 100 m of major roads. The third layer was developed to exclude areas of open water (e.g., Shark River Slough) that separates the ENP from the BCNP. Although tracking data have shown that panthers will cross water channels such as the Shark River Slough at low water levels, deep-water areas provide unsuitable den sites and cannot support a reproducing population. This layer was developed by overlaying aerial photography from the South Florida Water Management District with U.S. Geological Survey elevation maps in ArcGIS, which determined that areas of open water correlated with elevations lower than 1 m. Using the three landscape feature layers, we delineated discrete populations using the Spatial Data program of RAMAS GIS (Akçakaya 2005). The habitat suitability function required a habitat patch to consist of protected lands not within 100 m of a major road or highway, and at an elevation greater than 1 m. Using these requirements, we reproduced the four populations that represent the current panther metapopulation.

Density dependence is an important component of Florida panther demography and affects population size through limited habitat and food availability (Beier et al. 2003). Most recent analysis of panther tracking data estimated a density of one panther per 129 km^2 (Kautz et al. 2006). Using this density, a ceiling-type density dependence based on adult abundance was set in RAMAS GIS to affect survival rates only. Dispersal among disjunct populations is hindered by natural and man-made landscape features. In our model, the two features that separate panther populations are areas of open water and major roads. Male panthers have been noted to cross these barriers more frequently than do females and have average dispersal distances of 40–68 km from the natal range (Maehr et al. 2002b; Comiskey et al. 2002). Female panthers are reluctant to cross major roads and have not been document crossing the Shark River Slough that separates BCNP and ENP (Comiskey et al. 2002). Young female panthers develop home ranges that overlap their natal range and may inherit them as their mother moves to another part of her range; thus, juveniles may either lack any "dispersal event" (Comiskey et al. 2002) or disperse as little as 20 km from their natal range (Maehr et al. 2002b). For our female-based model, we therefore set dispersal as a function of density dependence and limited the range that females can disperse to 20 km.

A three-stage female-based matrix model was developed for the panther populations based on Land et al. (2004), which is the most recent analysis of 23 years of panther tracking data (table 3.1). The life stages were (1) kittens, 0–6 months old; (2) juveniles, 6–12 months old; and (3) adults, >12 months old. Although panthers of Texas descent had higher survival and birth rates immediately following the introgression, the lineage of recently born panthers is becoming increasingly more difficult to determine as the southern Florida population increases (Land et al. 2004). For model development, the number of kittens per female and kitten survival were estimated as an average determined from panthers of Florida and Texas descent (Land et al. 2004). To remain consistent with the duration of the first life stage, the time step for the model was six months. Females become reproductive as young as 18 months of age and produce an average of 2.5 kittens on average every 19 (± 9) months (Land et al. 2004). The interbreeding period was rounded down to 18 months to simplify calculations associated with adjusting reproduction to reflect the average number of offspring in a time step. Reproduction was adjusted to represent the average number of female

Table 3.1. Demographic parameters of female Florida panther (*Puma concolor coryi*) used for the metapopulation model developed in RAMAS GIS based on Land et al. (2004).

Parameter	Parameter value	Standard deviation
Kitten survival—0 to 6 months	0.62	0.12
Juvenile survival—6 months to adult	0.982	NA
Adult female survival (annual)	0.895	0.103
Average litter size (18 months)	2.5	0.92
Average female offspring (6 months)[a]	0.42	0.15
Litter sex ratio	1:1	NA
Birth interval	19.8	9.0
Carrying capacity	1/129 km^2[b]	NA

[a] Adjusted; see text for details.
[b] Kautz et al. (2006).

offspring produced at each time step, assuming a 1:1 sex ratio. Reproduction for the female-based model was estimated as 2.5 kittens/2 = 1.25 female offspring every 18 months, 1.25 female offspring/3 time steps = 0.42 female offspring every six months. The square root of annual adult survival was used to estimate six-month time survival, and model fecundity was the product of reproduction rate and adult survival, yielding the following matrix model, A:

$$A = \begin{pmatrix} 0 & 0 & 0.397 \\ 0.620 & 0 & 0 \\ 0 & 0.982 & 0.946 \end{pmatrix}$$

To account for model variation, standard deviations of the vital rates estimated by Land et al. (2004) were used as demographic stochasticity. Standard deviation of fecundity was estimated as the variance of the product of reproduction and survival variance, assuming that both variables are independent (Goodman 1960). The model had the following assumptions that were necessary to isolate the effects of Hg stress on panthers: (1) Vital rates were stable throughout the simulations, and there is no further genetic depression that may alter vital rates; (2) land management did not alter the available habitat during the simulations; and (3) there were no introductions of extrapopulation panthers during simulations.

Mercury exposure simulations

An initial simulation was run to determine population viability under baseline (unaltered) conditions. Baseline simulations began with all populations near carrying capacity with a stable age distribution and had no catastrophes or epidemics. Simulations ran for 200 time steps (100 years) for 1,000 iterations. Reproduction and survival effects of Hg stress were simulated in RAMAS GIS by multiplying fecundity and survivorship by 1—(% impairment/100). For simulations that determined metapopulation viability in the complete absence of the ENP population, BCNP population, and both

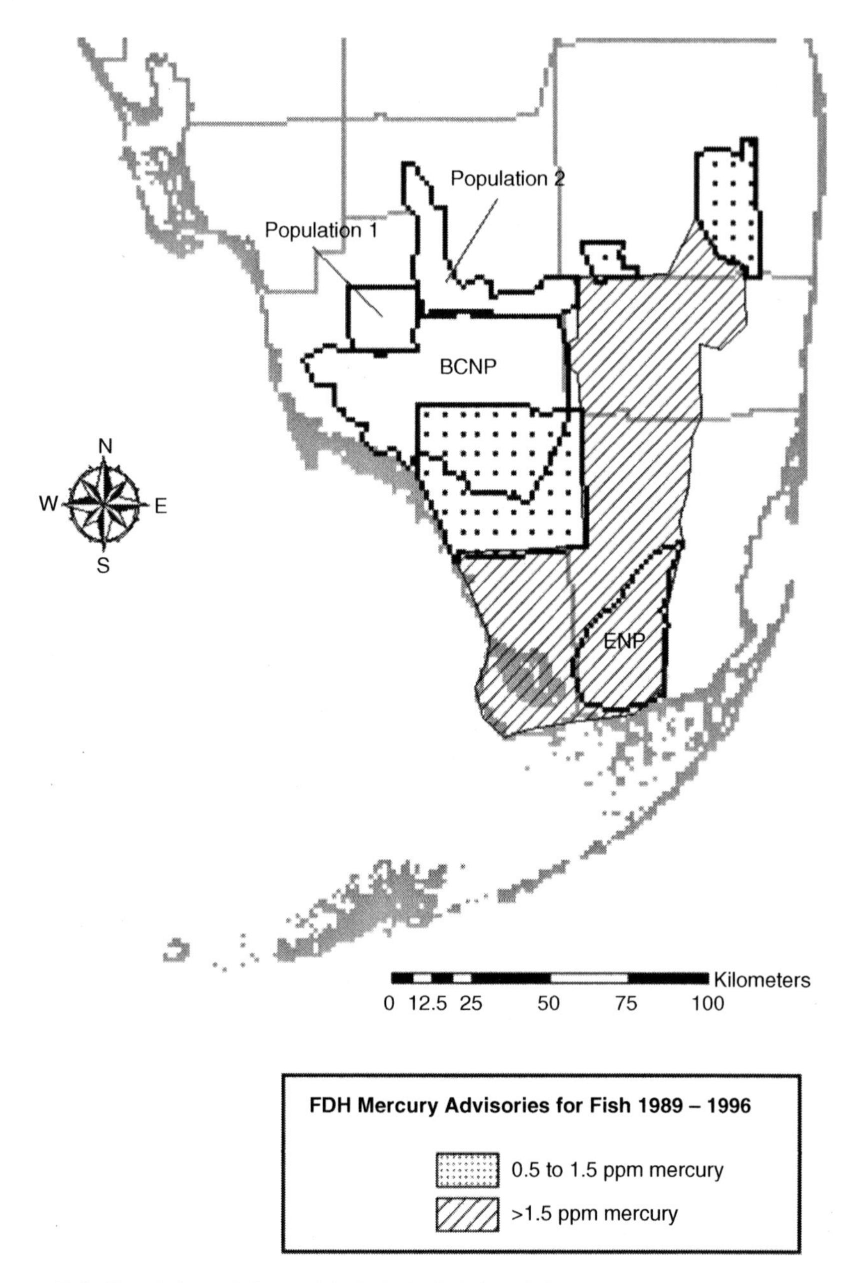

Figure 3.2. Populations delineated in RAMAS GIS and fish mercury advisories in southern Florida. Mercury advisories are adapted from the Florida Department of Health (FDH; http://sofia.usgs.gov/sfrsf/rooms/mercury/cht7.html, accessed November 2005).

ENP and BCNP populations, relative survival and fecundity were adjusted to 0 for the population(s) removed in each simulation.

Mercury exposure was modeled as spatially explicit impairments in reproduction and survival because chronic dietary MeHg exposure is known to increase the incidence of abortions and fetal abnormalities, neurological impairments, ataxia, convulsions, and death in toxicity studies with domestic cats (Khera 1973; Charbonneau et al. 1976). The spatial distribution of Hg within the range of each panther population was determined from general patterns of current and historical levels of Hg in southern Florida. Historically, ENP has had higher Hg levels due to its proximity to industrial, metropolitan, and agricultural areas, including the Everglades Agricultural Area (Beier et al. 2003). The spatial distribution of Hg in the panther's food web was modeled according to the pattern of fish consumption advisories established by the Florida Department of Health. These patterns include the highest level of bioaccumulated Hg surrounding ENP and moderate levels of Hg within the southern portions of BCNP (figure 3.2).

We simulated four scenarios of chronic Hg exposure to panthers in the southern Florida metapopulation. In the first two scenarios, Hg affected the ENP population only. In the first scenario, reproduction in the ENP population was systematically reduced to determine a population-level dose response of reduced fecundity on the viability of both the ENP population and the metapopulation. Similarly, the second scenario comprised systematic reductions in survival of ENP panthers to develop a dose response of reduced survival on ENP and metapopulation viability. In both scenarios, the probability of metapopulation extinction and the abundance of adult panthers in the ENP population were determined. Comparison of ENP population dynamics as part of the metapopulation and in the absence of the other three populations (i.e., relative survival and reproduction were set to 0 for populations 1, 2, and BCNP) showed identical dynamics of the ENP population under both scenarios. Thus, we removed the three northern populations from a separate set of simulations to obtain the probability of extinction for the ENP population at the levels of reduced of reproduction and survival used to generate the dose–response curves. The third scenario involved exposure of the BCNP population to determine how impairments to BCNP survival and reproduction affected metapopulation viability. In that scenario, we simulated low-level Hg stress on the BCNP population by reducing reproduction by only 25% and high-level stress by reducing both reproduction and survival by 50%. The fourth scenario included impacts to both ENP and BCNP populations to determine how metapopulation dynamics are affected by high levels of environmental MeHg that result in the local extinction of both populations. For this simulation, survival and relative reproduction of both ENP and BCNP populations were set to 0.

Results

Baseline model dynamics

RAMAS GIS delineated four populations representing the four centers of activity that comprise the current Florida panther metapopulation (figure 3.2). Populations 1 and 2 correspond to the areas east of State Road 29 and north of I-75, respectively, of

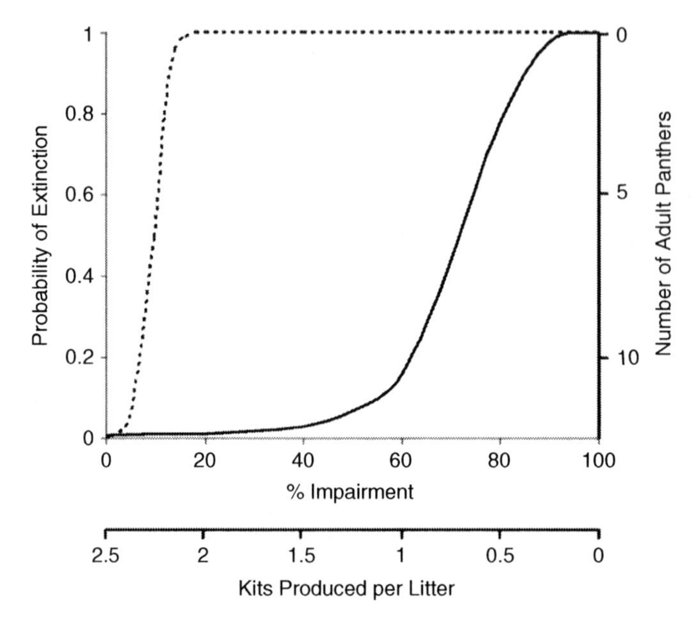

Figure 3.3. Dose–response curves of effects of impaired survival (dashed curve) and reproduction (solid curve) on ENP population abundance and probability of extinction.

Comiskey et al. (2002). BCNP represents the area of BCNP south of I-75, and ENP represents the southern area of ENP supporting a reproducing panther population. The area represented by our metapopulation was approximately 9,800 km^2, which is comparable to an estimated 10,000 km^2 of panther habitat currently recognized by the MERIT panther sub-team (Kautz et al. 2006). Based on the area of the four populations and a density of one panther per 129 km^2, the carrying capacity of the metapopulation modeled in RAMAS GIS was approximately 76 panthers, which is consistent with the current estimate of fewer than 90 panthers in this area (McBride 2003). Although the population estimates of McBride (2001, 2002, 2003) exceeded our estimate of carrying capacity, tracking data estimates of panther abundance included panthers on private lands and transient males that migrated to areas north of the Caloosahatchee River, which is considered a sink area due to lack of den sites. Simulations of panther dynamics under baseline (unaltered) conditions showed that all populations were stable for the duration of the simulations and that the probability of metapopulation extinction was less than 0.001. Population growth rate of the stage-structured matrix model under baseline conditions was 1.14, indicating that the population will increase at abundances less than carrying capacity.

Mercury stress

The dose response for the ENP population showed that at the probability of extinction had a sharp increase at approximately 60% reproductive impairment. At all levels of reproductive impairment of the ENP population, metapopulation risk was negligible (probability of extinction <0.01), and there was no change in the dynamics of

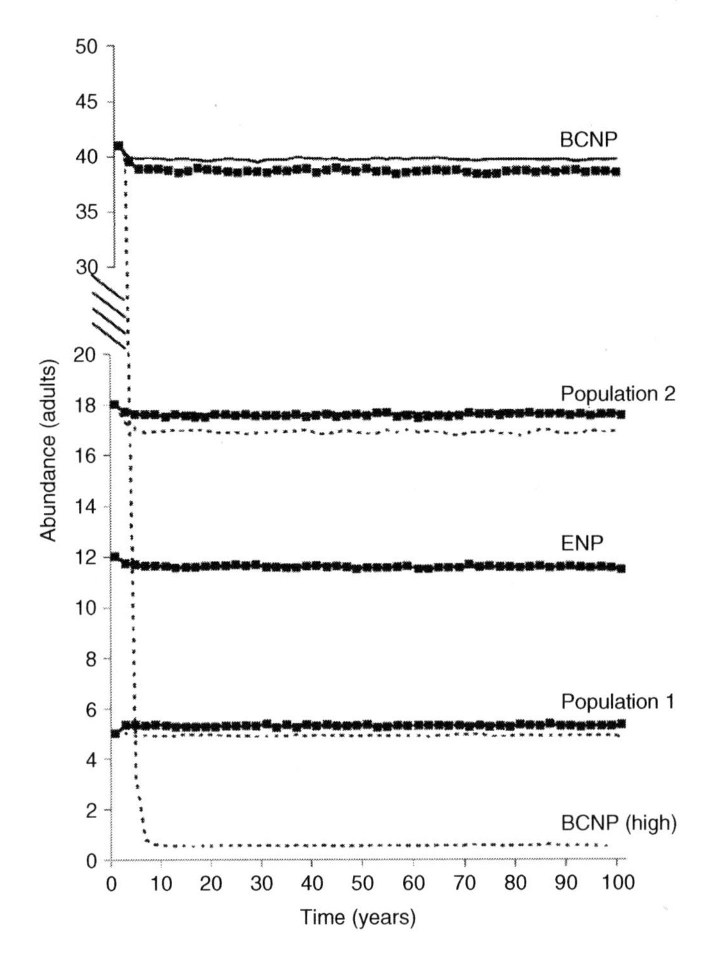

Figure 3.4. Times series of populations 1, 2, and ENP when abundance of BCNP population is affected by Hg stress. Solid lines represent baseline dynamics of each population, lines with square data points represent low Hg exposure (25% reduction in reproduction), and dashed lines represent the dynamics of each population in the presence of high Hg stress (50% reduction of survival and reproduction) in BCNP. Solid lines for populations 1, 2 and ENP are not visible under lines with square data points.

the other populations. The dose–response curve for ENP survival showed that at low levels of impairment of survival (<10%), the probability of extinction of the population was high, and at a 20% reduction in survival, the population had a 100% chance of extinction with the 100-year simulation (figure 3.3). In simulations where the ENP population rapidly became extinct (i.e., >50% reduction in survival), the risk of metapopulation extinction remained small (<0.01). In the simulations of Hg stress on the BCNP population, a 25% reduction of reproduction had negligible effects on the other three populations and reduced the BCNP population average by approximately two adult panthers (figure 3.4). Simulated high Hg stress on the BCNP population resulted in that population rapidly becoming extinct and lowered abundance of panthers in populations 1 and 2. High simulated Hg stress in BCNP did not, however, affect the ENP population (figure 3.4). Reducing vital rates of the BCNP population did not affect the probability of extinction of the metapopulation, which remained below 0.001 for all simulations. Similarly, reduced abundance in both ENP and BCNP populations, including extinction of both populations, only slightly lowered panther abundance in the remaining two populations, and the extinction probability of the metapopulation remained below 0.001 (data not shown).

Discussion

The baseline metapopulation model developed in RAMAS GIS provided a spatially explicit approach to assess Florida panther population viability that incorporated recent research on panther distribution and reproduction in southern Florida. The baseline model was consistent with field observations of the spatial distribution of adult panthers, den locations, fecundity, and dispersal patterns. Simulations of this baseline model over 100 years showed a high probability of panther persistence in each of the four populations inhabiting southern Florida under unstressed conditions. The baseline model allowed an assessment of Hg stress on population dynamics and extinction risk in an endangered large mammal with low abundance over a mosaic of habitats.

Simulation modeling of Hg stress on the ENP population predicted that even at relatively low Hg levels, the panther population in ENP would be at risk of extinction from impairments resulting in either reduced reproduction or reduced survival. The reduction and extinction of the ENP population did not affect the viability of the remaining three populations or the metapopulation. While the ENP population appeared independent of the other three populations based on the female-only model used in these simulations, it is linked to the metapopulation through the migration of males to and from the northern populations (Land et al. 2004). Simulation of Hg stress on the ENP population with two-sex matrix models may show a slight impact on the three northern populations by reducing the number of males dispersing from the ENP population. However, given the isolation of ENP and the relatively small number of males dispersing between these areas, this difference may be negligible.

Simulated reductions in survival and reproduction of the BCNP population resulted in minor decreases to the other populations but did not increase the extinction risk of the metapopulation. The close proximity of the three northern populations to each other resulted in more dispersing females between BCNP and the two northern populations. Thus, a greater impact on populations 1 and 2 occurred from reductions in BCNP abundance than from reductions in ENP abundance. Model simulations showed that in the absence of both the ENP and BCNP populations, the remaining two populations would remain viable and metapopulation extinction risk remained low (<0.001). This model prediction is in disagreement with the Kautz et al. (2006) population viability analysis, which concluded that a panther population size of less than 50 had a high probability of extinction in less than 100 years. This disparity was likely because our Hg stress model assumed that vital rates remain constant throughout the simulation and did not incorporate the likely effects of inbreeding at low population sizes.

Mercury stress was modeled using a range of reproductive and survival impairments to account for variable impacts on individual panthers caused by differences in dietary exposure and a spectrum of potential adverse effects. For example, chronic dietary Hg exposure in feline species has been demonstrated to cause a diversity of adverse effects, including an increased incidence of abortions and fetal abnormalities, neurological impairments, ataxia, convulsions, and death (Khera 1973; Charbonneau et al. 1976). Although individual female panthers with high Hg concentrations in blood and hair samples were reported to have lower reproductive success (Roelke 1991), there were inadequate data to link specific dietary concentrations of Hg to specified levels of impaired survival and reproduction in wild panthers. Instead, modeled Hg stress percentage was linked to specified reductions in survival and/or reproduction based on vital rates and Hg toxicosis. For example, a 40% reduction in fecundity equated to one less kitten per female every breeding event (18 months).

The baseline metapopulation model assumed that no land management changes altered the available habitat during the 100-year simulation period, either in number of populations or in patch size of the current populations, and that no extrapopulation panthers were introduced. The U.S. Fish and Wildlife recovery plan for the Florida panther involves range expansion and reintroduction of panthers to establish three viable populations (>240 animals) within the historic range of the subspecies, one of which should occur in southern Florida (U.S. Fish and Wildlife Service 2006). Within this proposal is the expansion of the southern Florida population modeled here to areas north of the Caloosahatchee River, currently considered a population sink. Success of land management plans would likely reduce the risk of Hg exposure on panther populations by increasing dispersal among the isolated ENP population and the remaining populations. Increased population tolerance to stress may also be achieved through increases in either the number of populations in the metapopulation or the carrying capacity of existing populations.

The primary concerns in the conservation of Florida panthers have been decreasing habitat and genetic depression associated with low population abundance (U.S. Fish and Wildlife Service 1999; Beier et al. 2003). Past assessments have also implicated Hg contamination in the Everglades as a causative factor in panther deaths, lower reproductive success, and elevated ecological risks (e.g., Roelke 1991; Barron et al. 2004). Although no definitive link can be made between Hg levels in panther tissue and survival and reproduction, tissue analysis of recently necropsied panthers (Land et al. 2004) found Hg levels in some panthers at levels reported by Roelke (1991) in dead panthers. Other recent reports also indicate high levels of Hg in the Everglades food web (e.g., Frederick et al. 2004; Porcella et al. 2004). Mercury risks to other wildlife species in the Everglades is location dependent (Duvall and Barron 2000), and our present findings indicate that Hg exposure and potential impacts on panther populations are also spatially heterogeneous. Simulations of Hg stress indicate that the isolated population of Everglades panthers would experience the greatest impacts from Hg exposure because of food web contamination but that the current panther metapopulation is unlikely to be affected.

Acknowledgments We thank Jane Comiskey for advice, assistance, and providing technical information; Anthony DiGirolamo and Yvonne Allen for assistance with maps; and Steve Walters for peer review. This chapter is contribution no. 1260 of the Gulf Ecology Division of the U.S. Environmental Protection Agency Office of Research and Development's National

Health and Environmental Effects Research Laboratory. The manuscript has been reviewed by the U.S. EPA's National Health and Environmental Effects Research Laboratory, Gulf Breeze, Florida. The information in this document does not necessarily reflect the views and policies of the U.S. EPA. Citation of products does not imply U.S. EPA endorsement.

References

Akçakaya, H. R. 2005. *RAMAS GIS: Linking Spatial Data with Population Viability Analysis* (version 5.0). Applied Biomathematics, Setauket, New York.

Barron, M. G., Duvall, S. E., and Barron, K. J. 2004. Retrospective and current risks of mercury to panthers in the Florida Everglades. *Ecotoxicology* 13: 223–229.

Beier, P., Vaughan, M. R., Conroy, M J., and Quigley, H. 2003. An Analysis of Scientific Literature Related to the Florida Panther. Final Report. Florida Fish and Wildlife Conservation Commission, Tallahassee, Florida.

Beissinger, S. R., and Westphal, M. I. 1998. On the use of demographic models of population viability in endangered species management. *Journal of Wildlife Management* 62: 821–841.

Charbonneau, S. M., Munro, I. C., Nera, E. A., Armstrong, F. A. J., Willes, R. F., Bryce, F., and Nelson R. F. 1976. Chronic toxicity of methylmercury in the adult cat: Interim report. *Toxicology* 5: 337–349.

Comiskey, E. J., Bass, O. L., Jr., Gross, L. J., McBride, R. T., and Salinas, R. 2002. Panthers and forests in south Florida: An ecological perspective. *Conservation Ecology* 6: 18. Available: www.consecol.org/vol6/iss1/art18.

Comiskey, E. J., Eller, A. C., Jr., and Perkins, D. W. 2004. Evaluating impacts to Florida panther habitat: How porous is the umbrella? *Southeastern Naturalist* 3: 51–74.

Duvall, S. E., and Barron, M. G. 2000. A screening-level probabilistic ecological risk assessment of mercury in Florida Everglades food webs. *Ecotoxicology and Environmental Safety* 47: 298–305.

Florida Legislature Office of Economic and Demographic Research. 2006. http://edr.state.fl.us/population.htm.

Frederick, P. C., Hylton, B., Heath, J. A., and Spalding, M. G. 2004. A historical record of mercury contamination in southern Florida (USA) as inferred from avian feather tissue. *Environmental Toxicology and Chemistry* 23: 1474–1478.

Goodman, L. A. 1960. On the exact variance of products. *Journal of the American Statistical Association* 55: 708–713.

Kautz, R., Kawula, R., Hoctor, T., Comiskey, J., Cansen, D., Jennings, D., Kasbohm, J., Mazzotti, F., McBride, R., Richardson, L., and Root, K. 2006. How much is enough? Landscape-scale conservation for the Florida panther. *Biological Conservation* 130: 118–133.

Khera, K. S. 1973. Teratogenic effects of methylmercury in the cat: Note on the use of this species as a model for teratogenicity. *Teratology* 8: 293–304.

Land, D., Cunningham, M., Lotz, M., and Shindle, D. 2004. Florida Panther Genetic Restoration and Management Annual Report 2003–2004. Florida Fish and Wildlife Conservation Commission, Tallahassee, Florida.

Maehr, D. S., and Cox, J. A. 1995. Landscape features and panthers in Florida. *Conservation Biology* 9: 1008–1019.

Maehr, D. S., and Deason, J. P. 2002. Wide ranging carnivores and development permits: Constructing a multi-scale model to evaluate impacts on the Florida panther. *Clean Technologies and Environmental Policy* 3: 398–406.

Maehr, D. S., Belden, R. C., Land, E. D., and Wilkins, L. 1990. Food habits of panthers in southwest Florida. *Journal of Wildlife Management* 54: 420–423.

Maehr, D. S., Lacy, R. C., Land, E. D., Bass O. L., Jr., and Hoctor, T. S. 2002a. Evolution of population viability assessments for the Florida panther: A multi-perspective approach. Pages 284–311 in S. R. Beissinger and D. R. McCullough (eds.). *Population Viability Analysis*. University of Chicago Press, Chicago, Illinois.

Maehr, D. S., Land, E. D., Shindle, D. B., Bass, O. L., and Hoctor, T. S. 2002b. Florida panther dispersal and conservation. *Biological Conservation* 106: 187–197.

McBride, R. T. 2000. Current Panther Distribution and Habitat Use: A Review of Field Notes, Fall 1999–Winter 2000. Report to Florida Panther Subteam of MERIT, U.S. Fish and Wildlife Service, South Florida Ecosystem Office, Vero Beach, Florida.

McBride, R. T. 2001. Current Panther Distribution and Habitat Use: A Review of Field Notes, Fall 2000–Winter 2001. Report to Florida Panther Subteam of MERIT, U.S. Fish and Wildlife Service, South Florida Ecosystem Office, Vero Beach, Florida.

McBride, R. T. 2002. Current Panther Distribution and Habitat Use: A Review of Field Notes, Fall 2001–Winter 2002. Report to Florida Panther Subteam of MERIT, U.S. Fish and Wildlife Service, South Florida Ecosystem Office, Vero Beach, Florida.

McBride, R. T. 2003. Documented panther population and its current distribution. Pages 63–73 in D. Shindle, M. Cunningham, D. Land, R. McBride, M. Lotz, and B. Ferree. Florida Panther Genetic Restoration Annual Report 2002–2003. Florida Fish and Wildlife Conservation Commission, Tallahassee, Florida. Available: www.panther.state.fl.us/news/pdf/FWC2002–2003PantherGeneticRestorationAnnualReport.pdf.

Meegan, R. P., and Maehr, D. S. 2002. Landscape conservation and regional planning for the Florida panther. *Southeastern Naturalist* 1: 217–232.

Newman, J., Zillioux, E., Rich, E., Liang, L., and Newman, C. 2004. Historical and other patterns of monomethyl and inorganic mercury in the Florida panther (*Puma concolor coryi*). *Archives of Environmental Contamination and Toxicology* 48: 75–80.

Porcella, D. B., Zillioux, E. J., Grieb, T. M., Newman, J. R., and West, G. B. 2004. Retrospective study of mercury in raccoons (Procyon lotor) in south Florida. *Ecotoxicology* 13: 207–221.

Roelke, M. E. 1991. Florida Panther Biomedical Investigation. Annual Performance Report, July 1, 1990 to June 30, 1991. Study no. 7506. Florida Game and Fresh Water Fish Commission, Tallahassee, Florida.

Roelke, M. E., Schulz, D. P., Facemire, C. F., and Sundlof, S. F. 1991. Mercury contamination in the free-ranging endangered Florida panther (*Felis concolor coryi*). *Proceedings American Association of Zoo Veterinarians*. 277–283.

Root, K. V. 2004. Using models to guide recovery efforts for the Florida panther. Pages 491–504 in H. R. Akçakaya, M. Burgman, O. Kindvall, C. C. Wood, P. Sjogren-Gulve, J. Hatfield, and M. McCarthy (eds.). *Species Conservation and Management: Case Studies*. Oxford University Press, New York.

Shrader-Frechette, K. 2004. Measurement problems and Florida panther models. *Southeastern Naturalist* 3: 37–50.

U.S. Fish and Wildlife Service. 1999. Florida panther recovery plan. Pages 4-117–4-150 in *South Florida Multi-species Recovery Plan*. Atlanta, Georgia.

U.S. Fish and Wildlife Service. 2006. Florida Panther Recovery Plan. Draft Report. Available: www.fws.gov/verobeach/.

4

Raccoon (*Procyon lotor*) Harvesting on and near the U.S. Department of Energy's Savannah River Site

Utility of Metapopulation Modeling for Prediction and Management of Hunter Risk

---

KAREN F. GAINES
JAMES M. NOVAK

Understanding the toxicodynamics of wildlife populations in contaminated ecosystems is one of the greatest challenges in ecotoxicology today. The goal is to manage these populations to minimize risk to ecosystem integrity as well as human health. Ecological risk assessments (ERAs) in the United States are designed to meet the regulatory mandates of the Comprehensive Environmental Response, Compensation and Liability Act (CERCLA) and the Resource Conservation and Recovery Act. According to the U.S. Environmental Protection Agency, an ERA evaluates the potential adverse effects that human activities have on the flora and fauna that define an ecosystem (U.S. Environmental Protection Agency 1997). When conducted for a particular geographic location, the ERA process can be used to identify vulnerable and valued resources, prioritize data collection, and link human activities with their potential effects. Risk assessment results provide a common framework for comparing different management options, thus enabling decision makers and the public to make better informed decisions about the management of ecological resources. The ERA uses available toxicological and ecological information to estimate the occurrence of a specified undesired ecological event or endpoint. The types of endpoints targeted for investigation depend on the objectives and the constraints imposed upon the risk assessment process (Newman and Strojan 1998) based on all of the relevant stakeholders; therefore, multiple endpoints at different scales may be necessary but are not commonly used (Gaines et al. 2004). In this case, the stakeholders are the public who live near and hunt on and near the Department of Energy's (DOE's) Savannah River Site (SRS; figure 4.1). To date, there is a dearth of knowledge concerning how environmental risk can be managed at the population level when using wildlife as endpoint (receptor) species.

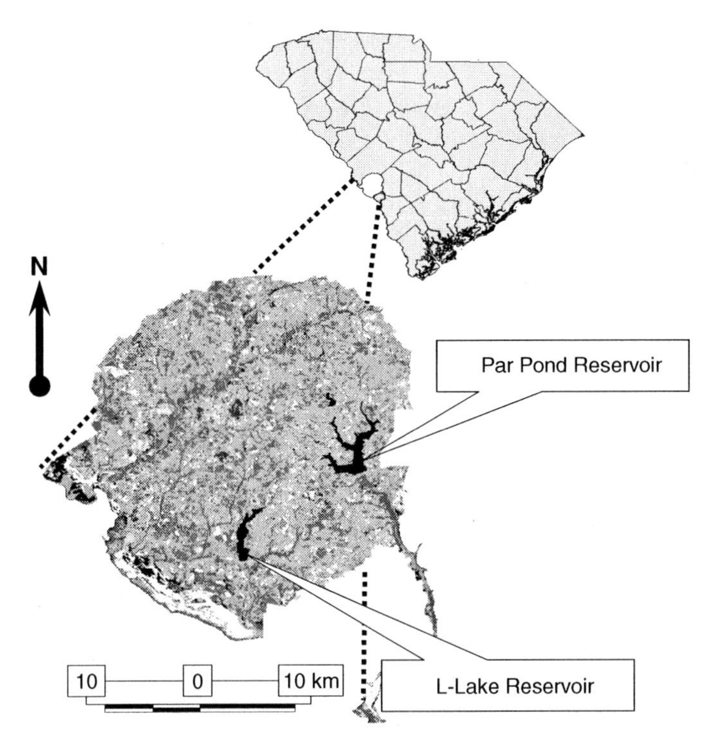

Figure 4.1. The Department of Energy's Savannah River Site located on the west-central border of South Carolina along the Savannah River Swamp. This habitat map shows the distribution of the major drainage systems of the Savannah River Site, including the former nuclear reactor cooling reservoirs, Par Pond and L-Lake (where raccoons are possible contaminant vectors), which are contaminated with ^{137}Cs.

Until recently, the landscape approach has been rarely used in ERAs when assessing wildlife receptor species, and especially on large federal facilities that would benefit from a landscape-level implementation. That is, contaminant exposure assessments have taken into account neither the spatial distribution of the pollutant nor the movements of groups of individuals over the landscape. Rather, fact gathering has remained biased toward lower levels of ecological organization, despite the acknowledged need for and relevance associated with information about effects at higher levels; for example, effects on higher trophic levels or populations (Taub 1989; Cairns 1996). Methods are rapidly changing due to the recognition that if a site is spatially heterogeneous with respect to either contamination or wildlife use, then models must be modified to include the dynamics imposed by those spatial constraints (Sample and Suter 1994). Although humans are often not considered a logical endpoint in an ERA, in many cases arguably they are the most appropriate. When considering the landscape structure of industrial sites such as the SRS (especially those that allow hunting) that are surrounded by rural areas, hunters are one of the main components influencing the population of many wildlife species and subsequently the structure of the ecosystem's food web. If hunters were not allowed to take game from these sites due to high consumption risks, it could have an impact on the population structure of the wildlife in those ecosystems and possibly contribute to new risks due to redistribution and movement of contaminants offsite. In this chapter, we describe how wildlife populations can be managed through harvest to minimize the bioavailability of toxicants in the environment for the SRS. We use raccoons as the focal species; however, many other ecologically important game species contribute to the environmental toxicodynamics of the SRS and similar landscape-level industrial sites that can also be managed to minimize both ecological and human risk. In the Southeast, raccoon hunting is extremely popular, and harvested individuals are used for both meat and fur (Gaines et al. 2000). In South Carolina, where the SRS is located, the raccoon-hunting season is usually from mid-September to mid-March, with no bag or possession limit. Thus, a diligent hunter who eats the meat could legally consume as much raccoon meat as desired.

Methods

Study species and area

The SRS is a 778-km^2 former DOE nuclear production and current research facility located in west-central South Carolina (33.1° N, 81.3° W; figure 4.1) that was closed to public access in 1952. On numerous occasions, both terrestrial and aquatic SRS ecosystems have been contaminated with radionuclides, metals, and organics, and areas have been affected by thermal effluents (White and Gaines 2001). In 1972, the entire SRS was designated as the nation's first National Environmental Research Park to provide tracts of land where the effects of human impacts upon the environment could be studied (Davis and Janecek 1997; White and Gaines 2001). Much of the suitable forested area of the SRS is managed primarily for commercial timber (pine) production by the U.S. Forest Service. More than 20% of the SRS is covered by wetlands, including bottomland hardwoods, cypress-tupelo swamp forests, creeks, streams, ponds, Carolina bays (which are natural elliptical depressions that vary in size and in the degree to which they retain water; Ross 1987), and two large former reactor cooling reservoirs (Par Pond, L-Lake) and associated floodplains and outflows that have been contaminated with the gamma-emitting radionuclide radiocesium (^{137}Cs). Hunting is allowed on the SRS proper. Both white-tailed deer (*Odocoileus virginianus*) and wild hogs (*Sus scrofa*) are hunted on site and are monitored for ^{137}Cs. Raccoons are hunted in the Crackerneck Wildlife Management Area located in the southwestern portion of the site, and individuals are not monitored for contaminant burden. Previous studies (authors' unpublished data) indicate that the raccoon population from Crackerneck Wildlife Management Area is not accumulating ^{137}Cs significantly above background levels. However, raccoons are hunted on the border of the SRS, and past studies (Gaines et al. 2000, Chow et al. 2005) have revealed elevated ^{137}Cs levels in individuals from populations that reside in contaminated areas on the SRS and along the SRS border. In this chapter, we focus on raccoon populations inhabiting two such border areas: the SRS border located west of the Par Pond reactor cooling reservoir, and the SRS border located south of the L-Lake reactor cooling reservoir along Steel Creek (figure 4.2). These two sites were chosen based on their juxtaposition relative to the SRS border, ^{137}Cs contamination, and the potential for harvest. Other border sites may also pose a risk of raccoons becoming contaminant vectors, but the toxicokinetics have not been well studied for those areas and therefore were not modeled.

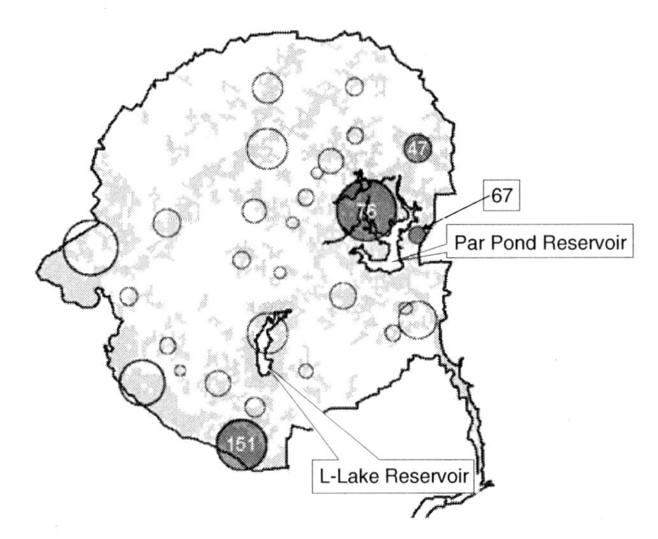

Figure 4.2. The population structure of Savannah River Site raccoons as estimated by the RAMAS Metapop program (circles). A probability surface of raccoon distribution was used to determine suitable habitat for raccoon population establishment (light gray pixels). Only populations with $n > 5$ were used in the analysis. The populations that were harvested (population ID #s 47, 67, 76, and 151) in the simulations are labeled and shaded.

Previous investigations (Arbogast 1999; Boring 2001; Gaines et al. 2000, 2005) have shown elevated levels of ^{137}Cs in these raccoon populations and have documented that raccoons move freely on and off the SRS. Thus, the raccoon is a useful species for both human risk assessments and ERAs. Since this species is extremely mobile, the likelihood of an animal's presence in specific microhabitats and the time spent in those habitats must be estimated to calculate reasonable risk estimates. For this study, home range habitat utilization information coupled with three years of harvest data and ^{137}Cs monitoring on and near the SRS supplied the data needed to develop a spatially explicit model to investigate population-level toxicodynamics.

Wildlife species are used as endpoints in the risk assessment process by the DOE, and raccoons in particular are a focal species. Specifically, raccoons have been used as a receptor species in both human risk assessments and ERAs for the SRS and other DOE sites using current information regarding home range, contaminant uptake, and food habits for populations both on and off the SRS. Several life-history characteristics of raccoons make them potential agents of contaminant distribution, including (1) high population levels with an extended range throughout North America in a variety of habitats, (2) their ability and proclivity to travel extended distances (Glueck et al. 1988; Walker and Sunquist 1997; Gehrt and Fritzell 1998), (3) a propensity to utilize human-altered habitats in combination with an ability to move freely in and out of most toxic waste sites (Hoffmann and Gottschang 1977; Clark et al. 1989; Khan et al. 1995), and (4) a broadly omnivorous diet that includes components of both terrestrial and aquatic food chains (Lotze and Anderson 1979; Khan et al. 1995).

Experimental methods

Toxicokinetics

Radiocesium uptake models were constructed from information collected for male raccoons from three consecutive harvests in the L-Lake corridor located near the border of the SRS (figure 4.2) adjacent to a private hunting preserve. This population was used because individuals spent 100% of their time in contaminated areas (as determined from radiotelemetry). Mean ^{137}Cs levels declined significantly from the first trap effort to the third trap effort (year 1, 127 Bq/kg; year 2, 63 Bq/kg; year 3, 29 Bq/kg; all activities are reported for wet weight; for analytical counting methods, see Arbogast 1999; Gaines et al. 2000; Boring 2001). The first two trapping efforts (Arbogast 1999; Gaines et al. 2000) harvested individuals from the population each year (n = 10, spring 1997; n = 13, spring 1998). Areas were trapped until no more individuals were caught after an additional two-week period. Therefore, it is assumed that the sample size represents the population of male raccoons for the immediate area, thus providing a baseline for the determination of the amount of ^{137}Cs that new recruits will accumulate in one year (for a detailed discussion of the ^{137}Cs dynamics in this raccoon population, see Gaines et al. 2000, 2005). For the first trapping effort, muscle was removed from raccoons and analyzed for ^{137}Cs. For the second trap effort, both muscle and whole-body ^{137}Cs burdens were determined, and a simple linear regression was performed to determine their predictive relationship. For the third trap effort, whole-body ^{137}Cs burdens were determined for all captured raccoons (n = 14). The muscle concentration was estimated using the simple linear regression model developed from the second trap effort (wet-weight muscle concentration [Bq/kg] = 1.7041 × whole body [Bq/kg] + 3.1; $r^2 = 0.9617$; Arbogast 1999). The aforementioned declining rates from year to year were used to estimate the proper "average weight" parameter in the Stages submenu of the model procedure in the RAMAS Metapop program described below. The maximum observed burden of 1,000 Bq/kg was used to allow conservative risk calculations.

Population model

Spatial structure A spatially explicit model of raccoon distribution for the SRS was developed using data from a raccoon radiotelemetry study and visualized with GIS. An inductive approach was employed to develop three submodels using the ecological requirements of raccoons studied in the following habitats: (1) man-made reservoirs, (2) bottomland hardwood/riverine systems, and (3) isolated wetland systems. Probabilistic resource selection functions were derived from logistic regression using habitat compositional data and landscape metrics (for further details, see Gaines et al. 2005). The final distribution model provides a spatially explicit probability (likelihood of being in an area) surface. This surface was used as the base map in the Spatial Data program. Specifically, the habitat suitability function used a probability threshold of 0.85 and a neighborhood distance of one 100 m × 100 m cell to derive the habitat suitability map used to determine the metapopulation spatial structure. Initial abundances for each population were calculated from the population density of raccoons based on the trapping efforts during and two years prior to the radiotelemetry study. Specifically, it was estimated that a maximum of eight individuals would occupy a 250-hectare (average home range) area (for further details, see Boring 2001). Therefore, the initial raccoon abundance was calculated based on the following equation:

$$\text{Initial abundance} = (\text{noc} \times 0.072)\ \text{ahs}, \tag{4.1}$$

where noc is the number of cells in the patch, ahs is the average habitat suitability, and 0.072 is the number of raccoons per cell. Populations with initial abundances of fewer

Table 4.1. Population parameters used for simulations of raccoon (*Procyon lotor*) populations residing in the Department of Energy's Savannah River Site.

Metric	Juvenile	Yearlings	2–3 years	4+ years
Fecundity[a]	0	1.4	1.55	1.0
Survival	0.8	0.8	0.8	0.64
Relative dispersal	1.0	1.0	0	0
^{137}Cs muscle concentration (Bq/kg)[b]	1.0	1.0	1,000	1,000

[a]Based on Zeveloff (2002).
[b]Using the average weight parameter in the Stages menu of the Metapop program.

than five individuals were not considered to be viable populations and therefore were removed from further analysis.

Population parameters Since natural resources required by raccoons are abundant in the Southeast, and both male and female raccoons have overlapping home ranges on the SRS, an exponential growth to carrying capacity (ceiling) density dependence was used. Moreover, the sex structure of the population was assumed to be mixed since there is little differentiation between how males and females use SRS habitats (although females may disperse less). As such, to maximize harvest rates, the populations were modeled at carrying capacity (K) with a 10% standard deviation, thus producing a conservative estimate for risk assessment purposes. For the purposes of harvest management, only a two-stage model is needed, juveniles and adults, since only adults are harvested. However, a five-stage model (juveniles, yearlings, 2-year-olds, 3-year-olds, and 4+ years) was needed to properly estimate contaminant burden and relative dispersal. Specifically, the average weight parameter in the Stages submenu of the model procedure in the RAMAS Metapop program (Akçakaya 2005) was used as a proxy for the concentration of ^{137}Cs in raccoon muscle tissue. For the purposes of this simulation, it was assumed that juveniles and yearlings have little to no contaminant body burden. Although past studies have shown that juveniles may inherit a burden from their mother (von Zallinger and Tempel 1998) and that yearlings will attain approximately one-third of their maximal burden (Boring 2000), to ensure that new recruits have background burden (the Metapop program will not differentiate weights on a per population basis) based on the toxicokinetic model described above, both juveniles and yearlings were modeled with a muscle concentration of 1 Bq/kg. To apply a conservative estimate for risk assessment purposes, the absolute maximum burden observed in the wild (1,000 Bq/kg) plus 10% was simulated for the 2- to 4-year stages. This absorbed the potential error of setting all juveniles and yearlings to have negligible burden. The dispersal-distance function used was based on movement data collected during the radiotelemetry study. The model assumed that only juveniles and yearlings disperse, and that the dispersal rate is distance dependent, with a negative exponential slope distance factor of 2 and a maximum dispersal distance of 25 km (the average diameter of the SRS). A 20% coefficient of variation was used for simulating environmental stochasticity in dispersal; due to the small sample size used to determine some population parameters, a 40% coefficient of variation was applied for sampling error of N under harvest management. It was also assumed that fecundity and survival were correlated, but carrying capacity was independent for the correlation structure of the stochastically varying parameters within each population. Table 4.1 shows additional population parameters used to construct model matrices. Fecundity was calculated by dividing the estimated litter size by 2 multiplied by the survival rate (0.8), assuming postbreeding census.

The harvest management scheme simulated yearly hunting of focal populations near the two border sites (figure 4.2). The proportion of individuals harvested was between 0.1 and 0.9 at 0.1 increments for all populations at both locations to determine if the ^{137}Cs burden would decrease at the population level over time as the harvest rate increased. Only yearlings through 4+ years were harvested in the simulation. The harvest started in year 4 to allow cohorts to equilibrate. Since highly suitable raccoon habitat extends off the SRS along the Steel Creek corridor, the estimated population size for that area most likely was underestimated using equation 4.1; therefore, the initial population size was doubled based on the area of the Steel Creek corridor that extends off the SRS. This was not done for the simulation around Par Pond since the habitats abruptly stop at the SRS border and change to habitats much less suitable for raccoon populations. However, similar forested and wetland habitats do extend offsite, which could affect dispersal (see discussion of isotropic scenarios below). Simulations were conducted using 1,000 replications over 50 years. The final-stage abundances and harvest summary were exported from the Metapop program for populations from each area (L-Lake, Par Pond). Two major scenarios were performed. The first assumed that recruitment rates of border populations would be less based on an anisotropic (asymmetric) migration trajectory and therefore used the initial population structure calculated from the Metapop program. To calculate the body burdens, the harvest totals (e.g., total ^{137}Cs Bq/kg) were divided by the total number of individuals in the harvested cohorts from the previous year (t–1) multiplied by the harvest rate. However, since border populations could have isotropic recruitment of noncontaminated individuals from offsite, the second scenario adjusted the numbers of individuals in the focal populations to assume 100% recruitment of each stage after an initial harvest. This was achieved by dividing the body burdens determined in scenario 1 by the proportion of individuals that were not recruited for each cohort. We also assumed that the harvested populations would receive new individuals from every age class via dispersal. This phenomenon was seen on the SRS during studies performed by Gaines et al. (2000).

Results

For the first set of simulations (scenario 1), populations were not adjusted to maximize new recruits. That is, the total number of individuals within each harvested cohort was summed, multiplied by the harvest rate, and used as the denominator to divide the total ^{137}Cs muscle concentration for determination of the toxicant concentration on a per individual basis. This revealed that the projected ^{137}Cs muscle concentration (range: Par Pond, 521–805 Bq/kg; L-Lake, 669–758 Bq/kg) will increase, with harvest rate peaking at 60%, and then decline as harvest rates increase (table 4.2). Nevertheless, the mean body burden diminishes only slightly below the European Economic Community (1986) standard of 600 Bq/kg. However, the second set of simulations (scenario 2) reveals that if the numbers of new recruits were maximized and assumed to come from neighboring populations that had no burden (isotropic

Table 4.2. Mean ^{137}Cs muscle concentration (Bq/kg) and associated lower and upper standard deviation (LSD, USD) estimates over 50 years for raccoon populations associated with the contaminated reservoir areas on the border of the Savannah River Site.

	L-Lake			Par Pond		
Harvest (%)	LSD	Mean	USD	LSD	Mean	USD
10%	258	717	1176	239	539	929
20%	307	698	1089	292	613	935
30%	351	751	1151	349	688	1027
40%	275	695	1114	231	521	812
50%	239	689	1140	245	623	1003
60%	325	758	1191	409	805	1201
70%	279	710	1142	329	697	1065
80%	287	702	1117	337	693	1050
90%	283	669	1054	331	664	998

The simulation assumes anisotropic recruitment of uncontaminated individuals after harvest due to border effects (e.g., no offsite recruitment).

recruitment), then the mean ^{137}Cs muscle concentration declines as the proportion of harvest increases (range over 50-year simulation: L-Lake, 164–813 Bq/kg; Par Pond, 156–654 Bq/kg; figure 4.3). Using the European Economic Community standard as a precautionary limit, a yearly harvest rate of at least 30% and 20% for the L-Lake and Par Pond populations, respectively, would need to be implemented to minimize risk to humans who may consume these animals.

To explore sensitivities in the model, a range of values for the movement and density dependence were used that would most likely affect the number of new recruits into harvested populations. Changing these parameters only altered the final number of raccoons in those populations for both simulations. However, the proportions of individuals within populations remained constant for the entire metapopulation, which is the most influential parameter regarding toxicant body burden.

Discussion

Most risk assessments that focus on the effects of toxicants to wildlife at the population level are concerned with impacts to population size due to such factors as reduced fecundity, acute toxification, or shifts in species composition. The population risk assessment presented here focuses on how managed populations through harvest may reduce contaminant mobility. Further, most ERAs do not use humans as an endpoint, assuming that they are independent from the trophic system of the focal environment. However, if hunting is not allowed due to risk concerns, wildlife may contain higher burdens than if harvests were implemented. The ^{137}Cs dynamics of the SRS is a typical

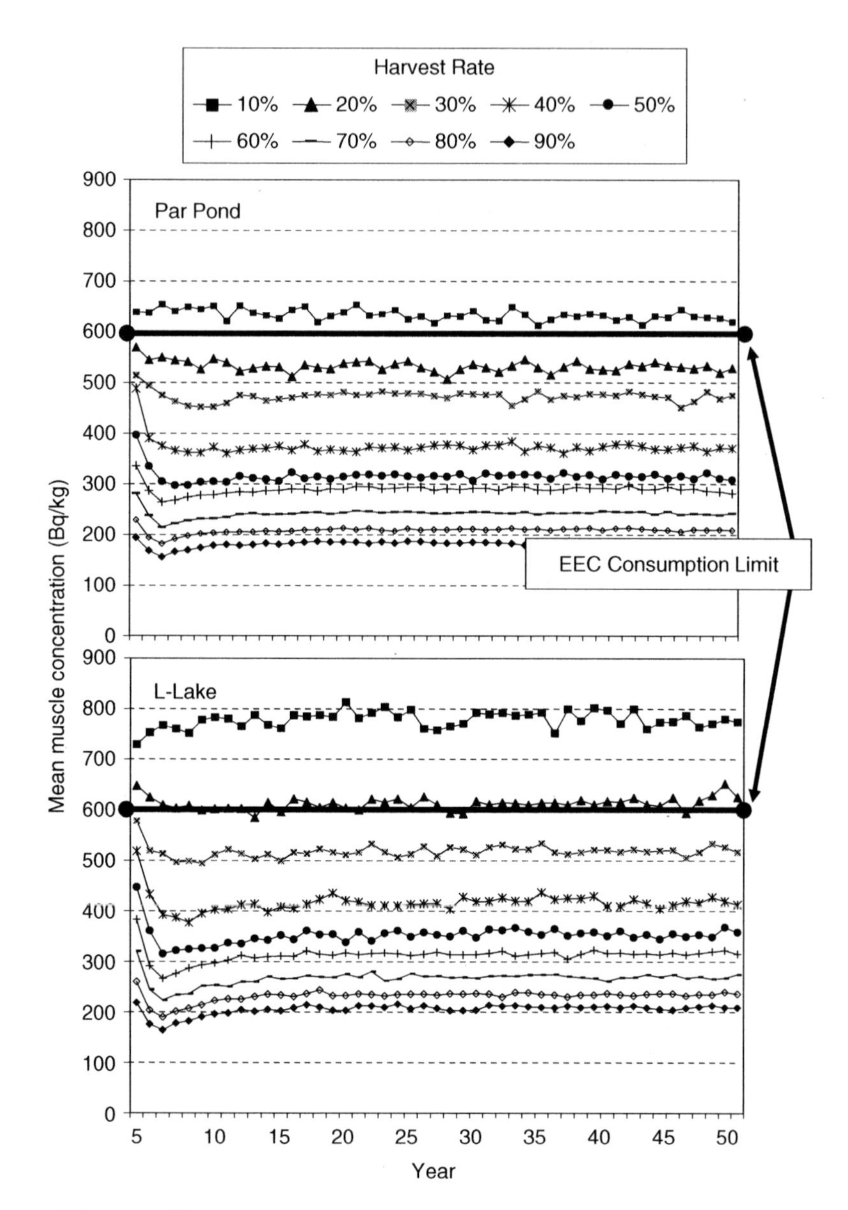

Figure 4.3. Mean ^{137}Cs muscle concentration (Bq/kg) estimates over 50 years for raccoon populations associated with the contaminated reservoir areas on the border of the Savannah River Site (Par Pond, L-Lake). The simulation assumes maximum (isotropic) recruitment of uncontaminated individuals after harvest. The European Economic Community (EEC) standard of 600 Bq/kg is referenced as a precautionary limit, showing that a yearly harvest rate of at least 20–30% would need to be implemented to minimize risk to humans who may consume these animals.

example of how a coupled human–natural system drives ecological risk. Ecosystem dynamics control the ecological half-life of ^{137}Cs, while hunting in and around the SRS influences receptor species population dynamics and thus the bioavailability of ^{137}Cs to humans and other consumers, as well as contaminant transport within the ecosystem.

The premise of this model is that the most important components of estimating the environmental toxicokinetics of contaminants in wildlife populations are residence time of different age cohorts of organisms and the residence time of the contaminant in the environment. We used values from the literature best suited for the U.S. Southeast to estimate many of the population parameters. Cohort survival and dispersal are regionally distinct (Zeveloff 2002) and will affect toxicant exposure to populations through their residence time in the contaminated environment. The physical half-life of ^{137}Cs is approximately 30 years. The biological turnover rates within a given organism are generally much shorter and are influenced by metabolism. Therefore, biological half-life should change based on biotic and abiotic parameters such as age, overall health, seasonality, and food availability and also depends on the sources and bioavailability of the contaminants within the animal's home range. However, bioavailability in these systems has additional complexities. When radioactive isotopes are released into ecosystems such as those associated with L-Lake or Par Pond on the SRS, the isotopes will theoretically also have an ecological half-life. This is the amount of time required for the level of an isotope (in this case, ^{137}Cs), once established and at equilibrium within a given ecosystem compartment, to decrease by 50%. This is a result of the isotope either becoming ecologically unavailable or being physically removed from a system (Brisbin 1991). The concept of ecological half-life is further constrained by the fact that most ecosystem compartments are extremely dynamic and rarely come to equilibrium. As the time required to achieve effective equilibrium increases, it becomes less likely that these conditions will remain constant (Peters and Brisbin 1996). Therefore, having a relative estimate of the ecological half-life of the toxicant as it relates to the species uptake and depuration rate is essential in modeling contaminant mobility. That is, once these parameters are known, the spatiotemporal patterns can be better predicted.

Specifically, in this model, the residence times of age cohorts within contaminated areas were a direct function of both home range and dispersal dynamics. The home range dynamics of raccoons are known from the SRS (Boring 2001), and raccoons have proved to be useful sentinel species in these and other contaminated ecosystems (Bigler et al. 1975; Smith et al. 2003). It is how raccoons disperse from contaminated populations that drives the risk process. The effect of dispersal on the demography of local populations is dependent on whether the populations are connected in a metapopulation spatial arrangement (Johnson et al. 2005) and whether the relationship between any two populations is in the context of source or sink (Pulliam 1988). Although dispersal contributes to spatiotemporal variation in population size (Nichols et al. 2000), its relative importance, compared to local recruitment, in any subpopulation is unclear due to a lack of empirical data (Bennetts et al. 2001; MacDonald and Johnson 2001). Until recently (Cam et al. 2004), much of metapopulation theory has been based upon simulations and has lacked rigorous empirical testing (Hanski 2001; MacDonald and Johnson 2001). Specifically, the anisotropy of dispersal caused by edge effects when modeling has not been addressed adequately. Since a potential study site boundary may or may not correlate to habitat boundaries, the effect on dispersal rates into peripheral populations can be greatly in error. That is especially why these effects were addressed in this study since peripheral subpopulations are important in terms of management considerations.

Two major simulations were performed due to the above considerations: One scenario assumed that there was no recruitment from offsite populations after harvest, and the other scenario assumed that "clean" raccoons from offsite would maintain these border populations at levels close to carrying capacity. High-quality raccoon habitat on the eastern border of the SRS near Par Pond does not continue offsite, which would make the former scenario more likely due to population isolation. However, suitable habitat does exist nearby, so both management scenarios should be considered. Conversely, since the southern border of the SRS where the L-Lake population resides has contiguous habitat that persists offsite as well as noncontiguous patches much like the SRS interior, the latter scenario is more biologically plausible. Additionally, previous investigations (Boring 2001; Gaines et al. 2005) have demonstrated that new recruits do move into the L-Lake border population after harvest and that the ecological half-life is at least twice the biological half-life. Specifically, based on this estimate, new recruits should become equilibrated within at least two years, which is why juveniles and yearlings were assumed to have no contaminant burden, and raccoons two or more years of age were assumed to have ^{137}Cs muscle concentration of 1,000 Bq/kg (the maximum observed in SRS raccoons). Federal facilities such as the DOE's SRS have sufficiently characterized the abiotic and biotic systems within their boundaries; therefore, in this case, understanding how ecosystem components offsite influence wildlife movement was more critical. To maintain isotropy for simulations assuming maximum recruitment, we used the population structure and trajectory of populations located on the interior of the SRS. Other population parameters used for this model are congruent with other studies investigating raccoon movement (Broadfoot et al. 2001).

Metapopulation dynamics can be extremely important in long-term management scenarios for species of human concern. Source–sink dynamics may influence the management of harvested (McCoy et al. 2005) or endangered (Kauffman et al. 2004) species or may be created by the harvest or management itself (Novaro et al. 2005). While studies are beginning to link metapopulation dynamics with physical resource transport between subpopulations as reciprocal food web subsidies (Nakano and Murakami 2001), the use of such models for contaminant transport studies within and between ecosystems (including human ecosystems) is a novel aspect of this study.

The simulations showed that recruitment type (local vs. dispersal) affects a population's estimated body burden. If clean individuals are recruited and harvested before they reach equilibrium with contaminants in the environment, then contaminant mobility into humans and other predators is truly minimized. Based on the harvest simulations, at least a 20% and 30% yearly harvest rate for Par Pond and L-Lake, respectively, would be required before the muscle concentrations would fall below the European Economic Community limit of 600 Bq/kg (figure 4.3). Managed hunting near L-Lake and along the adjacent SRS border is plausible since a private hunting preserve is already established outside but adjacent to the SRS. Since L-Lake most likely has new recruits coming from "clean" populations offsite, this would be an optimum management action to minimize the risk of contaminant movement into the human food chain.

Conversely, if edge populations have limited recruitment due to anisotropic migration, then the survivors in each cohort would more likely be original residents of the contaminated areas and thus have a higher burden. This is most likely the case for the Par Pond population. If hunting would continue in that area, then harvest success would be reduced due to lower population size; however, those individuals harvested would contain the highest burdens. Since the Par Pond SRS border is commercially owned, offsite hunting is unlikely to be encouraged as a management option. Given that the SRS is concerned with raccoons potentially moving contaminants offsite, the best risk management option would be to implement an onsite hunt to lower the population even though this would not reduce toxicant burden. However, caution should always be used in this type of scenario since reducing population size as a long-term management goal may have consequences on ecosystem integrity and function.

References

Akçakaya, H. R. 2005. *RAMAS GIS: Linking Spatial Data with Population Viability Analysis*, version 5.0. Applied Biomathematics, Setauket, New York.

Arbogast, M. L. 1999. The Use of Raccoons as Bioindicators of Radiocesium Contamination. M.S. Thesis, Rutgers University, Piscataway, NJ.

Bennetts, R. E., Nichols, J. D., Lebreton, J.-D., Pradel, R., Hines, J. E., and Kitchens, W. M. 2001. Methods for estimating dispersal probabilities and related parameters using marked animals. Pages 3–17 in J. Clobert, E. Danchin, A. Dhondt and J. D. Nichols (eds.). *Dispersal*. Oxford University Press, Oxford, UK.

Bigler, W. J., Jenkins, J. H., and Cumbie, P. H. 1975. Wildlife and environmental health: Raccoons as indicators of zoonoses and pollutants in the southeastern United States. *Journal of the American Veterinary Medical Association* 167:592–597.

Boring, C. S. 2001. The Interactions of Home-Range, Habitat Preference and Contaminant Burdens in Relation to Ecological Risk for Raccoon Populations Living near the Borders of the Department of Energy's Savannah River Site. M.S. Thesis, Rutgers University, Piscataway, NJ.

Brisbin, I. L., Jr. 1991. Avian radioecology. *Current Ornithology* 8:69–140.

Broadfoot, J. D., Rosatte, R. C., and O'Leary, D. T. 2001. Raccoon and skunk population models for urban disease control planning in Ontario Canada. *Ecological Applications* 11:295–303.

Cairns, J., Jr., and Niederlehner, B. R. 1996. Developing a field of landscape ecotoxicology. *Ecological Applications* 6:790–796.

Cam, E., Oro, D., Pradel, R., and Jimenez, J. 2004. Assessment of hypotheses about dispersal in a long-lived seabird using multistate capture-recapture models. *Journal of Animal Ecology* 73:723–736.

Chow, T. E., Gaines, K. F., Hodgson, M. E., and Wilson, M. D. 2005. Habitat and exposure modelling for ecological risk assessment: A case study for the raccoon on the Savannah River Site. *Ecological Modelling* 189:151–167.

Clark, D. R., Jr., Ogasawara, P. A, Smith, G. J., and Ohlendorf, H. M. 1989. Selenium accumulation by raccoons exposed to irrigation drainwater at Kesterson National Wildlife Refuge, California. *Archives of Environmental Contamination and Toxicology* 18:787–794.

Davis, C. E., and Janecek, L. L. 1997. DOE Research Set-Aside Areas of the Savannah River Site. Publication SRO-NERP-25. Savannah River Ecology Laboratory, Aiken, SC.

European Economic Community. 1986. Derived Reference Levels as a Basis for the Control of Foodstuffs Following a Nuclear Accident: A Recommendation from the Group of Experts Set Up under Article 31 of the Euratom Treaty. EEC Regulation 170/86. Commission of the EEC Printing Office, Brussels, Belgium.

Gaines, K. F., Lord, C. G., Boring, C. S., Brisbin, I. L., Jr., Gochfeld, M., and Burger, J. 2000. Raccoons as potential vectors of radionuclide contamination to human food chains from a nuclear industrial site. *Journal of Wildlife Management* 64:199–208.

Gaines, K. F., Porter, D. E., Dyer, S. A., Wein, G. R., Pinder, J. E., III, and Brisbin, I. L., Jr. 2004. Using wildlife as receptor species: A landscape approach to ecological risk assessment. *Environmental Management* 34:528–545.

Gaines, K. F., Boring, C. S., and Porter, D. E. 2005. The development of a spatially explicit model to estimate radiocesium body burdens in raccoons (*procyon lotor*) for ecological risk assessment. *Science of the Total Environment* 341:15–31.

Gehrt, S. D., and Fritzell, E. K. 1998. Resource distribution, female home range dispersion and male spatial interactions: Group structure in a solitary carnivore. *Animal Behaviour* 55:1211–1227.

Glueck, T. F., Clark, W. R., and Andrews, R. D. 1988. Raccoon movement and habitat use during the fur harvest season. *Wildlife Society Bulletin* 16:6–11.

Hanski, I. 2001. Population dynamic consequences of dispersal in local populations and metapopulations. Pages 283–298 in J. Clobert, E. Danchin, A. Dhondt, and J. D. Nichols (eds.). *Dispersal*. Oxford University Press, Oxford, UK.

Hoffmann, C. O., and Gottschang, J. L. 1977. Numbers, distribution, and movements of a raccoon population in a suburban residential community. *Journal of Mammalogy* 58:623–636.

Johnson, C. N., Vernes, K., and Payne, A. 2005. Demography in relation to population density in two herbivorous marsupials: Testing for source–sink dynamics versus independent regulation of population size. *Oecologia* 143:70–76.

Kauffman, M. J., Pollock, J. F., and Walton, B. 2004. Spatial structure, dispersal, and management of a recovering raptor population. *American Naturalist* 164:582–597.

Khan, A. T., Thompson, S. J., and Mielke, H. W. 1995. Lead and mercury levels in raccoons from Macon County, Alabama. *Bulletin of Environmental Contamination and Toxicology* 54:812–816.

Lotze J, and Anderson S. 1979. *Procyon lotor. Mammalian Species.* 119:1–8.

MacDonald, D. W., and Johnson, D. D. P. 2001. Dispersal in theory and practice: Consequences for conservation biology. Pages 359–372 in J. Clobert, E. Danchin, A. Dhondt, and J. D. Nichols (eds.). *Dispersal*. Oxford University Press, Oxford, UK.

McCoy, J. E., Hewitt, D. G., and Bryant, F. C. 2005. Dispersal by yearling male white-tailed deer and implications for management. *Journal of Wildlife Management* 69:366–376.

Nakano, S., and Murakami, M. 2001. Reciprocal subsidies: Dynamic interdependence between terrestrial and aquatic food webs. *Proceedings of the National Academy of Sciences of the USA* 98:166–170.

Newman, M.C. and Strojan C.L. 1998. *Risk Assessment: Logic and Measurement.* Ann Arbor Press, Chelsea, MI, USA

Nichols, J. D., Hines, J. E., Lebreton, J.-D., and Pradel, R. 2000. The relative contribution of demographic components to population growth: A direct estimation approach based on reverse-time capture-recapture. *Ecology* 81:3362–3376.

Novaro, A. J., Funes, M. C., and Walker, R. S. 2005. An empirical test of source-sink dynamics induced by hunting. *Journal of Applied Ecology* 42:910–920.

Peters, E. L., and Brisbin, I. L., Jr. 1996. Environmental influences on the ^{137}Cs kinetics of the yellow-bellied turtle (*Trachemys scripta*). *Ecological Monographs* 66:115–136.

Pulliam, H. R. 1988. Sources, sinks, and population regulation. *American Naturalist* 132:652–661.

Ross, T. E. 1987. A comprehensive bibliography of the Carolina bays literature. *Journal of the Elisha Mitchell Scientific Society* 103:28–42.

Sample, B. É., and Suter, G. W. II. 1994. Estimating Exposure of Terrestrial Wildlife to Contaminants. ES/ER/TM-12.5. Environmental Science Division, Oak Ridge National Laboratory, Oak Ridge, TN.

Smith, P. N., Johnson, K. A., Anderson, T. A., and McMurry, S. T. 2003. Environmental exposure to polychlorinated biphenyls among raccoons (Procyon lotor) at the Paducah gaseous diffusion plant, western Kentucky, USA. *Environmental Toxicology and Chemistry* 22:406–416.

Taub, F. B. 1989. Standardized aquatic microcosms. *Environmental Science and Technology* 23:1064–1066.

U.S. Environmental Protection Agency. 1997. Ecological Risk Assessment Guidance for Superfund: Process for Designing and Conducting Ecological Risk Assessments. EPA/630/R-021011. Washington, DC.

von Zallinger, C., and Tempel, K. 1998. Transplacental transfer of radionuclides. A review. *Journal of Veterinary Medical Science* A45:581–590.

Walker, S., and Sunquist, M. 1997. Movement and spatial organization of raccoons in north-central Florida. *Florida Field Naturalist* 25:11–21.

White, D. L., and Gaines, K. F. 2000. The Savannah River Site: Site description, land use and management history. Pages 8–17 in J. B. Dunning and J. C. Kilgo (eds.). *Avian Research at the Savannah River Site: A Model for Integrating Basic Research and Long-Term Management.* Studies in Avian Biology 21. Cooper Ornithological Society, Camarillo, CA.

Zeveloff, S. I. 2002. *Raccoons: A Natural History.* Smithsonian Institutional Press, Washington, DC.

5

Earthworms (*Lumbricus rubellus*) in Northwestern Europe

Sublethal Effects of Copper on Population Viability

CHRIS KLOK

In a large number of industrialized countries, soils are polluted with a mixture of contaminants such as heavy metals (Bengtsson and Tranvik 1989; Eeva and Lehikoinen 1996). In the Netherlands, for instance, as many as 175,000 sites in rural areas are classified as highly polluted on the basis of soil quality criteria for single pollutants (Swartjes 1999). Remediation of these sites would require an enormous effort and lead to an extreme high cost for society. Additionally, the extent of the risk these polluted soils actually pose to below- and above-ground ecosystems remains unclear. Earthworm populations are an important study organism to gain more insight in these risks. Earthworms form the largest part of the animal biomass in most temperate soils (Lavelle and Spain 2001); they strongly influence soil fertility (Edwards and Lofty 1977) and are a major food source for many species (Lee 1985). Given their link in many food chains, earthworms play a key role in transporting pollutants from the soil into the above-ground ecosystem.

Pollutant effects on populations can be studied under field conditions, but this approach can be time consuming and costly, and results are often difficult to interpret due to fluctuations in other factors. Effects of pollutants on organisms are traditionally studied under controlled laboratory conditions using standard protocols (e.g., Organisation for Economic Co-operation and Development 1984). To interpret laboratory-derived effects at the population level, mathematical models can be used. The software package RAMAS GIS (Akçakaya 2005) provides a user-friendly framework that can be applied to achieve this goal.

In this chapter, the focus is on the effect of sublethal copper concentrations on the population growth rate, equilibrium density, and risk of quasi extinction for *Lumbricus rubellus* (Hoffm.). The RAMAS GIS modeling environment is used to investigate these

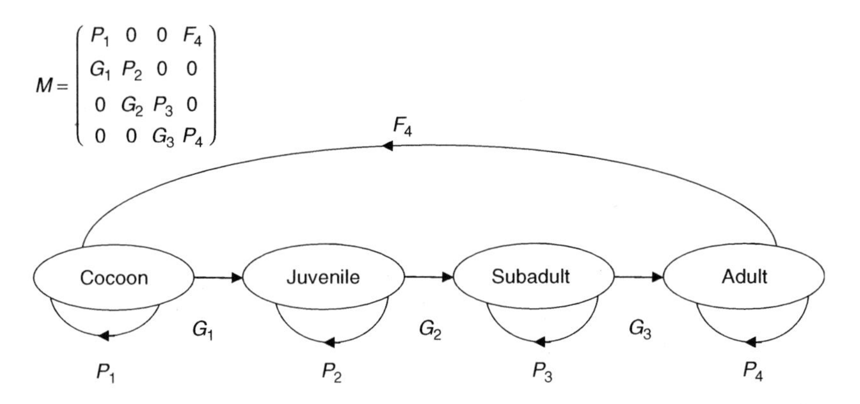

Figure 5.1. Life cycle graph of *Lumbricus rubellus.*

population-level effects. Time-invariant and more realistic density-dependent models, including stochasticity and catastrophes, are parameterized with literature data (Ma 1984; Klok and de Roos 1996; Klok et al. 1997). Pollutants can have an influence on different vital rates; however, these are not independent but are inherently linked by energy acquisition and utilization of the individual (Kooijman 2000). The dynamic energy budget (DEB) theory, developed by Kooijman (2000), offers a solid mechanistic general theory that integrates different individual-level endpoints. The theory is based on closed energy and mass balances; it describes the acquisition and use of energy by individuals in a mechanistic way and is of much value to ecotoxicology. The theory, as well as empirical evidence, is given in (Kooijman 2000; Kooijman and Bedaux 2000).

To assess the possible impact of copper at the population level, three parameterization scenarios are used: an integrative scenario using DEB (Klok and de Roos 1996; Kooijman 2000), an additive scenario assuming that effects on different vital rates are independent, and a single-effect scenario focusing on reproduction effects only.

Methods

Study species

Earthworms belong to the phylum Annelida, order Oligochaeta, class Clitellata. The group consists of more than 1,800 species distributed all over the world. Their role in soil processes had already been identified by Aristotle, who called them the "intestine of the earth." Charles Darwin contributed much to the knowledge of earthworms with a publication in 1881 titled "The formation of vegetable mould through the action of worms, with observations of their habitat." The species *Lumbricus rubellus* is common in most temperate soils and one of the dominant species in grassland soils (Didden 2001). This species lives in the upper 10 cm of the soil profile (Nordström and Rundgen 1973) and feeds primarily on litter. Juvenile, subadult, and adult developmental stages are distinguishable by visual characteristics. The age and weight at which the subadult and adult stages are reached depend on environmental variables (Klok et al. 2006). The life cycle of *L. rubellus* is shown in figure 5.1. Individuals are hermaphrodites; cocoon production is relatively high and can continue throughout the year if soil conditions are optimal, but usually halts during dry periods such as in summer and winter.

Model structure

Time-invariant model

I used a stage-classified model with four stages based on the life cycle of the species (figure 5.1) and a time step of one week. Since individuals are hermaphrodites, the model includes all individuals (type mixed in RAMAS GIS). A metapopulation structure is not considered. In the model, the life cycle of an individual is divided into four distinct developmental stages: cocoon, juvenile, subadult, and adult. Given the number of individuals in each of these developmental stages at some time t, the population model determines the size and composition of the population one time step later, $t + 1$. Let $c(t)$, $j(t)$, $s(t)$, and $a(t)$ be the number of cocoons, juveniles, subadults, and adults constituting the population at time t. On the basis of the life cycle graph (figure 5.1) the number in the different stages at $t + 1$ can be derived by

$$c(t+1) = P_1 c(t) + F_4 a(t),$$
$$j(t+1) = P_2 j(t) + G_1 c(t),$$
$$s(t+1) = P_3 s(t) + G_2 j(t),$$
$$a(t+1) = P_4 a(t) + G_3 s(t).$$

These equations specify that the number of cocoons at time $t + 1$ [$c(t + 1)$] is equal to the number of cocoons at time t that survived but did not hatch during the time interval t to $t + 1$ [$P_1 c(t)$] plus the number of viable cocoons produced by adults present at time t during the time interval t to $t + 1$ [$F_4 a(t)$].

The following formulas from Klok and de Roos (1996) were used to calculate the probability of remaining in a stage (P_i), growing to the next stage (G_i), and producing a certain number of cocoons (F_i):

$$P_i = \frac{\int_{A_1}^{A_2-\Delta} e^{-(r+\mu)\cdot a}\,da}{\int_{A_1}^{A_2} e^{-(r+\mu)\cdot a}\,da}\, e^{-\mu\Delta}$$

The numerator in this equation is proportional to the number of individuals that stay in stage i during a time step, and the denominator is proportional to the total number in stage i at t–Δ. Multiplying this fraction with the probability of surviving during the time step ($e^{-\mu\Delta}$) yields the complete expression for P_i:

$$G_i = \frac{\int_{A_1-\Delta}^{A_2} e^{-(r+\mu)\cdot a}\,da}{\int_{A_1}^{A_2} e^{-(r+\mu)\cdot a}\,da}\, e^{-\mu\Delta}$$

The proportion of individuals that survive and move to the next stage (G_i) is equivalent to P_i with the exception of the denominator, which now gives the number of individuals that move in one time step to the next stage. And the equation for the number of cocoons produced per time step (F) is equal to the product of the average adult reproductive rate (i.e., the total reproductive output of all adults in the population divided by the total number of adults present):

$$F_4 = \frac{\int_{A_{ad}}^{A_{max}} e^{-(r+\mu)\cdot a} \cdot m(a)da}{\int_{A_{ad}}^{A_{max}} e^{-(r+\mu)\cdot a} da} \Delta$$

In these three equations, i represents a stage, A_2–A_1 the duration of stage i, A_{ad} the adult age, A_{max} the maximum age, $m(a)$ the number of cocoons produced per worm of age a per day, r the intrinsic rate of population growth per time unit, μ the mortality rate, and Δ the time step in the model.

The entries of the matrix are based on laboratory data on the effect of copper in sandy loam soil on cocoon production (Ma 1984) and development (Klok and de Roos 1996). Data on earthworm growth pattern show that development is delayed with increasing copper concentration, whereas the weight at which individuals reach a developmental stage does not change (Klok and de Roos 1996). Cocoon production and development data under background copper conditions (13 mg/kg) were used to assess population growth in the absence of copper stress (Cu13 scenario). Since copper is a vital element for many enzymes, absence of copper will result in growth inhibition; therefore, a realistic background level is above zero copper. To assess the population-level impact of copper stress at 145 mg/kg, I used three parameterization scenarios to integrate the laboratory-derived effects of copper on reproduction and development: a DEB scenario (Cu145DEB), an additive scenario (Cu145add), and a single-effect scenario (Cu145single). The Cu145DEB scenario is equivalent to the best-fit scenario reported in Klok and de Roos (1996), which assumes that enhanced copper concentrations induce an increase in maintenance requirements for detoxification and an increase in the energy requirements to produce a single cocoon. The Cu145add scenario assumes that effects on reproduction and development are independent, and the Cu145single scenario uses effects on reproduction only.

All matrix entries and the standard deviation matrix entries of the unstressed population (Cu13) and the stressed populations (Cu145DEB, Cu145add, Cu145single) are based on data in table 5.1. The actual values of the entries can be found in the files L.rubellusCu13scramble.mp, L.rubellusCu145DEBscramble.mp, L.rubellusCu145addscramble.mp, and L.rubellusCu145singlescramble.mp in the CD-ROM included with this book.

Density dependence and stochasticity

Under field conditions, earthworm populations do not show unlimited growth, but tend to fluctuate around certain density levels, where levels depend on biotic and abiotic

Table 5.1. Life history data of *Lumbricus rubellus* under different copper levels in sandy loam soil.

Factor	Copper level in soil	
	13 mg/kg	145 mg/kg
Reproduction (cocoon/worm/day) average (CV%)	0.130 (30)	0.050 (50)
Duration in stage, in days		
Cocoon	42	42
Juvenile	63	91
Subadult	28	35
Adult[a]	619	584
Daily survival average (CV%)		
Cocoon[b]	0.999 (10)	0.999 (10)
Juvenile[c]	0.993 (10)	0.994 (10)
Subadult[c]	0.995 (10)	0.996 (10)
Adult[c]	0.997 (10)	0.997 (10)

[a] Based on a maximal life span of hatched individuals of 710 days (Klok and de Roos 1996).
[b] Based on the percentage of viable cocoons (Klok and de Roos 1996).
[c] Based on $S(t)=\left(\frac{(1-at)}{(1+bt)}\right)^k$, with a = 0.0014, b = 0.02, k = 0.369, and t based on the duration in stage (Klok and de Roos 1996).

environmental conditions. In particular, the influence of abiotic conditions on earthworms has received much attention in the literature. Soil properties have long been known to affect abundance and biomass of earthworms (Guild 1948; Edwards and Lofty 1977). Among these are pH (Edwards and Bohlen 1996) and organic matter content (Curry 1976; Edwards and Lofty 1977; Edwards and Bohlen 1996). Biotic factors, on the other hand, have received much less attention. However, only biotic factors such as predation and competition for food or space can be responsible for population regulation (Klok and de Roos 1996). Curry (2004) suggested that, in general, food limitation is a more important regulating factor than predation, given the fact that earthworm numbers strongly increase following organic amendment. However, predation can also have a dramatic impact on earthworm densities, as illustrated by predation by the golden plover *Pluvialis apricaria* that was reported to reduce earthworm abundance by 50% (Bengston et al. 1976).

In nonpolluted sandy loam grasslands, *L. rubellus* is reported to reach densities of 150 individuals per square meter, whereas under copper stress, at a copper level of 136 mg/kg, this density reduces to about 40 individuals per square meter (Ma 1988). Field measurements of earthworm density give a reasonable estimate of subadult and adult densities. Juveniles, however, are usually underrepresented because of their small size, which makes them difficult to detect. I therefore assumed that the densities reported by Ma (1988) are based on subadults and adults only and excluded cocoons and juveniles. I furthermore assumed a variation of 20% in these densities. Given ambiguity of what actually limits earthworm densities under field conditions, I applied two extreme forms of density dependence in the model to assess the range of possible density dependent effects: a compensatory density dependence (Beverton-Holt) and an overcompensatory density dependence (Ricker). Furthermore, I assumed that density dependence acts only on the subadult and adult stages.

Both demographic and environmental (lognormal) stochasticity was implemented in the models.

Catastrophes

Earthworms are very sensitive to droughts. Prolonged droughts may strongly reduce population densities, and recovery to predrought numbers can take up to two years. This is probably an effect of the strong decrease in cocoon production (Gerard 1967). Besides a halt in cocoon production, survival may also be reduced under low moisture conditions. Prevention of water loss is a major problem of earthworm survival, since water constitutes 75–90% of their body weight and their skin is permeable (Grant 1955). However, earthworms have evolved considerable ability to survive adverse moisture conditions by moving deeper into the soil, or even going into a quiescent stage (Edwards and Lofty 1977). Droughts usually take place in summer, but low temperatures in winter also reduce the humidity of the soil. I implemented droughts as catastrophes in the model assuming that a drought would completely halt cocoon production but would not affect survival. These catastrophes can occur twice a year; since the time step of the model is one week, this results in a probability equal to 0.019 (1/52) for both events.

Results

Time-invariant model

The largest eigenvalues of the matrices are given in table 5.2. This value decreases with an increase in copper and is lowest for the DEB scenario (Cu145DEB). The largest eigenvalue is directly related to the intrinsic rate of increase r ($\lambda = e^r$), which equals 0.13/day in unpolluted sandy loam soil and reduces to 0.004/day with 145 mg/kg copper under the assumption that vital rates are integrated (Cu145DEB). These values are in the range of those calculated in an earlier model application (Klok and de Roos 1996). Adult survival (entry P_4) has the highest elasticity, both without and with copper stress in all scenarios (see table 5.2). The ranking of the elasticities changes with the parameterization scenario. For Cu13, Cu145add, and Cu145single, this ranking equals $P_4 > P_2 > P_1 > P_3 > G_1 = G_2 = G_3 = F_4$, whereas for Cu145DEB, P_1 is replaced by P_3. Table 5.2 also shows that with an increase in copper, the older stages constitute a larger part of the population. In unpolluted soil, the cocoon stage makes up almost 50% of the population. With copper at 145 mg/kg, the percentage in the adult stage strongly increases (see table 5.2).

Density dependence, stochasticity, and catastrophes

Under density dependence, population growth is no longer unlimited but fluctuates around an equilibrium density. Figure 5.2 shows the trajectory summaries obtained from 500 simulation runs of the unpolluted (figure 5.2a,b) and the three polluted (figure 5.2c–h) model populations. The population density declined in the integrative scenario (Cu145DEB), whereas in the single-effect and additive scenarios (Cu145single

Table 5.2. Time-invariant model results of *L. rubellus* under sublethal copper stress in sandy loam soil.

Model	λ	Rank elasticities	Stable stage distribution (%)			
			Cocoons	Juveniles	Subadults	Adults
Cu13	1.094	$P_4>P_2>P_1>P_3>G_1=G_2=G_3=F_4$	47	34	7	12
Cu145DEB	1.034	$P_4>P_2>P_3>P_1>G_1=G_2=G_3=F_4$	30	36	13	21
Cu145single	1.056	$P_4>P_2>P_1>P_3>G_1=G_2=G_3=F_4$	37	34	8	21
Cu145add	1.048	$P_4>P_2>P_1>P_3>G_1=G_2=G_3=F_4$	35	39	8	19

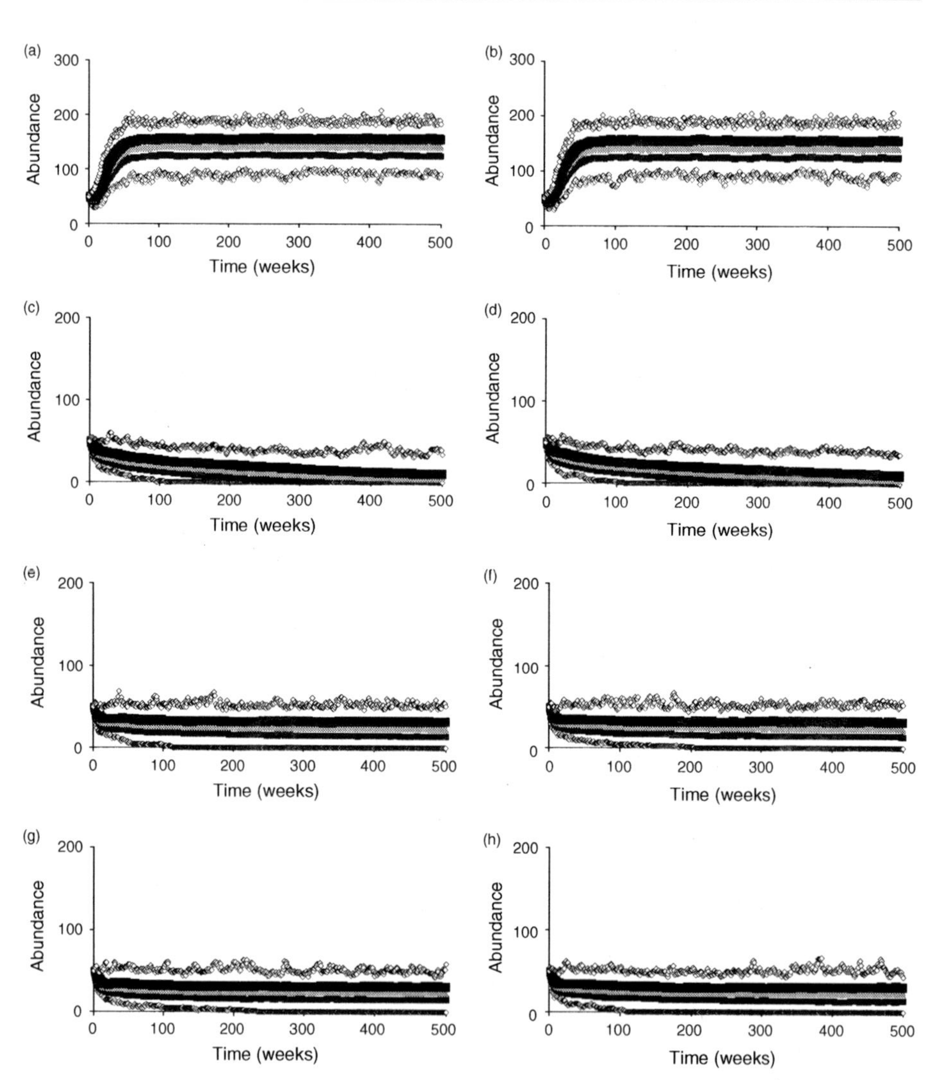

Figure 5.2. Mean population size of *L. rubellus* (adults and subadults; gray lines) with bounds (± one standard deviation; black lines) and population maxima and minima (diamonds) as a function of time, based on 500 replicates. Left column, Scramble competition; right column, Contest competition. a and b, Cu13; c and d, Cu145DEB; e and f, Cu145single; g and h, Cu145add.

Table 5.3. Density-dependent model results of *L. rubellus*: Final stage abundance [number/m^2, mean ± SD (composition in %)], based on 500 time steps and 500 replicates.

	Scramble (Ricker)				Contest (Beverton-Holt)			
Scenario	Cocoons	Juveniles	Subadults	Adults	Cocoons	Juveniles	Subadults	Adults
Cu13	330 ± 92 (47)	246 ± 46 (34)	39 ± 10 (7)	92 ± 10 (12)	337 ± 84 (47)	200 ± 46 (34)	49 ± 10 (7)	81 ± 10 (12)
Cu145DEB	3.6 ± 7.1 (31)	3.9 ± 7.4 (34)	1.2 ± 2.3 (10)	2.9 ± 5 (25)	3.2 ± 6 (30)	3.6 ± 6.5 (34)	1.1 ± 2.3 (10)	2.7 ± 4.7 (26)
Cu145single	29 ± 19 (35)	28 ± 14 (35)	6 ± 3 (7)	18 ± 7 (22)	29 ± 19 (36)	27 ± 14 (34)	5 ± 3 (6)	18 ± 7 (23)
Cu145add	28 ± 16 (33)	33 ± 15 (40)	6 ± 4 (7)	17 ± 6 (20)	27 ± 16 (33)	33 ± 16 (40)	2 ± 3 (7)	10 ± 7 (20)

Table 5.4. Density-dependent model results of *L. rubellus*: Quasi-extinction duration, number of successive weeks subadult and adult worms are absent from the population, based on 500 time steps and 500 replicates.

	Scramble (Ricker)		Contest (Beverton-Holt)	
Scenario	Mean	SD	Mean	SD
Cu13	0	0	0	0
Cu145DEB	98	58	102	63
Cu145single	6	31	6	24
Cu145add	2	15	4	21

and Cu145add) the population fluctuated around equilibrium. Trajectory summaries are relatively similar for both types of density dependence (Scramble, left column; Contest, right column). The final stage abundance (table 5.3) is equivalent to the stable stage abundance in the density-independent analysis, with the exception of the integrative scenario (Cu145DEB), where the percentage of adults in the population is higher in the density-dependent case (see table 5.2). The population composition under density dependence is insensitive to the type of competition.

The quasi-extinction duration (table 5.4) was zero with background copper levels (Cu13) and highest for the integrative scenario (Cu145DEB). The type of density dependence seems to play a role in the quasi-extinction duration, with longer quasi-extinction durations for Contest competition compared to Scramble competition; however, standard deviations are quite large, so these differences are not significant. Figure 5.3 shows the probability that the population abundance (adults plus subadults) will fall below a certain size. As can be inferred from figure 5.3, the probability that the unpolluted population will decline is relatively low. The polluted populations were highly likely to decline; this probability was largest for the integrative scenarios (Cu145DEB) where the likelihood of falling below 10 individuals/m^2 exceeded 80% (figure 5.3). For

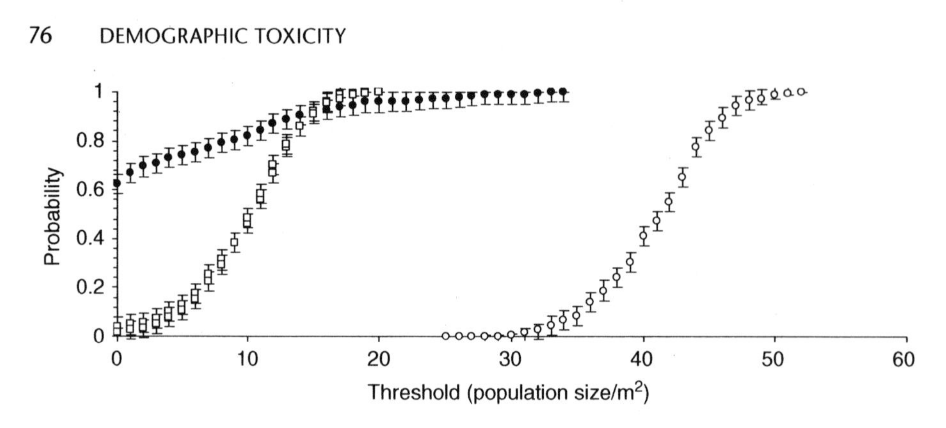

Figure 5.3. Interval quasi-extinction risk curve of *L. rubellus*: probability of falling below a particular population size (adults and subadults) within 500 time steps (from 500 replicates): Scramble competition results only (means and 95% confidence intervals). Solid circles, Cu145DEB; open squares, Cu145single and Cu145add; open circles, Cu13.

the single-effect and additive scenarios (Cu145single and Cu145add), the probability was lower (40%).

Discussion

Density-independent effects

Common practice in risk assessment is to apply standardized tests (e.g., Organisation for Economic Co-operation and Development 1984). These include tests on reproduction and survival of adult organisms with a limited duration. Generally, development is not tested, since for many species this may strongly extend the test duration. At the individual level, reproduction often is the most sensitive parameter, as is shown here for *L. rubellus* (see table 5.1). The question is whether populations will be protected if we set soil quality criteria based on effect concentrations for reproduction. The decrease in reproduction from 0.13 cocoons/worm/d to 0.05 cocoons/worm/d (see table 5.1) equals 62%. The intrinsic rate of increase, however, is reduced by 97% (0.13–0.004). The most sensitive endpoint at the individual level does not seem to reflect the risk for the population per se. Therefore, a population-level interpretation based on individual-level effects with structured models is a useful exercise to gain insight on population-level risk.

The integrative scenario has a stronger effect than the single and the additive scenarios. The integrative scenario works out as more than additive, because in the DEB model, reproduction is intimately linked to growth through the energy budget of the individual. An increase in maintenance costs for detoxification thus results not only in slower growth but also in lower cocoon production since smaller individuals produce fewer cocoons. Clearly, using only the reproductive data will strongly underestimate the population-level effect. Using reproduction and development data as independent and additive effects results in an overoptimistic estimate of consequences at the population level.

The elasticity analysis shows that adult survival (entry P_4 in the matrix) has the largest influence on the population growth rate. The experimental data on the growth pattern of *L. rubellus* show that with copper stress, the size at which individuals becomes adults is fixed, whereas the time to reach this size increases (Ma 1984). Given the fact that the life span of the species is not expected to change with copper (the tested levels induced only sublethal effects), the duration of the adult stage consequently decreases with the copper level. At extreme copper levels, this can result in individuals not reaching adult stage at all (Klok and de Roos 1996). Therefore, if population-level consequences of sublethal pollutant levels are of interest, studies only on individual reproduction do not seem to be the appropriate measure. Studies on the duration of the preadult stages will be a more informative indicator of the population-level consequences and therefore should be included in toxicity tests.

Density-dependent effects and the influence of droughts

Obviously, the time-invariant results are bound to give a conservative estimate of the effect of contaminants at the population level, since they are based on laboratory data that show the impact of the tested contaminant under optimal conditions. In the field, environmental conditions are not expected to be optimal; food usually is a limiting factor, and other factors tend to fluctuate. To assess the influence of copper under more natural conditions, I included density dependence, demographic stochasticity, and droughts in the model. Information on what actually limits density in earthworms is relatively unknown, as is the form of density dependence. Based on field data of earthworm densities in unpolluted sandy loam soil and in this soil with copper (Ma 1988), I implemented the two most extreme forms of density dependence, the compensatory Beverton-Holt function and the overcompensatory Ricker function, to gain insight in the influence of the form of density dependence on the population behavior. The results show that the actual form of density dependence does not have a large influence on this behavior. Under these more natural conditions, the polluted populations are highly likely to decline, a result that was not obvious from the density-independent results.

Integration of effects on vital rates

In ecotoxicological studies, endpoints at the individual level—growth, reproduction, and survival—are often treated as independent (e.g., Forbes and Calow 1999, and references therein). However, they are actually strongly linked (Kooijman and Metz 1984; Klok and de Roos 1996; Kooijman 2000). Individual growth can influence reproductive output. If individual growth is retarded, maturation is delayed. Treatment of growth and reproduction as independent endpoints may therefore result to an overestimation of population growth rate (as shown in this chapter).

Within the DEB framework, the impact of toxicants on vital rates works through changes in energy allocation, for example, maintenance through repair. Therefore, food levels can have a large impact on the perceptibility of toxic effects. Using the DEB framework exemplifies that an effect at high food levels does not necessarily result in the same effect at low food levels. If a toxicant increases the maintenance costs and food is abundant, effects may not be apparent, whereas with low food abundance this increase may have a dramatic influence on, for example, reproduction. This complicates assessment of population behavior under density dependence, for example, when density dependence results from competition for food. Physiologically structured population models (Kirkilionis et al. 2001; De Roos et al. 2004) can cope with these difficulties.

References

Akçakaya, H.R. 2005. *RAMAS GIS: Linking Spatial Data with Population Viability Analysis* (version 5.0). Applied Biomathematics, Setauket, New York.

Bengston, S.A., Nilsson, A., Nordström, S., and Rundgren, S. 1976. Effects of bird predation on lumbricid populations. *Oikos* 27: 9–12.

Bengtsson, G., and Tranvik, L. 1989. Critical metal concentrations for forest soil invertebrates. A review of the limitations. *Water Air and Soil Pollution* 47: 381–417.

Curry, J.P. 1976. Some effects of animal manures on earthworms in grassland. *Pedobiologia* 16: 425–438.

Curry, J.P. 2004. Factors affecting the abundance of earthworms in soils. Pages 91–113 in C.A. Edwards (ed.), *Earthworm Ecology*, 2nd ed. CRS Press, London.

Darwin, C. 1881. *The Formation of Vegetable Mould through the Action of Worms, with Observations of Their Habitats.* Murray, London.

De Roos, A.M., Persson, L., and McCauley, E. 2004. The influence of size-dependent life-history traits on the structure and dynamics of populations and communities. *Ecology Letters* 6: 473–487.

Didden, W.A.M. 2001. Earthworm communities in grasslands and horticultural soils. *Biology and Fertility of Soils* 33: 111–117.

Edwards, C.A., and Bohlen, P.J. 1996. *Biology and Ecology of Earthworms.* Champan and Hall, London.

Edwards, C.A., and Lofty, J.R. 1977. *Biology of earthworms*, 2nd ed. Chapman and Hall, London.

Eeva, T., and Lehikoinen, E. 1996. Growth and mortality of nesting tits (*Parus major*) and pied flycatchers (*Ficedula hypoleaca*) in a heavy metal pollution gradient. *Oecologia* 108: 631–639.

Forbes, V.E., and Calow, P. 1999. Is the per capita rate of increase a good measure of population-level effects in ecotoxicology? *Environmental Toxicology and Chemistry* 18: 1544–1556.

Gerard, B.M. 1967. Factors affecting earthworms in pastures. *Journal of Animal Ecology* 36: 235–252.

Grant, W.C. 1955. Temperature relasionships in the megasco-earthworm, Pheretima hupeiensis. *Ecology* 36: 412–417.

Guild, W.J.M. 1948. Studies on the relationship between earthworms and soil fertility. III. The effect of soil type on the structure of earthworm populations. *Annals of Applied Biology* 35: 181–192.

Kirkilionis, A.M., Diekmann, O., Lisser, B., Nool, M., Sommerijer, B., and De Roos, A.M. 2001. Numerical continuation of equilibria of physiologically structured population models. I. Theory. *Mathematical Models and Methods in Applied Sciences* 11: 1101–1127.

Klok, C., and de Roos, A.M. 1996. Population level consequences of toxicological influences on individual growth and reproduction in *Lumbricus rubellus* (Lumbricidae, Oligochaeta). *Ecotoxicology and Environmental Safety* 33: 118–127.

Klok, C., de Roos, A.M., Marinissen, J.C.Y., Baveco, J.M., and Ma. W.C. 1997. Assessing the impact of abiotic environmental stress on population growth in *Lumbricus rubellus. Soil Biolology and Biochemistry* 29: 287–293.

Klok, C., Faber, J., Heijmans, G., Bodt, J., and van den Hout, A. 2006. Influence of clay content and acidity of soil on development of the earthworm *Lumbricus rubellus* and its population level consequences. *Biology and Fertility of Soils*.43: 549–556.

Kooijman, S.A.L.M. 2000. *Dynamic Energy Mass Budgets in Biological Systems.* Cambridge University Press, Cambridge.

Kooijman, S.A.L.M., and Bedaux, J.J.M. 2000. Dynamic effects of compounds on animal energetics and their population consequences. Pages 27–41 in J.E. Kammenga and R. Laskowski (eds.), *Demography in Ecotoxicology*. Wiley, Sussex.

Kooijman, S.A.L.M., and Metz, J.A.J. 1984. On the dynamics of chemically stressed populations: The deduction of population consequences from effects on individuals. *Ecotoxicology and Environmental Safety* 8: 254–274.

Lavelle, P., and Spain, A.V. 2001. *Soil Ecology*. Kluwer Academic Publishers, Dordrecht.

Lee, K.E. 1985. *Earthworms Their Ecology and Relationships with Soils and Land Use*. Academic Press, London.

Ma, W.C. 1984. Sublethal toxic effects of Cu on growth, reproduction and litter breakdown activity in the earthworm *Lumbricus rubellus*, with observations on the influence of temperature and soil pH. *Environmental Pollution* 33: 207–219.

Ma, W.C. 1988. Toxicity of copper to lumbricid earthworms in sandy agricultural soils amended with Cu-enriched organic waste materials. *Ecological Bulletins* 39: 53–56.

Nordström, S., and Rundgen, S. 1974. Environmental factors and lumbricid associations in southern Sweden. *Pedobiologia* 14: 1–27.

Organisation for Economic Co-operation and Development. 1984. *Guidelines for the Testing of Chemicals 207. Earthworm Acute Toxicity Tests*. OECD Bookstore, Paris.

Swartjes, F.A. 1999. Risk-based assessment of soil and groundwater quality in the Netherlands. Standards and remediation urgency. *Risk Analysis* 19: 1235–1249.

6

Stressor Impacts on Common Loons in New Hampshire, USA

A Demonstration Study for Effects of Stressors Distributed across Space

STEVEN WALTERS
ANNE KUHN
MATTHEW C. NICHOLSON
JANE COPELAND
STEVEN A. REGO
DIANE E. NACCI

Organisms, the resources upon which they depend, and the stressors that affect them typically exhibit heterogeneous distributions. Such heterogeneity results in spatially distributed populations or, at more extreme levels of heterogeneity, metapopulations (or "populations of subpopulations"; Levins 1970; Hanski 1998) whose broader scale regional dynamics are driven by localized dynamics (births, deaths, etc.) coupled with interactions among subpopulations (e.g., dispersal). Methods and modeling frameworks that explicitly incorporate spatial dimensions of population dynamics can therefore be essential for effective risk assessments. Knowledge of the demography of a wildlife species can be combined with spatially explicit habitat data and information on stressors, through the use of GIS. Such tools can be used to forecast the relative spatial distribution of organisms and to identify potential hotspots at which species and stressors significantly overlap (Dunning et al. 1995; Turner et al. 1995, 2001; Johnson 2002; Landis 2002). Theoretical studies suggest that spatial dependence in wildlife demography and stressor exposure levels can lead to dramatic differences in predictions derived from nonspatial versus spatial modeling approaches (Hiebeler 1997, 2000). In particular, localized stressors affecting source and sink subpopulations (i.e., subpopulations exhibiting net positive vs. negative growth, respectively; Pulliam 1988), interconnected via dispersal, can translate to regional level metapopulation impacts (Pulliam and Danielson 1991).

To illustrate the importance of spatial context, we simulated spatially explicit stressor effects on populations of common loons (*Gavia immer*) with summer breeding habitat in lakes of New Hampshire. Common loons are listed as threatened in New Hampshire, and there are concerns that local and long-range human activities may negatively affect loon populations. For example, increasing human development may encroach upon lakeshore nesting sites (Sutcliffe 1978; Jung 1991), and atmospherically deposited pollutants, including mercury, may affect loon fitness (Barr 1986; Meyer et al. 1998). Such stressors can exhibit distinct spatial distributions due to regional climatic and physiographic variability and patterns of human settlement (CEP-NWF 2003; Chen et al. 2005; Kamman et al. 2005; Kramar et al. 2005; Yates et al. 2005).

In this chapter, we describe methods of integrating demography, habitat preferences, and dispersal characteristics for loon populations in New Hampshire to illustrate risks posed by exposure to stressors having complex spatial distributions. The study serves as the preliminary foundation of a broader study demonstrating the utility of various spatially explicit approaches for realistically assessing risks of multiple stressors to wildlife populations. As such, the objectives of this study focused on exploring factors relevant both to general spatial dependence in population dynamics and to the metapopulation modeling approach in particular, with the ultimate goal of contributing to a more detailed risk assessment. We used RAMAS GIS (Akçakaya 2005) to estimate source–sink characteristics (Pulliam 1988; Pulliam and Danielson 1991) and lake-level occupancy under hypothetical unstressed conditions. We then explored metapopulation dynamics under conditions of stressor-induced reductions in fecundity in either source or sink subpopulations, or in distinct subregions within New Hampshire. Simulations of varying stressor exposure scenarios were used to examine the hypothesis that local increases in exposure levels may have broader effects on the state-level metapopulation in a habitat network that is highly connected through dispersal.

Methods

Study species

The common loon (*Gavia immer*) is a large, aquatic-dependent obligate piscivorous bird (McIntyre and Barr 1997). North American breeding populations use freshwater lakes and large riverine systems as their summer breeding habitat, from as early as late March to as late as December, establishing nests on nearshore wetland or grassy areas or, when available, floating nest platforms (Barr 1986; Piper et al. 2002). Nearby coastal marine and estuarine habitats typically serve as destinations for winter migrations, although birds have been observed migrating from the Great Lakes to as far south as the Gulf of Mexico (Kenow et al. 2002). Breeding pairs are territorial, defending regions that range from 40 to 600 hectares in extent (Evers 2001). Loon pairs are highly site fidelic, returning to the same breeding site annually except when the territory is usurped by conspecifics (Piper et al. 1997, 2000). Loon territories can span one or multiple lakes (Piper et al. 1997; Evers 2001). Lakes smaller than about 40 hectares are too limited to function as a territory but can contribute to a multilake territorial assemblage. Lakes of roughly 80–120 hectares have been found to serve as a whole territory for a single pair, and lakes larger than 400 hectares are of sufficient size to support territories for multiple breeding pairs.

Common loons are thought to have a life span of 25–30 years, similar to that of Arctic loons (Nilsson 1977). Reproductive maturity is believed to occur in the third or fourth year, but mate finding and territorial success may be suboptimal in these young adults. Clutch sizes typically range from two to four eggs, and hatchlings fledge in late summer. Little is known about the movements and fate of juveniles until they reappear near natal lakes by three to five years of age (McIntyre and Barr 1997; Piper et al. 1997; Evers 2001).

Though loon populations at the scale of New Hampshire do not necessarily fit the classic definition of a metapopulation, the construct may nonetheless be useful for representing broad-scale dynamics across the network of lakes. Loon populations can be thought of as existing as metapopulations in the broad sense of the term (Hanski 1998) in that (1) discrete breeding groups occupy spatially distributed "patches" (i.e., freshwater lakes) with a low level of exchange between patches from year to year, and (2) patches experience nonnegligible rates of localized extinction and colonization through lake territory vacancies and (re)establishments via juvenile dispersers and "floaters" (i.e., unpaired, nonbreeding loons), respectively (Piper et al. 1997; Evers 2001). For simplicity's sake, we thus refer here to groups of territories on a lake (or cluster of lakes—see below) as a "subpopulation" and the collection of lakes in New Hampshire as a "metapopulation."

Study region

Located in the northeastern United States, New Hampshire is largely a forested state, with approximately 81.1% of its land area composed of deciduous and evergreen forest cover (NH GRANIT 2002). Of specific relevance to loon populations, however, is the presence of a considerable network of freshwater lakes, particularly in the southeastern part of the state (figure 6.1).

New Hampshire has been experiencing steady declines in forest cover over the past two decades (CEP-NWF 2003), which are attributable principally to a rapidly growing human population (U.S. Department of Commerce 2000) and the accompanying conversion of land to developed uses. The consequences are particularly severe for loons because a significant percentage of this human development has involved the construction of summer recreational housing around lake shores (Sutcliffe 1978; CEP-NWF 2003). The impacts for loons stem from both a decreased availability of suitable nest sites (Barr 1986; Jung 1991) and altered runoff regimes and water chemistry within the lakes (Kamman et al. 2005; Kramar et al. 2005).

Model components

Our model as implemented in RAMAS GIS (Akçakaya 2005) consisted of two primary subcomponents: a model of habitat suitability (HS; Morrison et al. 1998) and a stage-structured matrix population model (Caswell 2001). The models were derived from preliminary analyses of loon census data collected in New Hampshire (Loon Preservation Committee, unpublished data), New Hampshire environmental data (N.H. Department of Environmental Services), and banding studies conducted at multiple North American locations (Mitro et al. 2005). The HS model used in the present study is a simplified version of a series of HS modeling approaches currently being evaluated for common loons in New Hampshire (Kuhn et al. 2005). The model was developed using a binary logistic stepwise regression analysis relating lake-level loon presence/absence to the following attributes (NH GRANIT 2002): lake area,

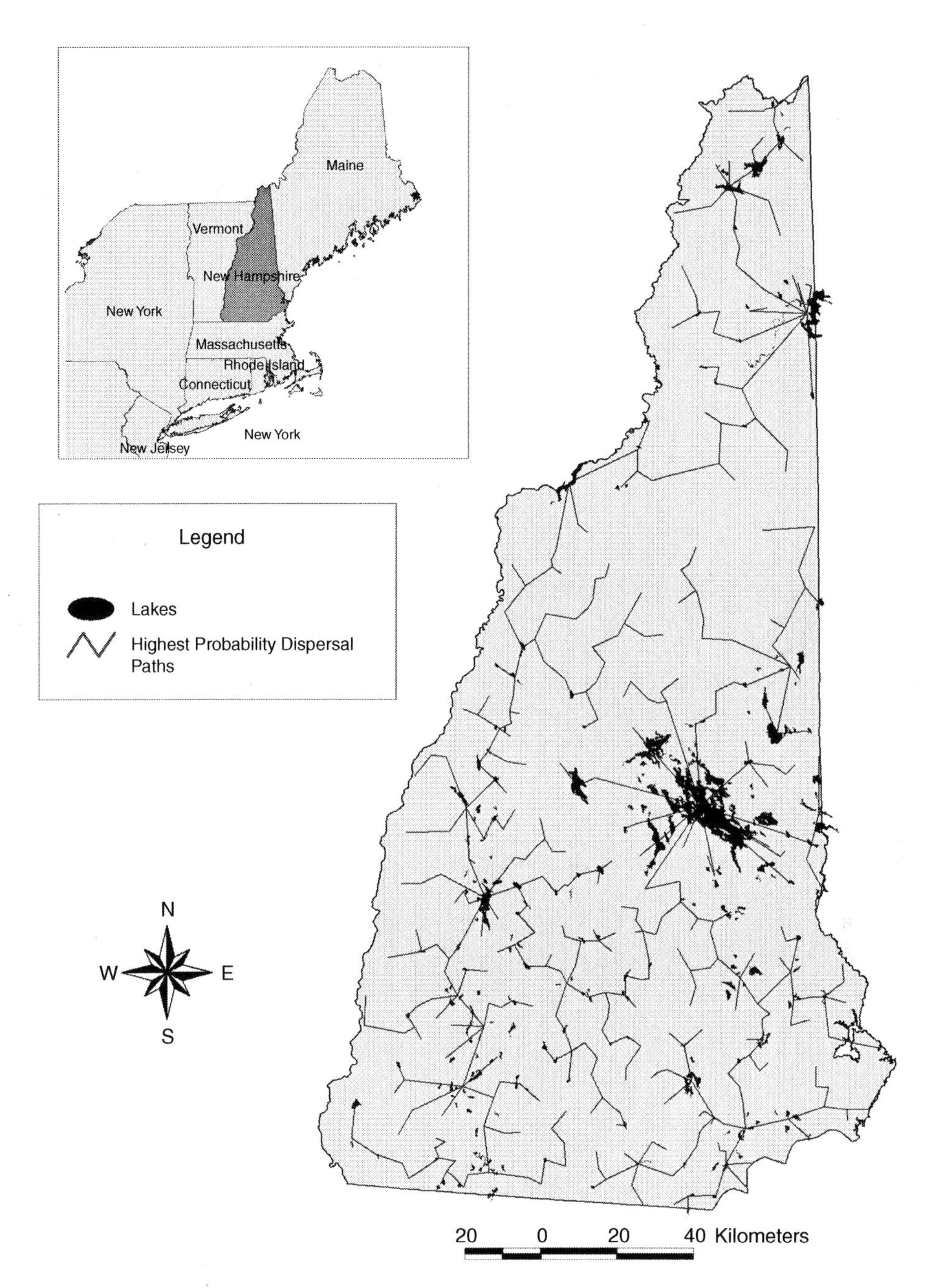

Figure 6.1. Lakes of New Hampshire, USA, serving as potential breeding habitat for common loons. Lines indicate the shortest interlake distances, and hence most probable dispersal routes assuming a negative exponential dispersal function, between all lakes within New Hampshire.

shoreline length, perimeter:area ratio, number of islands, depth, clarity, trophic status (oligo-, meso-, or eutrophic), and the areal extent of land cover classes (deciduous, evergreen and mixed forest, agriculture, developed, wetland, disturbed, and cleared classes) within 500 m of the shoreline. Of these variables, only lake area (in km^2; $\chi^2 = 10.21$, $p = 0.0014$), shoreline perimeter (in km; $\chi^2 = 25.46$, $p < 0.0001$), and deciduous forest area near the shoreline (in hectares; $\chi^2 = 21.42$, $p < 0.0001$) were significant and hence included in the final simplified HS model. The resultant logistic regression model took the following form:

$$HS = \frac{\exp(\hat{\pi})}{1 + \exp(\hat{\pi})}, \tag{6.1}$$

where

$$\begin{aligned}\hat{\pi} = {} & (2.33 \times 10^{-2} \cdot \text{deciduous}) - (1.40 \times 10^{-4} \cdot \text{perimeter}) \\ & + (2.27 \times 10^{-6} \cdot \text{area}) - 1.50.\end{aligned} \tag{6.2}$$

The result is a continuous variable, ranging from 0 to 1, that can be used as an index of HS for a given lake. Based on a comparison of presence/absence data with predicted HS values, we considered lakes with HS of 0.5 or greater to comprise habitat sufficient for one or more loon breeding pairs; the end result was a total of 224 out of 611 lakes being incorporated as subpopulation sites in the model.

Preliminary analyses of banding studies from several North American locations (Mitro et al. 2005) and loon census data collected in New Hampshire (Loon Preservation Committee, unpublished data) were then used to construct the matrix model subcomponent. The females-only, postbreeding-census, birth-pulse matrix model (Caswell 2001) consisted of two life stages: fledged juveniles, encompassing approximately the first three years of life prior to maturity when loons return to freshwater breeding territories, and adults (approximately three or more years of age; McIntyre and Barr 1997). Like the habitat model, the transition matrix was also an approximation derived from an ongoing evaluation of demographic modeling and estimation techniques (Grear et al. 2005). The basic matrix took the following form:

$$\mathbf{A} = \begin{bmatrix} 0.6 & 0.2 \\ 0.2 & 0.9 \end{bmatrix} \tag{6.3}$$

Diagonal terms indicate the probability that individuals survive and remain in their respective stage classes, the lower left term is the probability that juveniles survive and transition to adulthood in a given year, and the upper right term represents adult fecundity. The per capita deterministic rate of increase, λ, computed as the dominant eigenvalue of **A**, is 1.0, reflecting a growth rate of approximately zero. Given this growth rate, we assumed for the present study that the above base matrix describes a population at or near carrying capacity. Because productivity is considered to be the key factor limited by the amount of habitat suitable for breeding territories (McIntyre and Barr 1997; Evers 2001), we adjusted the matrix to simulate growth in subpopulations below carrying capacity (see below) by increasing adult fecundity from 0.2 to 0.4 (assuming a doubling of adult breeding propensity from 50% to 100%), resulting in an estimated maximum growth rate R_{max} of 1.07. Lastly, to simulate hypothetical effects of environmental stochasticity, we assumed a lognormal distribution for the demographic rates and arbitrarily estimated their standard deviations (SDs) as 10% of the mean parameter estimates in the transition matrix in equation 6.3.

The matrix population model was applied either to individual lakes or to clusters of lakes based on lake size. Because carrying capacity of lakes depends on the number of loon territories they can support (McIntyre and Barr 1997; Evers 2001), the model incorporated a Beverton-Holt (i.e., contest-based) form of density dependence (Morris and Doak 2002) by reducing fecundity as subpopulations approached saturation. We established the following rules in the spatial data subprogram of RAMAS GIS (Akçakaya 2005) for calculating carrying capacity as a function of lake size and HS (Barr 1986; Piper et al. 1997; Evers 2001):

1. Lakes smaller than 4 hectares are not used as part of a loon territory.
2. Lakes from 4 to 40 hectares and within 2 km of similar-sized lakes are used as components of multilake territories by single breeding pairs.
3. Lakes from 40 to 120 hectares are occupied by single breeding pairs.
4. Lakes greater than 120 hectares are composed of multiple territories, with 60–600 hectares per territory as a function of HS.

Data on loon dispersal were derived from Evers (2001) and Piper et al. (1997): Mean observed dispersal distance between a bird's natal territory and a newly colonized territory was 4.5 km, with an SD of ±4.6 km; the maximum observed dispersal distance was 20.5 km. We parameterized the negative exponential dispersal function in RAMAS GIS using these observations to estimate the probability of dispersal among lakes. The model included dispersal only within the subadult classes, as is common in loons and many other species (Greenwood 1980; McIntyre and Barr 1997; Piper et al. 1997; Evers 2001). The collective network of subpopulations occupying the lakes, when connected via dispersal among lakes, composed a hypothetical loon metapopulation for the State of New Hampshire.

Simulation scenarios

The first series of three simulation scenarios was designed to examine stressor effects on metapopulation dynamics as a function of demographic characteristics of the stressed subpopulations. We compared the results from these scenarios to examine the hypothetical effects of heightened stressor exposure at key spatial locations as they might affect the broader metapopulation. Stressors were incorporated in the metapopulation submodel of RAMAS GIS by altering the transition matrix (equation 6.3) for specific subpopulations, as described below. The scenarios were as follows:

1. Unstressed scenario—the model incorporated optimal fecundity levels for all lakes and lake assemblages, as reflected in equation 6.3.
2. Increased stressor effects in source subpopulations—fecundity was reduced by 20% in source subpopulations (i.e., sites that remained occupied through all time steps of the simulation and hence are net exporters of individuals) identified in scenario 1 (figure 6.2).
3. Increased stressor effects in sink subpopulations—fecundity was reduced by 20% in sink subpopulations (i.e., sites that were unoccupied more frequently than occupied through all time steps of the simulation and hence are net importers of individuals) identified in scenario 1 (figure 6.2).

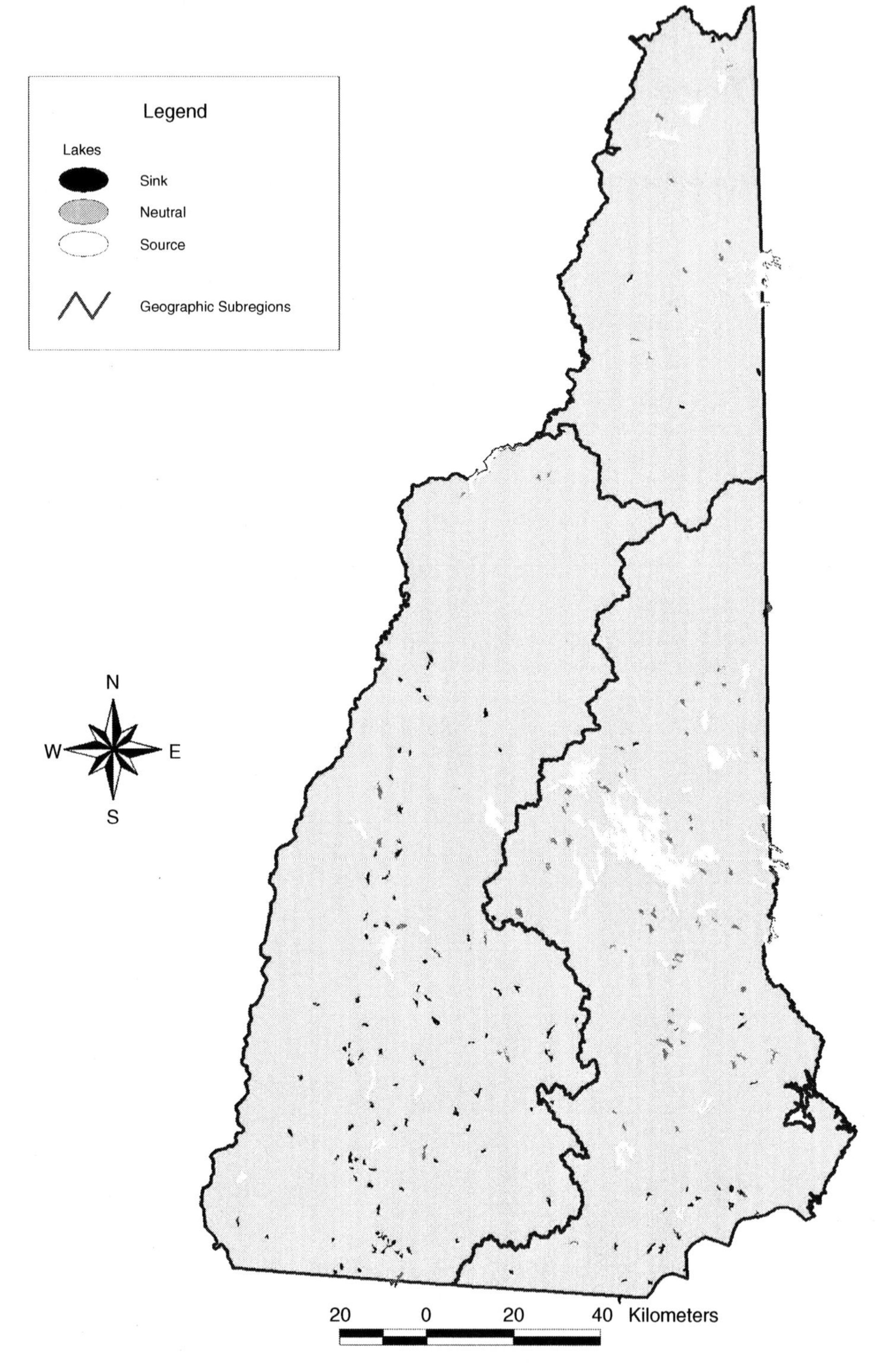

Figure 6.2. Delineation of lake subpopulations in New Hampshire for spatially explicit stressor exposure scenarios. Source, sink, and neutral populations were defined as a function of net occupancy rates in baseline simulations incorporating no stressor effects. Lines indicate the subregions used for simulations of geographically distinct stressor exposure effects.

Table 6.1. Number of source, sink, and neutral subpopulations (associated carrying capacities), by New Hampshire subregion.

	Demographic status			
Geographic subregion	Source	Sink	Neutral	Total by subregion
Northern	4 (105)	4 (5)	10 (12)	18 (122)
Southwestern	9 (113)	88 (101)	9 (22)	106 (236)
Southeastern	21 (564)	27 (31)	52 (73)	100 (668)
Total by demographic status	34 (782)	119 (137)	71 (107)	224 (1,026)

We conducted another set of simulations explicitly to examine stressor effects occurring solely as a function of geographic location, regardless of lake quality. The scenarios consisted of the following:

1. Heightened stress in northern subpopulations—fecundity was reduced by 20% in subpopulations located in northern New Hampshire (figure 6.2); such declines might represent increased stress due to heightened lake acidification effects, an increase in anthropogenic lakeshore development, or a combination of the two.
2. Heightened stress in southeastern subpopulations—fecundity was reduced by 20% in subpopulations located in southeastern New Hampshire (figure 6.2); this scenario could simulate conditions in which increasing human populations, already highest in this portion of New Hampshire (U.S. Department of Commerce 2000; CEP-NWF 2003), might impose greater stresses on the subpopulations.
3. Heightened stress in southwestern subpopulations—fecundity was reduced by 20% in subpopulations located in southwestern New Hampshire (figure 6.2); this scenario could simulate declines due to hypothetical increases in lake acidification and/or human development.

Table 6.1 summarizes the number of lakes (i.e., subpopulations) and their associated carrying capacities for the above scenarios. In all instances, simulations were run with 10,000 replications, for a total of 200 years each to allow subpopulation sizes to stabilize. Initial subpopulation sizes were set at half the carrying capacity of the respective lakes. Endpoints of interest were the rates of metapopulation occupancy (in number of lakes) and the probability of decline in metapopulation size (in total number of individuals across all subpopulations) under unstressed versus stressed conditions; measures for probability of decline were compared using the Kolmogorov-Smirnov two-sample test (Sokal and Rohlf 1981).

Results

Effects of stressor exposure in sources versus sinks

Incorporating stressor effects in distinct subpopulations within the metapopulation had clear effects on model endpoints, depending on the source–sink status of the subpopulations (figure 6.3). Relative to simulations that included no stressor effects, simulations incorporating heightened stressor levels in source subpopulations resulted in a consistently lower mean metapopulation occupancy rate (figure 6.3a) across all time steps: The number of occupied lakes declined by approximately 20% by the end of

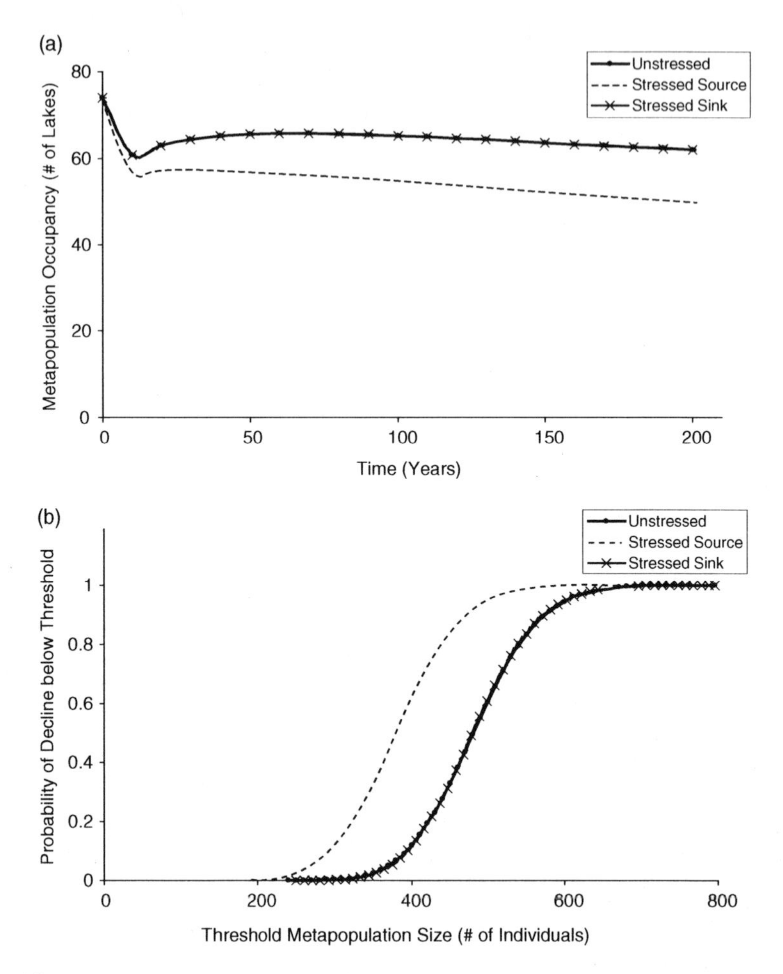

Figure 6.3. Metapopulation dynamics under different scenarios of stressor exposure: unstressed conditions, stressor effects in source subpopulations, and stressor effects in sink subpopulations. Trends are shown for metapopulation occupancy (a) and the probability of decline below a threshold metapopulation size (b) by the end of the 200-year simulations.

the 200-year simulations. The risk of metapopulation decline with increased stress in source subpopulations was also greater than observed under unstressed conditions: A decline in metapopulation size (i.e., the number of individuals across all subpopulations) of up to 480 individuals by the end of the 200-year simulations, occurring in about 50% of the unstressed scenario simulations, was observed in approximately 90% of simulations incorporating increased stressor levels in source subpopulations (figure 6.3b, table 6.2). Note that even in the unstressed case, population density never reached the estimated potential carrying capacity of 1,026 individuals or resulted in all 224 lakes being occupied, a model characteristic that is congruent with metapopulation theory (Levins 1970; Hanski 1998).

Results were quite different, however, with reduced fecundity on sink lakes. Trends in metapopulation occupancy (figure 6.3a) and risk of decline in metapopulation size

Table 6.2. Kolmogorov-Smirnoff test statistic (*D*) comparing risks of terminal threshold declines in total metapopulation size as a function of stressor effects in sources versus sinks (figure 6.3b).

	Kolmogorov-Smirnoff test relative to unstressed dynamics	
Stressor effects condition	*D*	*p*
Source subpopulations	0.535	<0.001
Sink subpopulations	0.014	NS[a]

[a] Not significant.

(figure 6.3b) did not differ from levels observed for unstressed conditions, despite such stressor effects occurring in more than half the subpopulations within the metapopulation.

Effects of geographically specific stressor distributions

Though less dramatic than simulated stressor effects in source subpopulations (figure 6.3), simulations incorporating hypothetical stressor-induced declines within geographic subregions of New Hampshire also exhibited metapopulation-level impacts relative to dynamics observed in the absence of stressor effects. Decreased fecundity in northern and southwestern subpopulations resulted in approximately a 3% decline in metapopulation occupancy by the end of the 200-year simulations, whereas similar demographic impairment in southeastern subpopulations resulted in more than a 10% decline (figure 6.4a). Differences in metapopulation dynamics under unstressed versus geographically explicit stressed conditions were also apparent in the risk of decline in metapopulation size (figure 6.4b, table 6.3). Effects of simulated stressors in northern and southwestern subpopulations were relatively similar: A decline to a threshold abundance of 480 individuals by the end of the simulation occurred in approximately 50% of unstressed simulations and approximately 60% of simulations in which stressors affected either northern or southwestern subpopulations. However, decline to the same threshold of 480 individuals occurred in approximately 79% of simulations that included decreased fecundity in southeastern subpopulations.

Discussion

Our demonstration study of hypothetical stressor effects on common loons in New Hampshire highlights the relevance of adopting a spatially explicit framework for wildlife risk assessment. The simplified models and assumptions incorporated in the present study preclude their use for a predictive risk assessment of regional decline in loon populations. Rather, the models were intended to provide the framework for ongoing research comprising more realistic, detailed assessment approaches and to explore both dynamics of spatial dependence in broad-scale populations and use of metapopulation models for examining such dependence. The use of a real-world system for

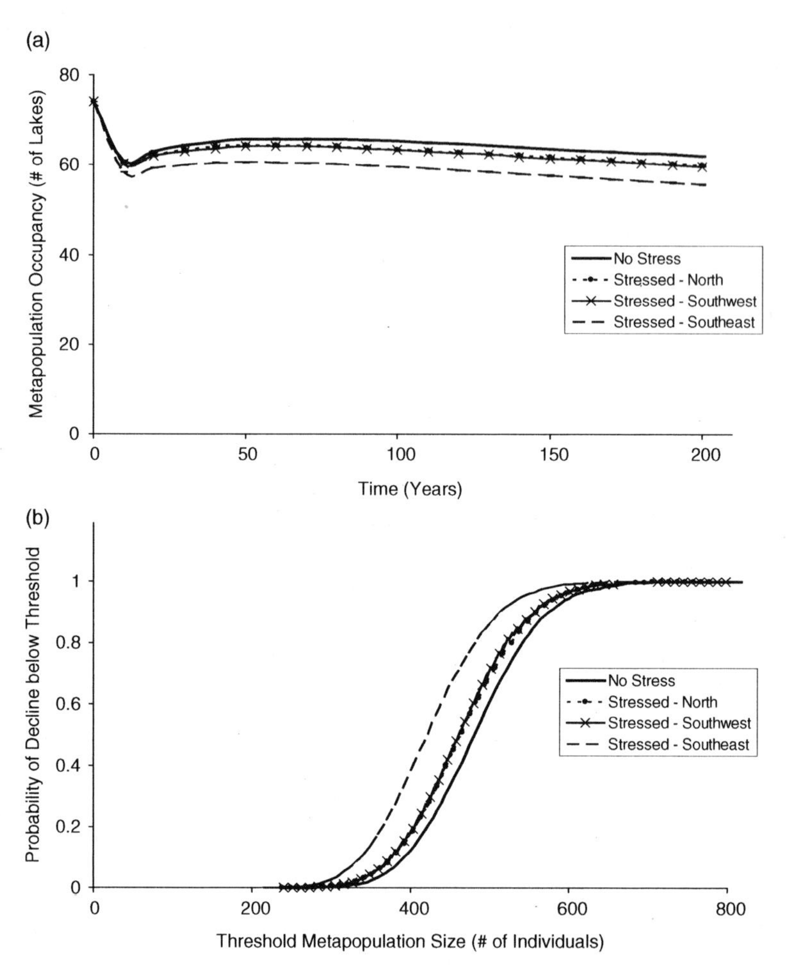

Figure 6.4. Metapopulation dynamics under scenarios of geographically specific stressor exposure: unstressed conditions, stressor effects in northern subpopulations, stressor effects in southwestern subpopulations, and stressor effects in southeastern subpopulations. Trends are shown for metapopulation occupancy (a) and the probability of decline below a threshold metapopulation size (b) by the end of the 200-year simulations.

Table 6.3. Kolmogorov-Smirnoff test statistic (*D*) comparing risks of terminal threshold declines in total metapopulation size as a function of stressor exposure effects in specific geographic subregions (figure 6.4b).

	Kolmogorov-Smirnoff test relative to unstressed dynamics	
Stressor effects condition	*D*	*p*
Northern subpopulations	0.094	<0.001
Southwestern subpopulations	0.108	<0.001
Southeastern subpopulations	0.334	<0.001

illustrating relative impacts of scenarios of potential stressor exposure provided a more concrete example emphasizing that knowing precisely where loon (or other wildlife) populations are expected to experience greater stressor levels can affect conclusions of ecological risk assessments.

Most notable of model outcomes were the relative differences in results from simulations imposing stressor effects in source versus sink subpopulations. Theories of metapopulation dynamics suggest that sufficiently high colonization rates can compensate for localized extinction events to allow regional populations to persist in the face of localized declines (Levins 1970; Gotelli 1991; Hanski 1998). One consequence of such dynamics is that through dispersal, more productive source populations can "rescue" less productive sinks from localized extinction events (Pulliam 1988; Gotelli 1991; Pulliam and Danielson 1991). The relationship between sources and sinks suggests that altered dynamics at one set of sites can translate into impacts at others; results from our simulations reflected this relationship. Increased stressor effects at 34 of the strongest source lakes—little more than 15% of the total number of populations in the metapopulation—translated into statewide declines in the total number of estimated loons as well as the number of occupied lakes. The result occurred despite the fact that populations still persisted at these source locations, which illustrates that the metapopulation-level declines, particularly in occupancy, were not merely a result of localized declines at the stressed lakes. The outcome corroborates our original hypothesis that increased localized stressor effects in a well-connected metapopulation can have regional-level impacts. The result is also congruent with metapopulation theory: With reduced productivity in source subpopulations, the likelihood of lakes being recolonized following local extinction events declines.

The observed outcome of reducing productivity in source populations highlights the necessity of identifying the most productive sites for wildlife populations. As reflected in the HS model (equations 6.1 and 6.2) and map of source and sink locations (figure 6.2), as well as in the territory usage patterns of loons (Piper et al. 1997; Evers 2001), high-quality sites in terms of loon presence and overall abundance consist of lakes that are the largest and/or have the most convoluted perimeter. We assumed in our model that such sites consequently are also the most productive. However, relating habitat quality to actual wildlife productivity can often be problematic (Garshellis 2000), which may translate into difficulties in identifying source locations. When accurate information on source locations is integrated with geographically referenced estimates of stressor levels, models can then be used to estimate consequences of increased stressor effects at sites important to wildlife populations, and hence can identify landscape-level hotspots that may require particular remediation and management effort.

The more unexpected results from our study were from scenarios including reduced productivity in sink populations. That metapopulation endpoints would remain unaffected by increased stress in sink populations relative to levels observed under unstressed conditions is at first glance counterintuitive: Given that more than half of all subpopulations experienced reduced fecundity in this scenario, we expected at least minor impacts at the metapopulation level. However, the explanation is tied in with the demographic characteristics of the sink subpopulations and, to a lesser extent, the level of connectivity among subpopulations. Because of the simplified relationships between lake size and habitat quality, carrying capacity for sink lakes was no more than two breeding pairs. As a consequence, the subpopulations had a strong likelihood of becoming unoccupied even in the absence of stressor effects. Furthermore, demography in such small subpopulations contributes little to the average occupancy and persistence rates of the metapopulation, relative to larger subpopulations with capacities up to 321 pairs. Neither artificially increasing the carrying capacity on sinks to five pairs (the mean carrying capacity across all subpopulations) nor increasing the number of subpopulations labeled as "sinks" to include larger lakes resulted in dynamics of stressed sink subpopulations affecting the metapopulation as a whole (unpublished data). Thus, the dynamics in the largest source subpopulations played a dominant role in driving the dynamics at the metapopulation level within this system. Connectivity also played a likely role in the results: Given the dispersal estimates included in the model, the network of lakes in New Hampshire is highly interconnected (figure 6.1), allowing for a high rate of recolonization when subpopulation densities decline or, in extreme cases, go extinct. Test simulations incorporating lower dispersal probabilities and distances did, in fact, result in relatively lower metapopulation occupancy rates and increased risks of decline with increased stressor effects in sink as well as source subpopulations (unpublished data). The degree of connectivity among subpopulations in the simulations allows any declines within sinks to be quickly compensated for by dispersal from sources.

The lack of metapopulation-level impacts from our simulations of localized stressor effects in sink subpopulations does not suggest that such subpopulations are unimportant within a larger regional framework. Macdonald and Johnson (2001) describe conditions in which sink subpopulations may effectively act as "sources" for recolonization if the typically more productive source subpopulations experience temporary local extinction events. Such a "reverse rescue effect" could be precluded by heightened stressor effects at these sink locations. Furthermore, sink habitats that appear favorable to the organism but in fact encapsulate high stressor levels could effectively siphon off dispersers that could otherwise contribute to subpopulations in more favorable habitats (Delibes et al. 2001). Lastly, the discrete source–sink distinction can actually cover a continuum of differences in terms of habitat quality, demographic characteristics, and so forth. In our simulations, sources were generally composed of very large, high-capacity lakes while sinks were composed of only the smallest lakes (which arguably may be unlikely to be considered "subpopulations" in the strictest definition). In a system with less disparity between characteristics of sources versus sinks, the effects of heightened stressor impacts in sink subpopulations could potentially be quite different from those observed in our simulations.

The impact of larger, more productive subpopulations on metapopulation-level dynamics is further reflected in the scenarios incorporating geographically discrete stressor effects. Both the northern and southwestern subregions included large lakes that were well connected to other lakes in the network via dispersal (figures 6.1 and 6.2). Incorporating stressor effects in these subregions thus moderately reduced metapopulation occupancy and increased risks of decline relative to unstressed conditions, as would be expected based on the importance of large subpopulations. However, simulations that incorporated stressor effects in the southeastern subregion, which encompassed 21 of the 34 largest source subpopulations (table 6.1), resulted in even lower metapopulation occupancy rates and increased risks of decline. The results again emphasize the impact of spatial dependence in stressor distributions relative to locations that are demographically important in a regional (meta)population.

Our present approach, intended as a demonstration study and not a thorough risk assessment, makes a number of simplifying assumptions that, if altered, could lead to markedly different trends in output but would not likely change the important effects of spatial dependence evident in our results. Stressors were incorporated as chronic exposures that did not vary in strength or distribution throughout the course of the simulations. One consequence of this is that metapopulation size and occupancy consistently reached equilibrium levels, which might lead to the incorrect conclusion that, based on our estimates of loon demography and habitat preferences, their populations are unlikely to experience long-term detrimental effects from stressor exposure. Such simplified representations of stressor levels and distributions do not reflect the spatial and temporal variability inherent in natural systems that could have a dramatic, and potentially more deleterious, effect than indicated in our stressor exposure scenarios. Behavioral components of loon natural history, such as aggregative versus competitive interactions that can affect territory occupancy and colonization rates (Piper et al. 1997, 2000), may affect metapopulation dynamics in a manner not captured by the dispersal assumptions included in the model. Similarly, basing dispersal probabilities on the Euclidean distance between lakes may not be an adequate representation if loon dispersal is affected functionally by more complex landscape patterns (Wiens 2001; Walters 2007)—although available data for loon dispersal do not currently support a more detailed description. Lastly, a model of habitat quality based on lake productivity, rather than loon presence/absence, could affect identification of sources and sinks such that the largest lakes are not always the most productive relative to their size. Such a revised habitat model could significantly alter the dynamics between sources and sinks if, for instance, certain small, isolated lakes were found to be particularly productive, since such source subpopulations might not be sufficiently capable for rescuing sink populations from localized extinctions (Walters 2001).

Despite its simplifications, the present study illustrates the relevant factors that can play a role in determining risk levels for wildlife populations within a spatially explicit framework. The results demonstrate that heightened stressor levels at important sites within a metapopulation can lead to regional level impacts on wildlife populations: Nonspatial measures of stressor levels, or measures of average stressor exposure across a given spatial extent, would thus be likely to underestimate the actual (meta)population-level risks posed by those stressors (Hiebeler 1997, 2000). This study therefore highlights the importance of implementing a spatially explicit framework for wildlife risk assessments and of obtaining site-specific habitat data and stressor distributions necessary for such assessments. Though the life-history information and data needs for spatially explicit models pose nontrivial challenges in terms of model complexity and data availability, the approach is nonetheless of considerable utility in qualitatively, if not quantitatively, estimating the condition of wildlife populations across heterogeneous landscapes and at broad spatial scales.

Acknowledgments This is contribution AED-06-064 of the U.S. Environmental Protection Agency Office of Research and Development, National Health and Environmental Effects Research Laboratory, Atlantic Ecology Division. Although the research described in this contribution has been funded partially by the U.S. EPA, it has not been subjected to agency-level review. Therefore, it does not necessarily reflect the views of the agency. Mention of trade names, products, or services does not constitute endorsement or recommendation for use.

References

Akçakaya, H. R. 2005. *RAMAS GIS: Linking Spatial Data with Population Viability Analysis* (version 5.0). Applied Biomathematics, Setauket, New York.

Barr, J. F. 1986. *Population Dynamics of the Common Loon (*Gavia immer*) Associated with Mercury-Contaminated Waters in Northwestern Ontario.* Occasional Paper, Canadian Wildlife Service, Ottawa, Ontario, Canada.

CEP-NWF. 2003. *U.S. State Reports on Population and the Environment: New Hampshire.* Center for Environment and Population, Portsmouth, New Hampshire, and National Wildlife Federation, Montpelier, Vermont.

Chen, C. Y., R. S. Stemberger, N. C. Kamman, B. M. Mayes, and C. L. Folt. 2005. Patterns of Hg bioaccumulation and transfer in aquatic food webs across multi-lake studies in the northeastern US. *Ecotoxicology* 14:135–147.

Delibes, M., P. Gaona, and P. Ferreras. 2001. Effects of an attractive sink leading into maladaptive habitat selection. *American Naturalist* 158:277–285.

Dunning, J. B., D. J. Stewart, B. J. Danielson, B. R. Noon, T. L. Root, R. H. Lamberson, and E. E. Stevens. 1995. Spatially explicit population models: current forms and future uses. *Ecological Applications* 5:3–11.

Evers, D. C. 2001. Common Loon Population Studies: Continental Mercury Patterns and Breeding Territory Philopatry. Ph.D. Dissertation, University of Minnesota.

Garshellis, D. L. 2000. Delusions in habitat evaluation: measuring use, selection and importance. Pages 111–164 in L. Boitani and T. K. Fuller (eds.), *Research Techniques in Animal Ecology.* Columbia University Press, New York.

Gotelli, N. J. 1991. Metapopulation models: the rescue effect, the propagule rain, and the core-satellite hypothesis. *American Naturalist* 138:768–776.

Grear, J. S., D. E. Nacci, A. Kuhn, and S. Walters. 2005. Methods for Developing Water Quality Criteria Based on Population-Level Risks of Multiple Stressors to Aquatic Life and Aquatic-Dependent Wildlife: A Population Modeling Demonstration for the Common Loon. Internal Report, Contribution AED-05-119. U.S. Environmental Protection Agency, Office of Research and Development, National Health and Environmental Effects Research Laboratory, Atlantic Ecology Division, Narragansett, Rhode Island.

Greenwood, P. J. 1980. Mating systems, philopatry and dispersal in birds and mammals. *Animal Behaviour* 28:1140–1162.

Hanski, I. A. 1998. Metapopulation dynamics. *Nature* 396:41–49.

Hiebeler, D. 1997. Stochastic spatial models: from simulations to mean field and local structure approximations. *Journal of Theoretical Biology* 187:307–319.

Hiebeler, D. 2000. Populations on fragmented landscapes with spatially structured heterogeneities: landscape generation and local dispersal. *Ecology* 81:1629–1641.

Johnson, A. R. 2002. Landscape ecotoxicology and assessment of risk at multiple scales. *Human and Ecological Risk Assessment* 8:127–146.

Jung, R. 1991. Effects of human activities and lake characteristics on the behavior and breeding success of common loons. *Passenger Pigeon* 53:207–218.

Kamman, N. C., N. M. Burgess, C. T. Driscoll, H. A. Simonin, W. Goodale, J. Linehan, R. Estabrook, M. Hutcheson, A. Major, A. M. Scheuhammer, and D. A. Scruton. 2005. Mercury in freshwater fish of northeastern North America—a geographic perspective based on fish tissue monitoring databases. *Ecotoxicology* 14:163–180.

Kenow, K. P., M. W. Meyer, D. C. Evers, D. C. Douglas, and J. Hines. 2002. Use of satellite telemetry to identify common loon migration routes, staging areas and wintering range. *Waterbirds* 25:449–458.

Kramar, D., W. M. Goodale, L. M. Kennedy, L. W. Carstensen, and T. Kaur. 2005. Relating land cover characteristics and common loon mercury levels using geographic information systems. *Ecotoxicology* 14:253–262.

Kuhn, A., J. Copeland, J. Grear, S. Walters, M. Nicholson, and D. Nacci. 2005. Report on habitat suitability indices to support population models for projecting relative risk of multiple stressors including toxic chemicals and habitat alteration to Common Loons (*Gavia immer*). Internal Report, Contribution AED-05-118. U.S. Environmental Protection Agency, Office of Research and Development, National Health and Environmental Effects Research Laboratory, Atlantic Ecology Division, Narragansett, Rhode Island.

Landis, W. G. 2002. Uncertainty in the extrapolation from individual effects to impacts upon landscapes. *Human and Ecological Risk Assessment* 8:193–204.

Levins, R. 1970. Extinction. *Lecture Notes in Mathematics* 2:75–107.

Macdonald, D. W., and D. D. P. Johnson. 2001. Dispersal in theory and practice: consequences for conservation biology. Pages 358–372 in J. Clobert, E. Danchin, A. A. Dhondt, and J.D. Nichols (eds.), *Dispersal*. Oxford University Press, Oxford, UK.

McIntyre, J. W., and J. F. Barr. 1997. Common loon (*Gavia immer*). Volume 313 in A. Poole and F. Gill (eds.), *The Birds of North America*. Academy of Natural Sciences, Philadelphia, Pennsylvania, and Washington, DC.

Meyer, M. W., D. C. Evers, J. J. Hartigan, and P. S. Rasmussen. 1998. Patterns of common loon (*Gavia immer*) mercury exposure, reproduction, and survival in Wisconsin, USA. *Environmental Toxicology and Chemistry* 17:184–190.

Mitro, M. G., D. C. Evers, M. W. Meyer, and W. F. Piper. 2005. Mercury and common loon survival rates in New Hampshire and Wisconsin. Unpublished manuscript.

Morris, W. M., and D. F. Doak 2002. *Quantitative Conservation Biology: Theory and Practice of Population Viability Analysis*. Sinauer Associates, Sunderland, Massachusetts.

Morrison, M. L., B. G. Marcot, and R. W. Mannan. 1998. *Wildlife-Habitat Relationships: Concepts and Applications*, 2nd ed. University of Wisconsin Press, Madison, Wisconsin.

NH GRANIT. 2002. New Hampshire Land Cover Assessment: Final Report. New Hampshire Geographically Referenced Analysis and Information Transfer System, Durham, New Hampshire. Available: www.granit.sr.unh.edu.

Nilsson, S. G. 1977. Adult survival rate of the black-throated diver, *Gavia arctica*. *Ornis Scandinavica* 8:193–195.

Piper, W. H., J. D. Paruk, D. C. Evers, M. W. Meyer, K. B. Tischler, M. Klich, and J. J. Hartigan. 1997. Local movements of color-marked common loons. *Journal of Wildlife Management* 61:1253–1261.

Piper, W. H., K. B. Tischler, and M. Klich. 2000. Territory acquisition in loons: the importance of take-over. *Animal Behaviour* 59:385–394.

Piper, W. H., M. W. Meyer, M. Klich, K. B. Tischler, and A. Dolsen. 2002. Floating platforms increase reproductive success of common loons. *Biological Conservation* 104:199–203.

Pulliam, H. R. 1988. Sources, sinks, and population regulation. *American Naturalist* 132:652–661.

Pulliam, H. R., and B. J. Danielson. 1991. Sources, sinks, and habitat selection: a landscape perspective on population dynamics. *American Naturalist* 137:S50–S66.

Sokal, R. R., and F. J. Rohlf. 1981. *Biometry: The Principles and Practice of Statistics in Biological Research*. W. H. Freeman, New York.

Sutcliffe, S. A. 1978. Changes in status and factors affecting common loon populations in New Hampshire. *Transactions of the Northeastern Section of the Wildlife Society* 35:219–224.

Turner, M. G., G. J. Arthaud, R. T. Engstrom, S. J. Hejl, J. Liu, S. Loeb, and K. McKelvey. 1995. Usefulness of spatially explicit population models in land management. *Ecological Applications* 5:12–16.

Turner, M. G., R. H. Gardner, and R. V. O'Neill. 2001. *Landscape Ecology in Theory and Practice: Pattern and Process*. Springer-Verlag, New York.

U.S. Department of Commerce. 2000. *US Census Bureau: Census 2000, American FactFinder*. Washington, DC. Available: factfinder.census.gov.

Walters, S. 2001. Landscape pattern and productivity effects on source-sink dynamics of deer populations. *Ecological Modelling* 143:17–32.

Walters, S. 2007. Modeling scale-dependent landscape pattern, dispersal, and connectivity from the perspective of the organism. *Landscape Ecology* 22:867–881.

Wiens, J. A. 2001. The landscape context of dispersal. Pages 96–109 in J. Clobert, E. Danchin, A. A. Dhondt, and J. D. Nichols (eds.), *Dispersal.* Oxford University Press, Oxford, UK.

Yates, D. E., D. T. Mayack, K. Munney, D. C. Evers, A. Major, T. Kaur, and R. J. Taylor. 2005. Mercury levels in mink (*Mustela vison*) and river otter (*Lontra canadensis*) from northeastern North America. *Ecotoxicology* 14:263–274.

7

Population-Level Effects of PCBs on Wood Frogs (*Rana sylvatica*) Breeding in Vernal Pools Associated with the Housatonic River, Pittsfield to Lenoxdale, Massachusetts

W. TROY TUCKER
J. D. LITZGUS
SCOTT FERSON
H. REŞIT AKÇAKAYA
MICHAEL E. THOMPSON
DOUGLAS J. FORT
JOHN P. LORTIE

It is a well-established fact that amphibian populations are declining worldwide (Blaustein and Wake, 1990; Houlahan et al., 2000; Stuart et al., 2004). Several factors have been implicated in global amphibian losses, including habitat degradation, pollution, overexploitation, introduced species, emerging infectious diseases, and climate change (Kiesecker et al., 2001; Collins and Storfer, 2003). Chemical stressors, such as pesticides, heavy metals, and nitrogen-based fertilizers, can have lethal, sublethal, direct, and indirect effects on amphibians (e.g., Blaustein et al., 2003; Hayes et al. 2002). Some of these effects include death, decreased growth rates, developmental and behavioral abnormalities, decreased reproductive success, and weakened immune systems. There has been a recent increase and widespread occurrence of malformations in natural populations of amphibians (e.g., Sessions 2003), and this is perceived as a major environmental problem.

Population viability analyses are commonly used in environmental and conservation planning. Computer-based simulations allow scientists to make predictions about population changes as a result of exposure to various stressors, such as contaminants and climate change. The purpose of our study was to use population viability analyses to examine the effects of polychlorinated biphenyls (PCBs) on the population persistence of a metapopulation of wood frogs (*Rana sylvatica*) occupying the floodplain of the Housatonic River, Massachusetts, a site contaminated by operations at General Electric's Pittsfield facility transformer manufacturing division. Widespread PCB contamination downstream of the GE plant is due to transport of contaminated sediment and floodplain soil by river flow and flooding. The GE plant is the only known source of PCBs in the system, and the PCBs detected are predominantly Aroclor 1260, with a small amount of Aroclor 1254.

The U.S. Environmental Protection Agency is conducting an ecological risk assessment for the portion of the Housatonic River and its floodplain beginning at the confluence of the east and west Branches of the river in Pittsfield, Massachusetts, and continuing downstream (Weston Solutions, Inc. 2004a). The primary study area (PSA) for these investigations is the area between the confluence and Woods Pond Dam in Lenoxdale.

Breeding amphibians use portions of the river and temporary ("vernal") and permanent pools in the floodplain for courtship and egg laying. These areas then support larval amphibians for periods ranging from several weeks to more than a year, resulting in the exposure of developing amphibians to sediment and water contaminated by PCBs and other contaminants of concern. Wood frogs (*Rana sylvatica*) are the most abundant frog species breeding in the vernal pools within the PSA (~83% of counts of all breeding adult amphibians), and spotted salamanders (*Ambystoma maculatum*) are the most common salamanders (~4% of counts of all breeding adult amphibians) (WAI 2002, sec. III, ch. 4; WAI 2003).

Previous studies within the Housatonic River PSA have demonstrated that PCBs can have harmful effects on developing amphibians, including direct mortality and internal and external malformations (FEL 2002). A stochastic population model was developed to determine whether these effects on individual wood frogs influence the dynamics of the wood frog population within the Housatonic River PSA.

In this chapter, we present the results of an effort to model wood frog populations within the PSA (Weston Solutions, Inc. 2004b). A description of the study area encompassed by the model and of the amphibian community within the PSA can be found in WAI (2002, 2003).

Methods

Study species and area

Sixty-six temporary (i.e., vernal) and permanent pools have been mapped in the Housatonic River PSA (WAI 2002). Not all, however, are suitable as wood frog breeding habitat. Based on field surveys conducted in the PSA since 1998 (WAI 2002), field studies of specific pools within the PSA (WAI 2003), and field collections for laboratory-based studies (FEL 2002), 27 vernal pools within the floodplain of the PSA were identified as suitable wood frog breeding habitat.

The wood frog breeding population within the PSA was defined as those frogs breeding within the 27 vernal pools described above. For the purposes of the model, a closed population was assumed (i.e., no immigration or emigration). Although it is probable that a small number of wood frogs enter or leave the PSA, wood frogs are not known to migrate long distances, and they show fidelity to their natal pools (Berven and Grudzien 1990). It is reasonable to assume, therefore, that the population is closed for the purposes of this model.

Experimental methods

Laboratory data were used to estimate the impact of total PCB (tPCB) exposure on larval survival and on metamorph malformation rates. A summary of laboratory results

Table 7.1. Larval wood frog mortality in relation to spatially weighted mean tPCB concentrations in vernal pool sediment.

Pool name	SPATWGT mean tPCBs	End mean % mortality
23b-VP-1	0.21	89
23b-VP-2	0.3	83
46-VP-5	1.36	36
46-VP-1	0.8	87
18-VP-2	4.9	98
8-VP-1	24.6	67
38-VP-1	28.5	26
38-VP-2	32.3	52
WML-1	0	77
WML-2	0	87
WML-3	0	75

SPATWGT, Spatially weighted concentration.
tPCB, mg/kg sediment.

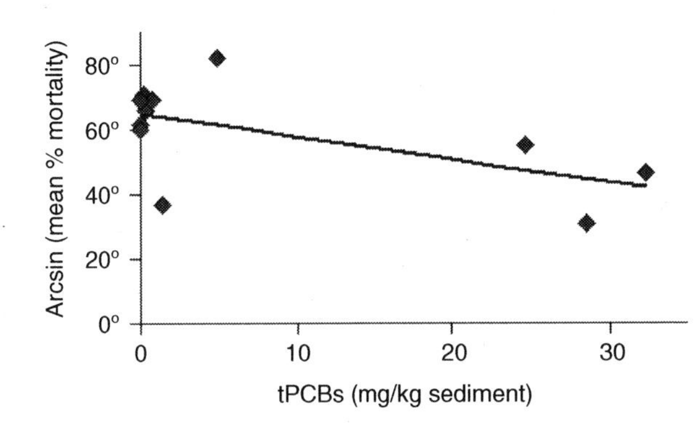

Figure 7.1. Relationship between spatially weighted tPCBs and larval wood frog mortality ($R^2 = 0.36$, df = 1,10, $F = 5.00$, $p = 0.052$; $y = 64.68 - 0.71x$).

and methods for adapting them for use in stochastic population modeling is presented in this section (for method details, see Weston Solutions, Inc. 2004a; FEL 2002).

Larval survival

A study of wood frogs collected from the PSA (FEL 2002) revealed a relationship between larval wood frog mortality and the spatially weighted concentration of tPCBs in vernal pool sediment (table 7.1). The percent mortality data in table 7.1 was arcsin transformed, and a linear regression was performed to describe the relationship between spatially weighted tPCBs and larval mortality (see figure 7.1). This relationship was used to predict the relative decrease in larval mortality associated with increased tPCBs in the 27 vernal pools (see table 7.2).

Metamorph malformation rates

Additional research (FEL 2002) included collecting wood frog metamorphs from 10 vernal pools in the PSA and reporting the percent malformed for each sex (see table 7.3).

Table 7.2. Relationship between spatially weighted tPCBs in vernal pool sediment and larval wood frog mortality rates.

Pool name	SPATWGT mean tPCBs (mg/kg sediment)	Mean larval mortality (tPCB = 0) (%)	Mean larval mortality given pond-specific tPCBs (%)	Decreased mean mortality/100	Increased survival proportion due to tPCBs
23b-VP-1	0.21	81.71	81.51	0.00	1.00
23b-VP-2	0.3	81.71	81.42	0.00	1.00
46-VP-1	0.76	81.71	80.98	0.00	1.00
19-VP-7	0.82	81.71	80.92	0.00	1.00
8-VP-4	0.95	81.71	80.80	0.00	1.00
46-VP-5	1.36	81.71	81.02	0.00	1.00
12-VP-1	1.72	81.71	80.04	0.01	1.01
23a-VP-1	3.04	81.71	78.73	0.01	1.01
40-VP-1	3.69	81.71	78.07	0.02	1.02
27b-VP-2	4.18	81.71	77.57	0.02	1.02
18-VP-2	4.9	81.71	76.82	0.02	1.02
66a-VP-1	5.31	81.71	76.39	0.03	1.03
18-VP-1	9.03	81.71	72.40	0.05	1.05
27b-VP-3	10.05	81.71	71.26	0.05	1.05
27-VP-1	10.21	81.71	71.09	0.05	1.05
38-VP-3	13.49	81.71	67.36	0.07	1.07
42-VP-3	20.12	81.71	59.50	0.11	1.11
49a-VP-1	24.34	81.71	54.35	0.14	1.14
8-VP-1	24.56	81.71	54.08	0.14	1.14
38a-VP-1	25.77	81.71	52.59	0.15	1.15
38-VP-1	28.54	81.71	49.18	0.16	1.16
19-VP-1	30.67	81.71	46.56	0.18	1.18
8-VP-5	31.86	81.71	45.10	0.18	1.18
38-VP-2	32.31	81.71	44.55	0.19	1.19
26-VP-1	38.81	81.71	36.69	0.23	1.23
39-VP-1	42.96	81.71	31.84	0.25	1.25
8-VP-2	54.98	81.71	19.03	0.31	1.31

These data were arcsin transformed, and regression relationships for females and males were derived (figure 7.2). Table 7.4 shows the predicted percent malformed at each of the 27 vernal ponds as a function of tPCBs using the relationships from figure 7.2.

To determine the effect of malformation on mortality, several sources were consulted. Glennemeier and Begnoche (2002) report increased mortality rates between 20% and 60% due to malformation caused by laboratory exposure of two species of frogs (*R. pipiens* and *R. utricularia*) to comparable levels of PCBs. Mortality rates between 70% and 100% are estimated among Housatonic *R. sylvatica* given the severity of malformations (FEL 2002). Based upon this information, simulations were run with models parameterized such that (1) 50% of malformed one-year-old frogs died and (2) 100% of malformed one-year-old frogs died. These values are intended to bracket the uncertainty regarding the mortality rate of malformed frogs.

In addition to gross malformations, the number of malformed frogs that exhibited gonadal abnormalities was tallied (FEL 2002; see table 7.5). Fifty-seven percent of malformed female metamorphs were observed to have gonadal abnormalities. There are

Table 7.3. Malformation rates in recently metamorphed wood frogs in relation to spatially weighted tPCB concentrations in vernal pool sediment.

Pool name	SPATWGT mean tPCBs (mg/kg sediment)	% Malformed metamorph		
		Male	Female	Total
WML-1	0	0	0	0
WML-3	0	0	5.9	2.9
23b-VP-1	0.21	3.9	5.9	4.9
23b-VP-2	0.3	5	6.5	5.9
46-VP-5	1.36	3	12.3	9.2
46-VP-1	0.8	8.2	8.9	8.6
18-VP-2	4.9	13.8	32.8	26.9
8-VP-1	24.6	0	66.7	66.7
38-VP-1	28.5	20	46.3	41
38-VP-2	32.3	42.1	53.8	51.5

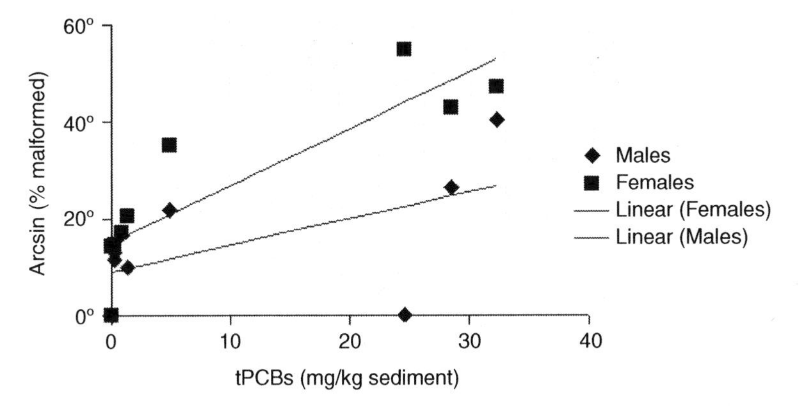

Figure 7.2. Relationship between spatially weighted tPCBs and the arcsin transform of the percent of metamorphs malformed (females: $R^2 = 0.78$, df = 1,9, $F = 28.05$, $p = 0.0007$; $y = 15.22 + 1.17x$; males: $R^2 = 0.33$, df = 1,9, $F = 3.9$, $p = 0.0837$; $y = 8.79 + 0.56x$).

no direct data, however, regarding the proportion of gonadal abnormalities leading to reproductive impairment. An estimated 70–100% of females with gonadal abnormalities are sterile, based upon the observed severity of the abnormalities (FEL 2002). Given this information, simulations were run with models parameterized such that (1) 50% of females with gonadal abnormalities were sterile and (2) 100% of females with gonadal abnormalities were sterile. These values are intended to bracket the uncertainty regarding the sterility rate of female frogs with gonadal abnormalities.

Impacts of tPCBs on vital rates were incorporated into the various affected parameterizations as proportions by which fertility and mortality were increased or decreased. The proportion by which fertility was affected is unique to each vernal pool and can be found in the Populations window in RAMAS Metapop (a part of RAMAS GIS, Akçakaya 2005). Separate values were used for the assumptions that 50% and 100%, respectively, of gonadally abnormal females are sterile. These values are factors by which fecundity values in the Leslie matrix were multiplied for

Table 7.4. Predicted malformation rates in wood frogs in relation to spatially weighted tPCB concentrations in vernal pool sediment.

Pool name	SPATWGT mean tPCBs (mg/kg sediment)	Predicted % female malformed	Predicted % male malformed
23b-VP-1	0.21	7.1	2.4
23b-VP-2	0.3	7.2	2.4
46-VP-1	0.76	7.7	2.6
19-VP-7	0.82	7.8	2.6
8-VP-4	0.95	7.9	2.7
46-VP-5	1.36	8.4	2.8
12-VP-1	1.72	8.8	2.9
23a-VP-1	3.04	10.4	3.4
40-VP-1	3.69	11.3	3.6
27b-VP-2	4.18	11.9	3.8
18-VP-2	4.9	12.9	4.1
66a-VP-1	5.31	13.5	4.3
18-VP-1	9.03	19.3	6.0
27b-VP-3	10.05	21.0	6.6
27-VP-1	10.21	21.3	6.6
38-VP-3	13.49	27.1	8.4
42-VP-3	20.12	40.2	12.7
49a-VP-1	24.34	49.0	15.8
8-VP-1	24.56	49.4	15.9
38a-VP-1	25.77	52.0	16.9
38-VP-1	28.54	57.7	19.1
19-VP-1	30.67	62.1	20.9
8-VP-5	31.86	64.5	21.9
38-VP-2	32.31	65.4	22.3
26-VP-1	38.81	77.6	28.2
39-VP-1	42.96	84.4	32.2
8-VP-2	54.98	97.6	44.4

Table 7.5. Gonadal malformation rates in wood frogs in the Housatonic River PSA.

Frequency	Male	Female	Total
Abnormal	24	124	148
Gonadal abnormality	11	71	82
Proportion	0.46	0.57	0.55

each pond. The pool-specific multipliers modifying the impact of tPCBs on mortality are found in *.SCH text files, referenced for each pool in the Populations window of Metapop. Impacts are specific to age class and sex, and half of all malformed frogs were assumed to die. Note that mortality in the zero age class decreases as tPCB concentrations increase and that mortality due to malformation also occurs in age class 0 (FEL 2002). The age class 0 multipliers are the sum of these positive and negative effects. The *.SCH files are named using the pool alphanumeric designations, and the first letter of the file name, either "A" or "B," indicates that the values assume 50% or 100%, respectively, of malformed frogs die.

Population model

A dynamic population model projecting wood frog population trends 10 years into the future, and computing the risk of population decline (Ginzburg et al. 1982), was constructed using vital rate information from the literature (Berven 1990) and initial abundances derived from studies of vernal pools in the PSA (WAI 2002, 2003). The model was age and sex structured and employed yearly time steps between age classes (Caswell 2001). In addition, both demographic and environmental stochasticity were incorporated (Burgman et al. 1993, ch. 4), as was density dependence. The model also considered the spatial location of pools and allowed migration between them as a function of distance based upon dispersal relationships described by Berven and Grudzien (1990).

The impact of tPCBs on the wood frog population was assessed by comparing population projections from a base population model (i.e., a wood frog population in the absence of tPCBs), with projections from population models that included the effect of tPCBs on population vital rates (see FEL 2002). Two projection comparisons were performed based on simulations of (1) a nondeclining base population, and (2) a declining base population. All models were constructed using RAMAS Metapop (Akçakaya 2002).

Population projections from four parameterizations of the model were analyzed for each of the two projection comparisons (i.e., nondeclining and declining base populations). The first parameterization was a base model of population change over 10 years in the absence of tPCB contamination. The other three parameterizations included the effect of tPCBs on initial population size and vital rates, each incorporating a slightly different combination of assumptions regarding the impact of tPCBs on fertility and mortality. These three parameterizations comprised combinations of low and high estimates of the proportion of malformed frogs that subsequently die or become reproductively incapacitated. The impact of tPCBs on initial population size and vital rates used in these parameterizations were derived from vernal pool studies conducted in the PSA (WAI 2002, 2003), from laboratory-based studies (FEL 2002), and from a literature study of the effect of tPCBs on amphibians (Glennemeier and Begnoche 2002).

The wood frog population modeling study was designed from the outset to examine the sensitivity of the population-level projections of tPCB impacts to the choices necessary to construct the models given incomplete data. For instance, long-term studies needed to establish the vital rates (survival and fecundity) of the local population are lacking. Data from the literature were analyzed to estimate these rates, and two sets of vital rates were developed. One set of rates described a nondeclining population; the other, a declining population. In the face of uncertainty over which set of literature-derived vital rates might best represent the local population, parallel simulations were performed using each set. Comparison of the population projections produced by each set of vital rates indicates the sensitivity of the model output to the choice of vital rates.

Similarly, although tPCBs are known to cause malformation of wood frog metamorphs, field studies necessary to relate the malformation rate to mortality and fertility are also lacking. Estimates vary on the proportion of malformed wood frogs that die and of the proportion of surviving malformed wood frogs that are rendered sterile.

Table 7.6. Base population model Leslie matrix for nondeclining base population projection[a]

		From					
		Females			Males		
		Age 0	Age 1	Age 2	Age 0	Age 1	Age 2
To Females	Age 0	0.05	2.9	2.9	0	0	0
	Age 1	0.29	0	0	0	0	0
	Age 2	0	0.13	0	0	0	0
To Males	Age 0	0.05	2.9	2.9	0	0	0
	Age 1	0	0	0	0.29	0	0
	Age 2	0	0	0	0	0.13	0

[a] In the Leslie matrix shown, the numbers in the "Age 0" *rows* represent fertility in terms of average number of eggs produced in each age class that survive to be censused. The numbers in the matrix on the *subdiagonal* are the proportion of wood frogs surviving from the previous age class (see Akçakaya 2002).

In the face of this uncertainty, parallel population models were constructed that project wood frog populations assuming that either 50% or 100% of malformed wood frogs die, and assuming that either 50% or 100% of gonadally malformed wood frogs are sterile. These values were chosen for simulation in order to bracket the range of estimates and opinions expressed by researchers in the literature. Comparing the results of models with different assumptions indicates the sensitivity of the model output to the level of tPCB-induced mortality and infecundity employed.

Additional sensitivity studies were performed to assess the robustness of the models to changes in other modeling assumptions. For both declining and nondeclining models, vital rates were increased and decreased by 5% and by 10%, and the robustness of the results to the particular vital rates was determined. A similar sensitivity analysis was performed on environmental correlation assumptions and on dispersal rates. Finally, to examine the sensitivity of the projections to the form of density dependence assumed, a series of models were developed using a Ricker-type density dependence function, and results were compared to the density ceiling function used in the other simulations.

Projection 1 (Nondeclining base population) Population matrix and vital rates

Projection matrix Three age classes were modeled in one-year steps for females and males, respectively, assuming a prebreeding census. Age class 0 spanned the period from egg to just less than one year of age, age class 1 spanned the period from one year to just less than two years, and so forth; no frogs were modeled as living past three years of age. Table 7.6 shows the Leslie matrix used in the base model.

For projection 1, fertility and survivorship for wood frogs were derived from one of the seasonal ponds in Maryland studied by Berven (1990). That study reported six yearly life tables for two ponds calculated from data collected between 1976 and 1980. Fertility and survivorship from the pond with the most years of data were averaged across years to

Table 7.7. Derivation of Leslie matrix survival rates (S_x) from survivorship data (l_x) reported in Berven (1990), by age class: Projection 1.

	l_x (from Berven 1990)			S_x		
Year	0	1	2	0	1	2
1976	0.0012	0.0002	0.0000	0.0012	0.1767	0.0755
1978	0.0298	0.0183	0.0016	0.0298	0.6141	0.0874
1979	0.0163	0.0035	0.0008	0.0163	0.2147	0.2286
1980	0.0220	0.0031	0.0004	0.0220	0.1429	0.1242
Mean				0.02	0.29	0.13
Standard Deviation				0.01	0.22	0.07

Table 7.8. Derivation of Leslie matrix fertility rates (F_x) from fertility data (m_x) reported in Berven (1990): Projection 1.

	Number of eggs per female of age x (reported in Berven 1990)			Fertility of females age x ($m_x \times S_0$)		
Year	m_0	m_1	m_2	F_0	F_1	F_2
1976	0	287	390	0	0.34	0.47
1978	3.34	373	344	0	0.07	0.09
1979	15.8	333	301	0.1	11.12	10.25
1980	1.04	286	352	0.26	5.43	4.91
Mean				0.095	5.793	5.84
Std Dev				0.116	4.412	4.194
Coefficient of Variation				1.226	0.762	0.718

produce the figures in the Leslie matrix in table 7.6.[1] The left half of table 7.7 shows the survivorship data reported in Berven (1990) for one pond across six years of study. The right half of the table shows the survival rates calculated from these data. The average survival rate for each age class was used as the survival rate in the Leslie matrix (table 7.6). Standard deviations from these averages were used to model environmental stochasticity.

The left half of table 7.8 shows the fertility data reported in Berven (1990) by age class and year. The right half of the table shows the fertility rate inputs to the Leslie matrix (table 7.6) calculated from the fertility data and the survival rate data. Each fertility rate in the right half of the table results from multiplying the fertility data to the left with the survival rate of age class 0 (shown in table 7.7); for example, for 1976, $F_0 = m_0 \times S_0$, $F_1 = m_0 \times S_1$, and $F_2 = m_0 \times S_2$. The fertility rates were averaged over years for each age class. A 50:50 sex ratio was assumed for eggs, and half of the multiyear average was assumed to be males, and half females, in the Leslie matrix (table 7.6).

[1] Berven (1990) reports survivorship (l_x) as the number surviving from birth to the beginning of each age class. This figure was transformed to produce the proportion surviving from the previous to the current age class by dividing l_x by l_{x-1} for each age class.

Table 7.9. Derivation of Leslie matrix survival rates (S_x) from survivorship data (l_x) reported in Berven (1990), by age class: Projection 2.

	l_x			S_x		
Year	0	1	2	0	1	2
1976	0.0012	0.0002	0.0000	0.0012	0.1767	0.0755
1977	0.0002	0.0000	0.0000	0.0002	0.1333	0.0109
1978	0.0298	0.0183	0.0016	0.0298	0.6141	0.0874
1979	0.0163	0.0035	0.0008	0.0163	0.2147	0.2286
1980	0.0220	0.0031	0.0004	0.0220	0.1429	0.1242
1981	0.0018	0.0005	0.0001	0.0018	0.2994	0.1509
Mean				0.01	0.26	0.11
Std Dev				0.01	0.20	0.06
CV				0.70	0.77	0.54

Table 7.10. Derivation of Leslie matrix fertility rates (F_x) from fertility data (m_x) reported in Berven (1990): Projection 2.

	Number of eggs per female of age x			Fertility of females age x ($m_x \times S_0$)		
Year	m_0	m_1	m_2	F_0	F_1	F_2
1976	0	287	390	0	0.344	0.468
1977	0	289	371	0	0.069	0.089
1978	3.34	373	344	0.10	11.12	10.25
1979	15.8	333	301	0.258	5.428	4.906
1980	1.04	286	352	0.023	6.283	7.733
1981	21.7	325	352	0.038	0.575	0.623
Mean				0.07	4.0	4.0
Std Dev				0.10	4.44	4.31
CV				1.42	1.12	1.08

Stochasticity Demographic stochasticity was incorporated in all vital rates. Environmental stochasticity was assumed to be distributed lognormally and was incorporated in vital rates using standard deviations around vital rates measured over four years at the pond studied by Berven (1990) (see tables 7.7 and 7.8). The demographic stochasticity standard deviation inputs to the population models are found in Metapop in the Standard Deviation Matrix window. Vital rates were independent from year to year but were assumed to be perfectly correlated within a single year.

Projection 2 (Declining base population) Population matrix and vital rates

Projection matrix In this projection, life table data from both of Berven's (1990) study pools were combined, resulting in a model population undergoing moderately rapid decline. Fertility and survivorship from the ponds were averaged across years and ponds to produce the figures in the Leslie matrix for this base model. Tables 7.9 and 7.10 show the derivation of the survival rates and fertility rates, respectively, from the raw data reported by Berven (1990).

Table 7.11. Breeding adult wood frog densities in four vernal pools of the PSA in relation to spatially weighted sediment tPCB concentration.

Pool	Pool volume (m^3)	SPATWGT mean tPCBs (mg/kg sediment)	Adults (frequency)			Density (per m^3)	
			Males	Females	Total	Male	Female
8-VP-1	72	24.56	104	112	216	1.44	1.56
8-VP-2	824	54.98	91	73	164	0.11	0.09
38-VP-2	202	32.31	318	245	563	1.57	1.21
46-VP-5	68	1.36	151	119	270	2.22	1.75

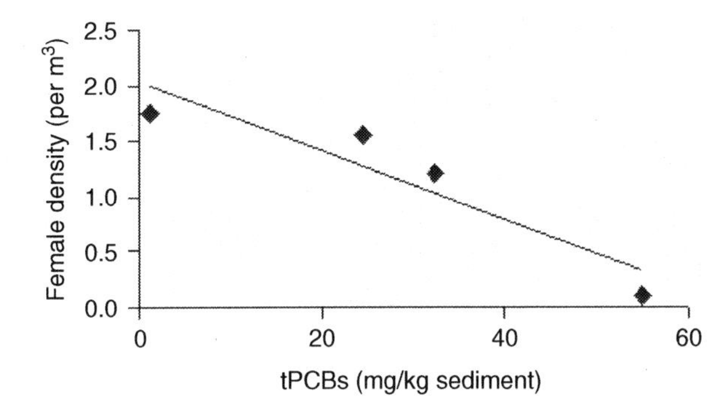

Figure 7.3. Relationship between spatially weighted tPCBs and breeding adult female wood frog density ($R^2 = 0.86$, df = 3, $F = 12.45$, $p = 0.072$; $y = 2.03 - 0.03x$).

Stochasticity Demographic stochasticity was also incorporated in all vital rates for the declining base population projection, and environmental stochasticity was assumed to be distributed lognormally. Because data from two ponds were used, the standard deviations include both temporal and spatial variability. To avoid overestimating the temporal variability input needed for the population model, however, the standard deviations were adjusted so that the coefficients of variation matched those calculated for the projection 1 model, which included only temporal variability. Vital rates were independent from year to year but were assumed to be perfectly correlated within a single year.

Population size and initial abundances

Base model initial abundance The initial number of adults was derived from data collected in 1999 at four vernal pools in the PSA (WAI 2003) (see table 7.11). The data in table 7.11 suggest a negatively linear relationship between spatially weighted tPCBs and the density of breeding adults entering the vernal pool. Therefore, for the base model parameterization, the relationship between male and female density and tPCBs was modeled using linear regression (figures 7.3 and 7.4).

Initial adult populations were calculated for the 27 pools corrected for the effect of tPCBs on population size (i.e., the regression equation was used to predict total adults given pool volume and assuming tPCBs = 0). The age structure of the initial population in each pool for the nondeclining base model is found in the Initial Abundances

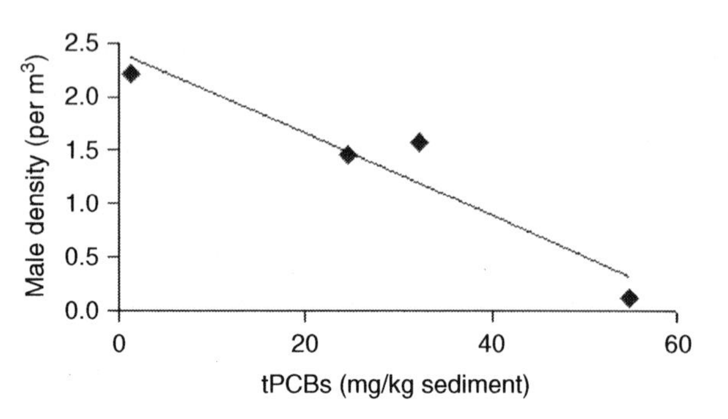

Figure 7.4. Relationship between spatially weighted tPCBs and breeding adult male wood frog density ($R^2 = 0.91$, df = 3, $F = 19.6$, $p = 0.047$; $y = 2.4 - 0.04x$).

window of Metapop. The total initial population estimate was distributed among age classes to approximate the numbers expected assuming the population exhibited a stable age distribution.

tPCB-affected models initial abundance For the nondeclining base model, initial abundances were calculated by controlling for the effect of tPCBs on adult density as observed in studies of vernal pools in the PSA (WAI 2003). The effect used is shown in figures 7.3 and 7.4. For each of the three tPCB-affected parameterizations of the projection 1 model, initial population size was recalculated to include the observed effect of tPCBs.

Density dependence

Berven (1990) notes that larval population densities fluctuate up to two orders of magnitude from minimum to maximum, and adults up to one order of magnitude. Some of these fluctuations are potentially related to density-dependent effects. For all simulations, therefore, a carrying capacity ceiling was calculated at 10 times the initial population size of each pool, and populations were not allowed to grow beyond this ceiling. This ceiling brackets the highest observed population fluctuations. A second density-dependent model based upon a Ricker-type function was applied as an analysis of the sensitivity of the model results to this ceiling-type density dependence assumption.

Environmental correlation

Correlation coefficients of 0.5 were assumed between all pools when calculating year-to-year demographic stochasticity due to environmental fluctuation. This is a relatively strong correlation intended to simulate ponds experiencing very similar, but not identical, environmental changes from year to year. A sensitivity analysis is presented below that examines the effect of varying the correlation coefficient (from 0.5 to 0.25) on model results.

Dispersal

The metapopulation models constructed for the population projection simulations were geographically explicit. Distances between ponds were calculated, and the dispersal equation reported for wood frogs by Berven and Grudzien (1990) was used to calculate the proportion of frogs dispersing from each pond that immigrate to each other pond. The equation used was negative exponential in form:

$$y = 0.4392 \times 10^{-0.000560x},$$

where x is distance (in meters) and y is the proportion of frogs dispersing that distance. Berven and Grudzien (1990) also report that, on average, 18.54% of wood frogs disperse in their first year, with no difference in dispersal rates between males and females. This percentage was used as the proportion dispersing from each pool in the simulation. A sensitivity analysis is presented below that examines the effect of varying dispersal rates between pools by ±5% and ±10%.

Results

The wood frog metapopulation in the PSA was projected over a 10-year period in one-year time steps based on 1,000 replications of each projection. Two population projections were run, each with three combinations of assumptions regarding the impact of malformation on fertility and mortality, respectively. The first projection used the life table data from Berven (1990) for a single pond monitored over four years as a base model. This population projection matrix results in a stationary or moderately declining population size over the 10 years of the simulation. This base model was compared to the following three impact assumptions: (1) impact 1—50% of malformed metamorphs die, 100% of gonadally abnormal metamorphs are sterile; (2) impact 2—100% of malformed metamorphs die; and (3) impact 3—50% of malformed metamorphs die, 50% of gonadally abnormal metamorphs are sterile.

The results of the first projection are shown in figure 7.5, where curves represent the means of the 1,000 simulations and indicate the probability that the population will fall at or below the population size on the x-axis. Higher probabilities of falling below a specified population size (toward the top of the graph) imply higher probability of extinction. All three impact scenarios show a significantly increased risk of population decline compared to the unaffected base model. Impact 3 (50% of malformed metamorphs die, 50% of surviving gonadally malformed frogs are sterile) has the smallest adverse effect on population size after 10 years. Impact 2 (100% of malformed metamorphs die) has the largest adverse effect. Impact 1 (50% of malformed metamorphs die, 100% of surviving gonadally malformed frogs are sterile) has an adverse effect intermediate to the other impacts. Under impact 3, the probability of the population size falling below 11,000 or 12,000 is increased by approximately 27%. When all malformed metamorphs are assumed to die, the probability of the population dropping below 5,000–6,000 increases by 44–45%. Note that the current population size estimated from PSA data was about 75,000. With no impact, the probability of falling below the current population size at the end of 10 years was estimated by the base model to be about 80%. With the impact of tPCBs, the probability of falling below the current population size increases to between 95% and 97% (maximum differences and p-values associated with Kolmogorov-Smirnov statistical tests between the base model and the various affected models are shown in table 7.12).

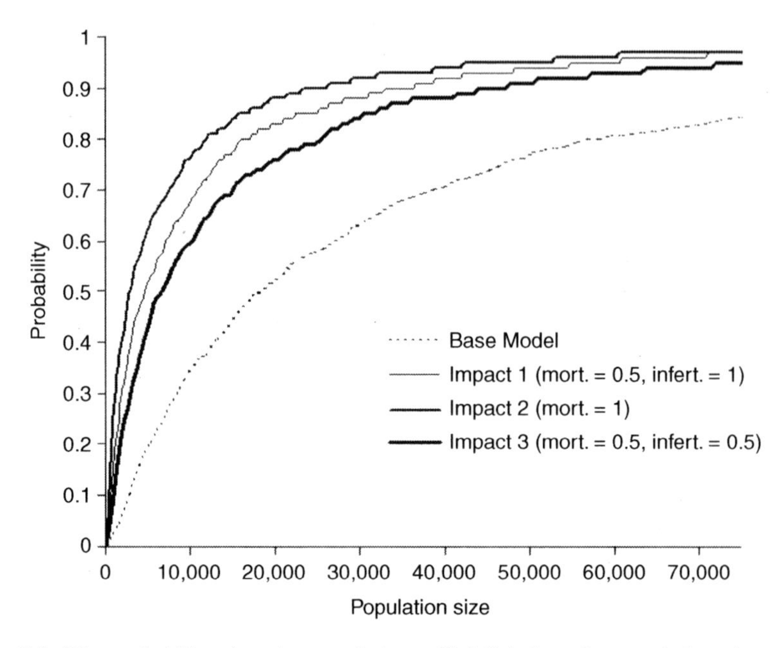

Figure 7.5. The probability that the population will fall below the population sizes on the x-axis at the end of 10 years: comparison of base model and PCB-affected wood frog population projections assuming a nondeclining population. The estimated initial adult population size for the 27 pools in the PSA is approximately 75,000.

Table 7.12. Maximum differences and Kolmogorov-Smirnov tests of significance comparing four PCB impact scenarios for modeled stable wood frog populations: Projection 1.

	Maximum difference (*D*)	*p*-Value
Impact 1	0.347	<0.0001
Impact 2	0.441	<0.0001
Impact 3	0.271	<0.0001

The base model for the second projection used the life table data from Berven (1990) for two ponds. Environmental stochasticity in this data set was adjusted to remove spatial variation while maintaining temporal variation. The resulting population projection matrix shows declining population size over the 10 years of the simulation. This base model was compared to the three affected projections (see figure 7.6).

Given the declining base model, the population projection indicates a 99% chance that the population will be smaller than it is now (i.e., smaller than 75,000) at the end of 10 years (not shown in figure 7.6). The probability that the population will be $\frac{1}{15}$th of its current size (about 5,000 on the graph) with no tPCB impact is about 75%. The effect of tPCB impacts is to increase this probability to as much as about 95%. Maximum differences and Kolmogorov-Smirnov tests of significance are shown in table 7.13.

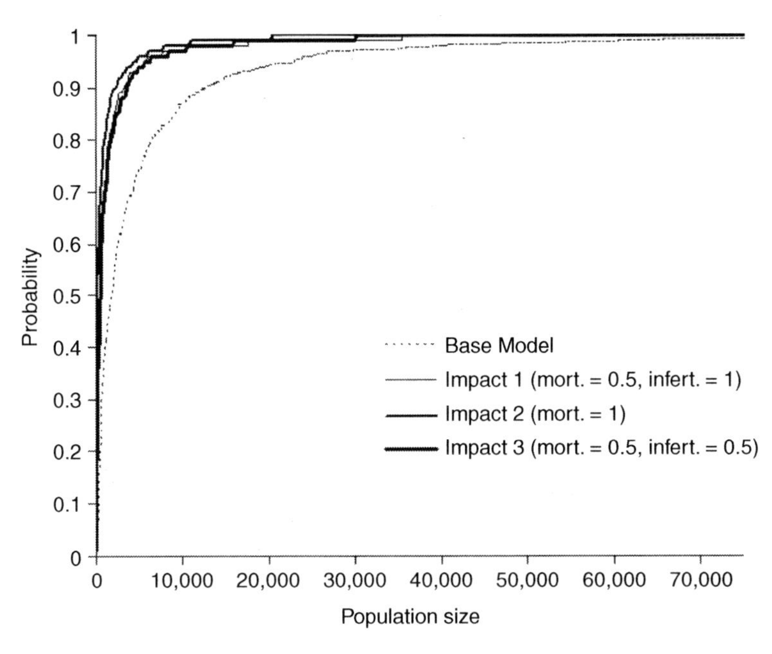

Figure 7.6. The probability that the population will fall below the population sizes on the x-axis at the end of 10 years: comparison of base model and PCB-affected wood frog population projections assuming a declining population. The estimated initial adult population size for the 27 pools in the PSA is approximately 75,000.

Table 7.13. Maximum differences and Kolmogorov-Smirnov tests of significance comparing four PCB impact scenarios for modeled declining wood frog populations: Projection 2.

	Maximum difference (*D*)	*p*-Value
Impact 1	0.336	<0.0001
Impact 2	0.442	<0.0001
Impact 3	0.314	<0.0001

Another series of simulations was run using the same two projections and parameterizations to assess the amount of time before the population faces *quasi extinction* (Ginzburg et al. 1982). Two thresholds were used to define quasi extinction: 50% or greater population decline from present levels and 95% or greater population decline from present levels. The 50% quasi-extinction threshold is commonly used to classify a species as endangered (e.g., IUCN 1994). The 95% quasi-extinction threshold is a more severe criterion. Under projection 1, the risk of a 50% population decline is less than about 82% over the next 10 years in the base model, and between 95% and 99% in the tPCB-affected parameterizations. For projection 2, where the base population is declining, the risk of a 50% population decline over the next 10 years is more than 98% in all parameterizations. In all cases, tPCB concentrations decrease the time to extinction. Note that the model assumes the 27-pool population in the PSA is closed to immigration

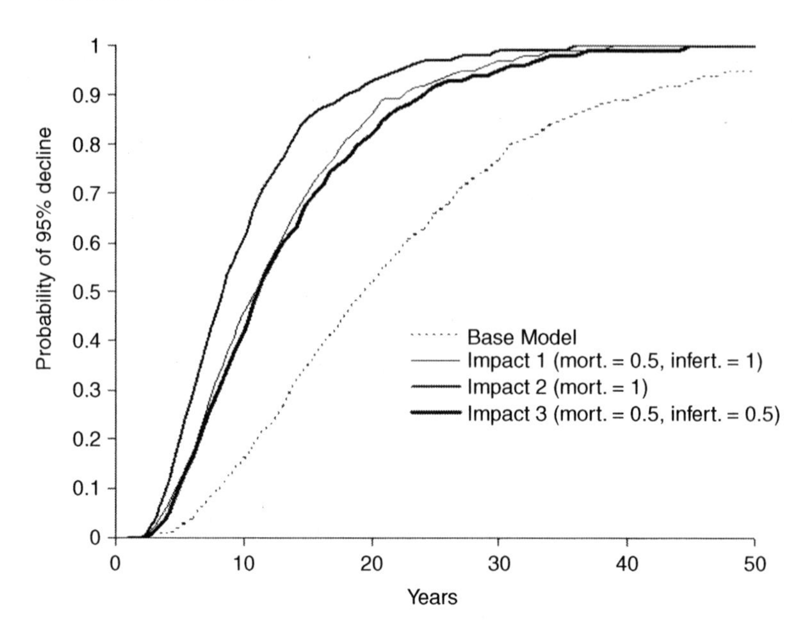

Figure 7.7. Time (in years) for the population to decline by 95%: comparison of base model and PCB-affected wood frog population projections assuming a nondeclining population.

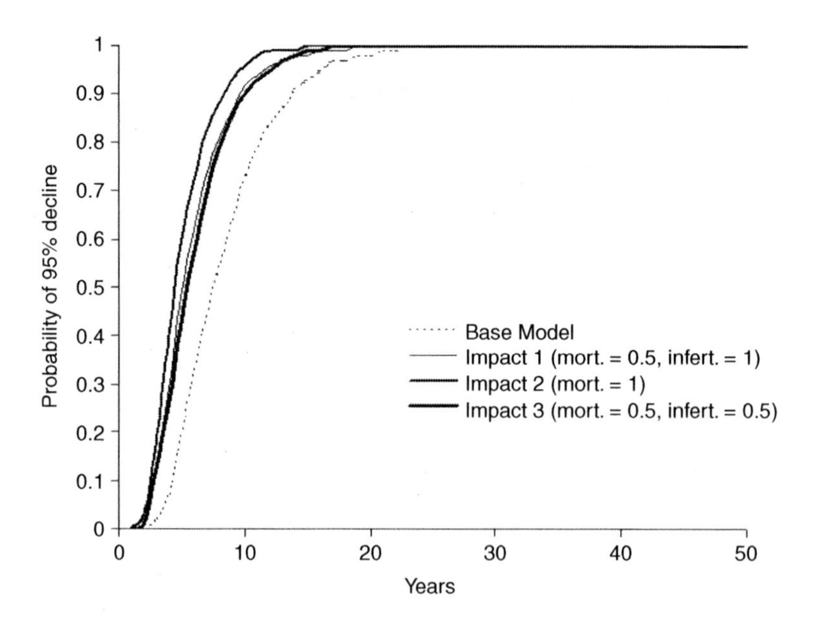

Figure 7.8. Time (in years) for the population to decline by 95%: comparison of base model and PCB-affected wood frog population projections assuming a declining population.

Table 7.14. Expected minimum abundance at the end of 10 years, average local extinction duration over 10 years, and the expected median time to metapopulation extinction.

	Base model	Impact 1	Impact 2	Impact 3	Difference from base (min.–max.)
Projection 1					
Expected minimum abundance	21,822	7,616	5,637	9,816	12,006–16,185
Average local extinction duration (years)	0.96	3.26	4.04	2.26	1.3–3.1
Median time to extinction (years)	20.0	11.0	8.5	11.5	8.5–11.5
Projection 2					
Expected minimum abundance	5,560	1,604	1,129	1,643	3917–4431
Average maximum local extinction duration (years)	3.63	5.67	5.89	5.52	1.9–2.3
Median time to extinction (years)	7.5	5.1	4.5	5.7	1.8–3

from the outside. This assumption allows for near certain population extinction. Were immigration to the PSA modeled, times to extinction would be lengthened.

Figures 7.7 and 7.8 show the results of the 95% quasi-extinction study for projections 1 and 2, respectively. In figure 7.7, the risk of a 95% population decline over the next 10 years is a mere 16% for the base model. In contrast, the affected models face quasi-extinction risks between 46% and 61% over that same time. In figure 7.8, the base model risk of 95% population decline over 10 years is 73%. The affected models face a 90–96% chance of quasi extinction. In all parameterizations of both projections, tPCB-affected populations are at higher risk of quasi extinction.

The population in the PSA can be expected to decline over the next 10 years for projection 1. The probability of a 75% reduction in abundance is just greater than 50% for the base model and 15–30% higher for the tPCB-affected parameterizations. For projection 2, with the declining base population, there is a near certainty (93–99% chance) of 75% decrease in abundance by the end of 10 years. The parameterizations with the tPCB impact are more likely to decline than with the base model.

Expected minimum abundance at the end of 10 years, average local extinction duration over 10 years, and the expected median time to metapopulation extinction are summarized in table 7.14. For projection 1, the impact of tPCBs reduces the minimum population size by at least half. A larger impact is seen with projection 2.

Average local extinction duration is the average time (in years) that individual vernal pools remain empty after a local extinction. In projection 1, the average local extinction duration in the base model is about one year. Local extinctions last two to four times longer when the impact of tPCBs is modeled. For projection 2, where the base population is already declining, tPCB impacts increase the average maximum local extinction duration by up to two additional years.

Median time to extinction (in years) is the median number of years by which population abundance decreases by 95%. In projection 1, the base model exhibits a 20-year median. tPCB impacts decrease this median by between 8.5 and 11.5 years, depending

Table 7.15. Impact of assumptions made in constructing wood frog population models for the Housatonic River PSA.

Impact	Assumption
?	Age-structured model with yearly time step
?	Density-dependent model with ceiling carrying capacity
C	Vital rates cross-correlated by 0.5 within each time step
C	Environmental correlation assumed to be 0.5
O	Omitted effect of PCBs on sex ratios
?	Time horizon of 10 years (terminal risk)
?	Housatonic data used to determine initial abundances
?	Data from Maryland population used to determine vital rates
?	Data from Virginia population used to determine dispersal
O	Assumed males could breed with up to 10 females
O	Modeled laboratory-observed effect showing tPCBs increase larval survival
?	Assumed gonadal abnormalities can cause sterility of 50–100%
?	Assumed malformation can cause death of 50–100%
C	Assumed no four-year-olds survive or breed
?	Assumed only zero-year-olds disperse
?	Corrected standard deviations of vital rates for declining population to remove spatial variability

C, conservative; O, optimistic; ?, unclear.

on the parameterization. In projection 2, the base median time to extinction is 7.5 years, and tPCBs are seen to reduce the median by two to three years.

Sensitivity Analyses

The wood frog population models presented here examined the sensitivity of the population-level projections of tPCB impacts to the data used to generate the vital rates (i.e., projection 1 vs. projection 2) and to the fertility and mortality rates associated with metamorph malformation (i.e., impacts 1, 2, and 3). In this section, we present the results of additional sensitivity studies performed to assess the robustness of the models to changes in other assumptions. Table 7.15 lists all of the major assumptions necessary to construct the population models. Of these, four were chosen for further study: vital rates, environmental correlations, dispersal rates, and the form of the density dependence assumption. These sensitivity studies assay both declining and nondeclining models for all impact and nonimpact parameterizations. In table 7.15, each assumption marked with "O" indicates that it is likely to be optimistic and may understate the effect of PCBs on population decline; "C," probably conservative and may overstate the effect of PCBs; "?," unclear whether the assumption is optimistic or conservative.

Vital rates

Vital rates were increased and decreased by 5% and 10% in order to determine the sensitivity of the models to the exact values chosen for each survival and fertility rate in the Leslie matrices. Increasing vital rates generally reduced the risk of extinction,

Table 7.16. Maximum difference in terminal extinction risk relative to the base model: Sensitivity analyses of vital rates.

	Maximum difference in terminal extinction risk		
	Vital rates increased by 10%	Vital rates unchanged	Vital rates decreased by 10%
(Original base model compared to impact models)			
Nondeclining (projection 1)			
Impact 1	0.07 ($p = 0.03$)	0.31 ($p < 0.0001$)	0.57 ($p < 0.0001$)
Impact 2	0.20 ($p < 0.0001$)	0.47 ($p < 0.0001$)	0.68 ($p < 0.0001$)
Impact 3	0.04 ($p = 0.29$ NS)	0.30 ($p < 0.0001$)	0.53 ($p < 0.0001$)
Declining (projection 2)			
Impact 1	0.08 ($p = 0.002$)	0.32 ($p < 0.0001$)	0.58 ($p < 0.0001$)
Impact 2	0.22 ($p < 0.0001$)	0.45 ($p < 0.0001$)	0.65 ($p < 0.0001$)
Impact 3	0.04 ($p = 0.29$ NS)	0.27 ($p < 0.0001$)	0.56 ($p < 0.0001$)
Base models with altered vital rates compared to impact models			
Nondeclining (projection 1)			
Impact 1	0.38 ($p < 0.0001$)	0.31 ($p < 0.0001$)	0.3 ($p < 0.0001$)
Impact 2	0.48 ($p < 0.0001$)	0.47 ($p < 0.0001$)	0.46 ($p < 0.0001$)
Impact 3	0.34 ($p < 0.0001$)	0.30 ($p < 0.0001$)	0.53 ($p < 0.0001$)
Declining (projection 2)			
Impact 1	0.35 ($p < 0.0001$)	0.32 ($p < 0.0001$)	0.29 ($p < 0.0001$)
Impact 2	0.49 ($p < 0.0001$)	0.45 ($p < 0.0001$)	0.39 ($p < 0.0001$)
Impact 3	0.31 ($p < 0.0001$)	0.27 ($p < 0.0001$)	0.26 ($p < 0.0001$)

NS, not significantly different from base model.

and decreasing vital rates increased the risk of extinction (table 7.16, top). When vital rates were increased by 10%, under both projections 1 and 2, Impact 3 (half malformed die, and half of survivors are infertile) became indistinguishable from the unaffected base models with the original vital rates (figures 7.9 and 7.10). Thus, the strength of the tPCB impact on reducing fertility and mortality when half of malformed metamorphs are assumed to die and half of the surviving gonadally abnormal metamorphs are assumed to be sterile is approximately equal in population-level effect to a 10% reduction in all vital rates. However, when compared to unaffected models with vital rates increased or decreased by the same percentage (table 7.16, bottom), the affected models display the same increased probability of population decline and extinction as models with the original vital rates. This means that the effect observed of tPCBs increasing the risk of population decline and extinction is robust to assumptions regarding vital rates over this range.

Environmental correlation

Environmental correlations between ponds were varied from 0.5 to 0.25 in order to determine the sensitivity of the models to the value chosen for this parameter. The 0.5

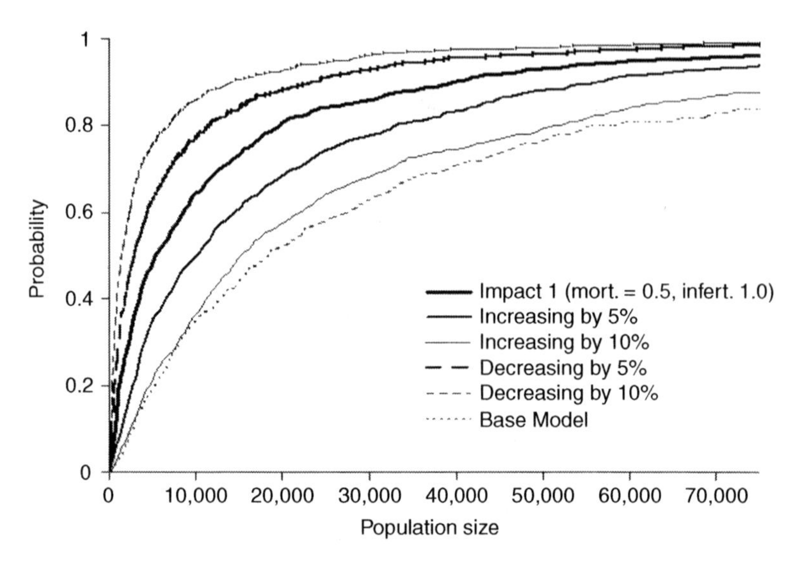

Figure 7.9. Effect of altering vital rates by ±5% and ±10% on terminal extinction risk for impact 1 (half of malformed metamorphs die, all gonadally abnormal metamorphs are sterile) of projection 1 (nondeclining population).

correlation assumption represents a fairly strong correlation between pools. Reducing this to 0.25 increases the chance that, although one pool may experience a bad year, some pools around it will not be so severely affected. Lowering the correlation value reduced the risk of extinction for both projections 1 and 2 (table 7.17, figures 7.11 and 7.12). By reducing the environmental correlation value, each pond experiences environmental conditions independent of neighboring ponds, such that a negative effect on one pond does not necessarily cause a negative effect on its neighbor. This pond autonomy in the face of environmental fluctuation allows greater overall persistence of the metapopulation because multiple pools are less likely to become simultaneously extinct. When environmental correlation is low, pools that do become extinct are more rapidly recolonized by their neighbors.

In figures 7.11 and 7.12, the effect of the strength of environmental correlation on the impact of tPCBs relative to the base models is clearly negligible. In figure 7.11, when the environmental correlation is reduced from 0.5 to 0.25, the projected extinction risk changes noticeably for both the unaffected and the affected models. The difference between the unaffected and the affected models remains about the same, however, with a significantly greater probability of population decline and extinction in the affected case. In figure 7.12, the effect of environmental correlation on the simulations under the declining population assumption is very small. Again, the difference between base model and affected model is comparable when either correlation is assumed.

Dispersal rates

Dispersal rates were generated as a function of distance between pools according to a relationship reported in the literature. To determine the sensitivity of the models to

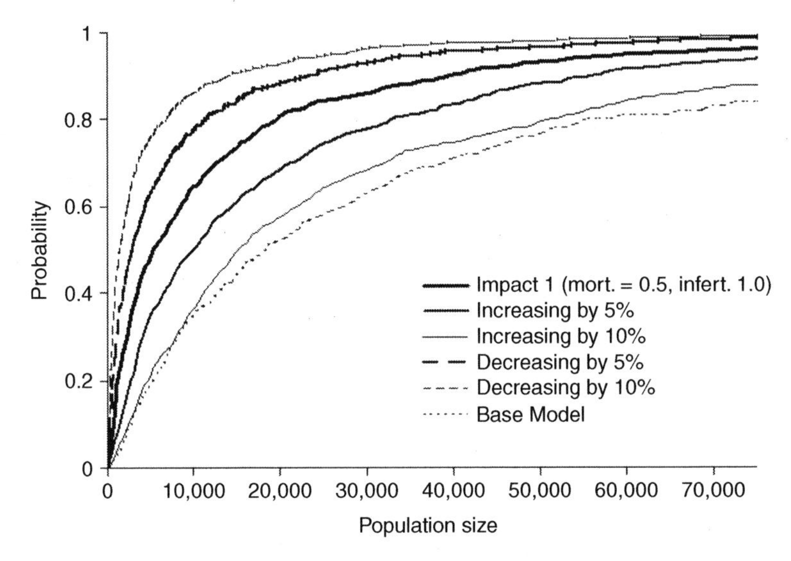

Figure 7.10. Effect of altering vital rates by ±5% and ±10% on terminal extinction risk for impact 1 (half of malformed metamorphs die, all gonadally abnormal metamorphs are sterile) of projection 2 (declining population).

the exact values chosen for probability of dispersal between pools, the values were increased and decreased by 5% and 10%. However, changing dispersal rates had no significant effect on terminal extinction risk in any of the models ($p > 0.05$ for all maximum differences). These results suggest that the models are highly robust to differences in dispersal rates within this range.

Density dependence

To examine the sensitivity of the projections to the form of density dependence assumed, a series of models were developed using a Ricker-type density dependence function, and results were compared to the density ceiling function used in the other simulations. Density dependence is the relationship between population parameters such as survival and fecundity and population size or density. This relationship can take many forms, but the two basic kinds are compensation (increase in survival and/or fecundity as density decreases because of more resources per individual) and depensation (decline in survival and/or fecundity as density decreases due to Allee effects). Compensation generally increases population stability and leads to lower risks of decline or extinction, whereas depensation generally leads to higher risks. In this study, depensation (Allee effects) are not explicitly modeled, but it is implicitly modeled by focusing on risk of decline to a small population instead of risk of extinction.

Incorporating compensation in a model requires knowledge that average vital rates (survival rates and/or fecundities) increase at low population sizes, and data on the magnitude of this increase. The average rate of population increase at low population sizes under density dependence is known as maximum growth rate, R_{max}. For wood frogs, there is no information on either the existence of density effects or their

Table 7.17. Maximum difference in terminal extinction risk relative to the base model for sensitivity analyses examining manipulation of environmental correlation between ponds (all comparisons significant at $p < 0.0001$).

	Maximum difference from base model	
	Correlation = 0.5	Correlation = 0.25
Nondeclining (projection 1)		
Impact 1	0.35	0.27
Impact 2	0.44	0.41
Impact 3	0.27	0.22
Declining (projection 2)		
Impact 1	0.34	0.27
Impact 2	0.44	0.39
Impact 3	0.31	0.23

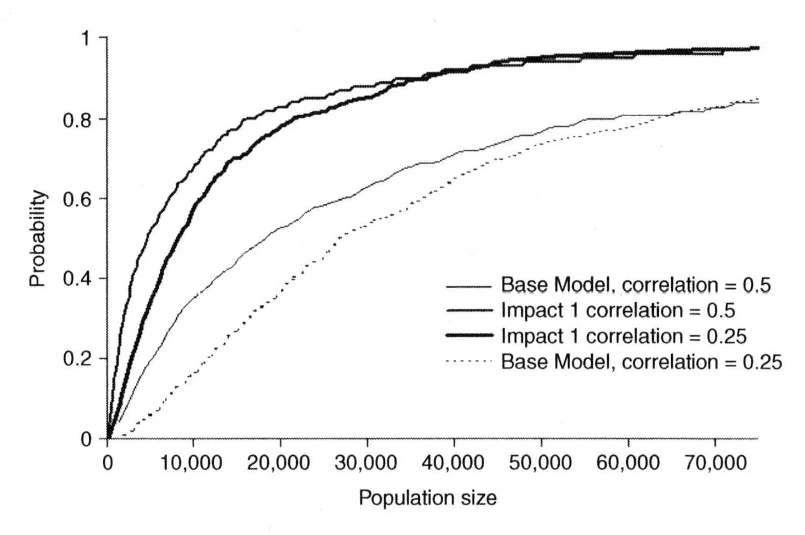

Figure 7.11. Effect of reducing environmental correlation from 0.5 to 0.25 on terminal extinction risk for impact 1 (half of malformed metamorphs die, all gonadally abnormal metamorphs are sterile) of projection 1 (nondeclining population).

magnitude. Therefore, we used the maximum annual population increase in the two observed population trajectories (Berven 1990) as a crude approximation of the maximum growth rate. The maximum rate of growth reported occurred when a population increased from 205 to 500 individuals, giving $R_{max} = 2.4$. This may overestimate R_{max} if the increase was in part due to an above-average year in terms of environmental conditions (e.g., rainfall) in addition to being a low-population year. If this is the case, this sensitivity analysis will likely overestimate the stabilizing effects of density dependence.

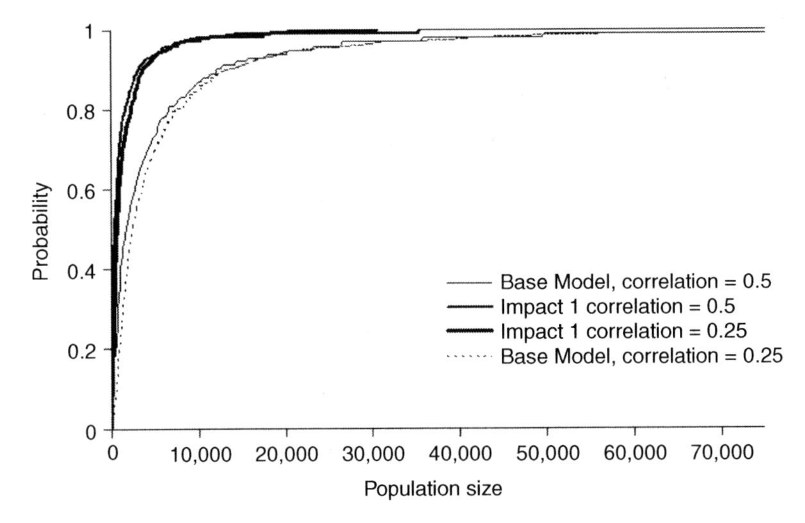

Figure 7.12. Effect of reducing environmental correlation from 0.5 to 0.25 on terminal extinction risk for impact 1 (half of malformed metamorphs die, all gonadally abnormal metamorphs are sterile) of projection 2 (declining population).

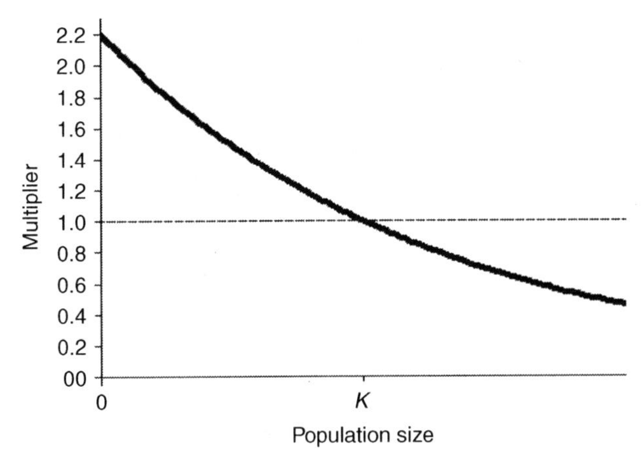

Figure 7.13. The density dependence function used in the sensitivity analysis. At each time step of a simulation, the stage matrix elements were multiplied by the value given by this function for the population size at that time step. The function is a Ricker equation with $\alpha = 2.2$. K is carrying capacity.

To incorporate a density dependence with $R_{max} = 2.4$, we multiplied all the survival and fecundity values of the base model by a constant of 2.2, which resulted in the base model having a growth rate of 2.4. In other words, when all elements of the stage matrix (of the base model) are multiplied by 2.2, the stage matrix has an eigenvalue of 2.4. We then used RAMAS Metapop to modify the stage matrix according to the size of the population at each time step: If population size was very low, the stage matrix was multiplied by 2.2; if the population size was equal to the carrying capacity, the stage matrix was unmodified (multiplied by 1.0); and if the population size was greater than the carrying capacity, the stage matrix was multiplied by a value less than 1.0. For this function, we used a Ricker equation with $\alpha = 2.2$ (see figure 7.13).

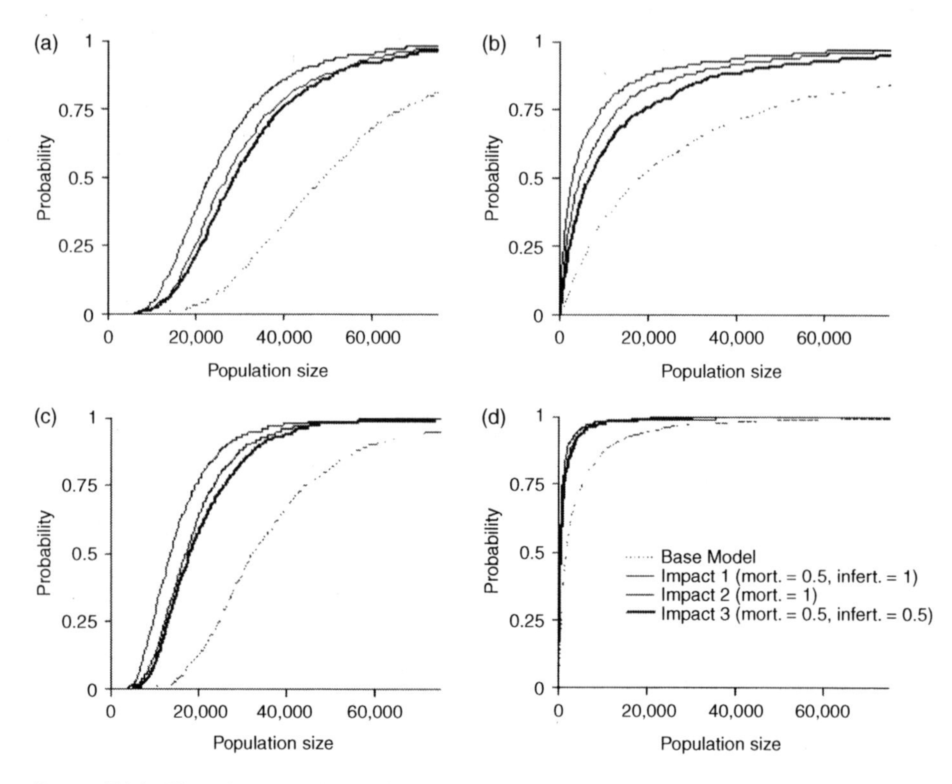

Figure 7.14. Time (in years) for the population to decline by 95%: Comparison of base model and PCB-affected wood frog population projections assuming a declining population, Ricker-type (*A* and *C*) and ceiling-type (*B* and *D*) density dependence, nondeclining (*A* and *B*) and declining (*C* and *D*) base model.

The projection 1 (nondeclining) and projection 2 (declining) simulations were reperformed using the Ricker-type density dependence function. Just as was done for the ceiling-type density dependence, the base model parameterizations were compared to affected parameterizations in order to determine the population-level effect of tPCB contamination. As expected, the compensatory density dependence resulted in decreased terminal extinction risk for all parameterizations. However, the magnitude of the negative impact of tPCBs on population decline was little different from the ceiling-type density dependence model simulations.

Figure 7.14 compares terminal extinction risks for the Ricker-type density dependence to the ceiling-type density dependence. For the nondeclining base model, figure 7.14A shows the Ricker density dependence results, and figure 7.14B reproduces the ceiling model results from figure 7.5. All of the projections in figure 7.14A (compensatory density dependence case) are more likely to have larger population sizes after 10 years compared to those in figure 7.14B (ceiling density case). This is because the compensatory density function retards population decline. Note, however, that the impact of tPCBs relative to the nonaffected base model is comparable for the two density assumptions. In every case, tPCB contamination hastens population decline and increases the likelihood of extinction relative to the uncontaminated base models.

Figure 7.14, C and D, shows the same comparison for the declining base model projections.

Discussion

The results of the population viability modeling exercise indicate an impact of tPCBs on wood frog population growth and abundance. tPCBs hasten population decline, reduce population numbers, and increase the likelihood of extinction. Data collected in the PSA provide field evidence supporting the population-level effects of tPCBs seen in the simulations. The relationship between sediment tPCB concentrations and adult male and female density shown in figures 7.3 and 7.4 indicate that increased tPCB concentration co-occurs with decreased density—particularly for adult females. It is notable that, even given the extremely small sample sizes, the negative relationships seen in the figures approach or are at the 0.05 level of statistical significance.

Sensitivity analyses were performed with the population viability model to assess the robustness of the model projections in the face of uncertainty regarding several key assumptions. The assumptions analyzed included vital rates, environmental correlation, dispersal rates, and density dependence. Projections of time to extinction and terminal extinction risk were sensitive to all of these assumptions except for dispersal rates. The impact of tPCBs on population decline was not sensitive to any of these assumptions. Instead, the increased risk due to tPCB contamination relative to the uncontaminated base models was maintained in each sensitivity study. The models are robust in projecting the increased risk of population decline and extinction of wood frogs due to tPCB contamination.

The process of deciding what remedial action should be taken to protect the wood frog metapopulation in the Housatonic River valley is ongoing. General Electric is preparing a corrective measures study proposal based on the ecological risk assessment and the population modeling results. The population modeling results presented here provide support for the more general results of the ecological risk assessment. The population model was accepted by a peer-review panel of international experts, and the results were presented and discussed in a series of public meetings. The modeling results make clear that, under a wide variety of assumptions, the mortality and morbidity caused by PCBs in the lab and measured in the field are having population-level ramifications.

References

Akçakaya, H. R. 2002. *RAMAS Metapop: Viability Analysis for Stage-Structured Metapopulations* (version 4.0). Applied Biomathematics, Setauket, New York.

Akçakaya, H. R. 2005. *RAMAS GIS: Linking Spatial Data with Population Viability Analysis* (version 5.0). Applied Biomathematics, Setauket, New York.

Berven, K. A. 1990. Factors affecting population fluctuations in larval and adult stages of the wood frog (*Rana sylvatica*). *Ecology* 71(4):1599–1608.

Berven, K. A., and Grudzien, T. A. 1990. Dispersal in the wood frog (*Rana sylvatica*): Implications for genetic population structure. *Evolution* 44(8):2047–2056.

Blaustein, A. R., and D. B. Wake. 1990. Declining amphibian populations—a global phenomenon. *Trends in Ecology and Evolution* 5:203–204.

Blaustein, A. R., Romansic, J. M., Kiesecker, J. M., and Hatch, A. C. 2003. Ultraviolet radiation, toxic chemicals and amphibian population declines. *Diversity and Distributions* 9:123–140.

Burgman, M. A., Ferson, S., and Akçakaya, H. R. 1993. *Risk Assessment in Conservation Biology.* Chapman and Hall, London.

Caswell, H. 2001. *Matrix Population Models: Construction, Analysis, and Interpretation.* Sinauer Associates, Sunderland, MA.

Collins, J. P., and Storfer, A. 2003. Global amphibian declines: Sorting the hypotheses. *Diversity and Distributions* 9:89–98.

FEL (Fort Environmental Laboratories, Inc.) 2002. *Final Report—Frog Reproduction and Development Study. 2000 Rana sylvatica Vernal Pool Study.* Study Protocol WESR01-RSTS03-1. Fort Environmental Laboratories, Inc., Stillwater, OK.

Ginzburg, L. R., Slobodkin, L. B., Johnson, K., and Bindman, A. G. 1982. Quasiextinction probabilities as a measure of impact on population growth. *Risk Analysis* 2:171–181.

Glennemeier, K. A., and Begnoche, L. J. 2002. Impact of organochlorine contamination on amphibian populations in southwestern Michigan. *Journal of Herpetology* 36(2):233–244.

Hayes, T., Haston, K., Tsui, M., Hoang, A., Haeffele, C., and Vonk, A. 2002. Herbicides: Feminization of male frogs in the wild. *Nature* 419:895–896.

Houlahan, J. E., Findlay, C. S., Schmidt, B. R., Meyer, A. H., and Kuzmin, S. L. 2000. Quantitative evidence for global amphibian population declines. *Nature* 404:752–755.

IUCN 1994. IUCN Red List Categories. Prepared by the IUCN Species Survival Commission as approved by the 40th Meeting of the IUCN Council, Gland, Switzerland.

Kiesecker, J. M., Blaustein, A. R., and Belden, L. K. 2001. Complex causes of amphibian population declines. *Nature* 410:681–684.

Sessions, S. K. 2003. What is causing deformed amphibians? Pages 168–186 in R. D. Semlitsch (ed.), *Amphibian Conservation.* Smithsonian Institution, Washington, DC.

Stuart, S., Chanson, J. S., Cox, N. A., Young, B. E., Rodrigues, A. S. L., Fishman, D. L., and Waller, R. W. 2004. Status and trends of amphibian declines and extinctions worldwide. *Science* 306:1783–1786.

WAI (Woodlot Alternatives, Inc.). 2002. *Ecological Characterization of the Housatonic River.* Prepared for U.S. Environmental Protection Agency. Woodlot Alternatives, Inc., Topsham, ME.

WAI (Woodlot Alternatives, Inc.). 2003. *Amphibian Reproductive Success within Vernal Pools Associated with the Housatonic River, Pittsfield to Lenoxdale, Massachusetts.* Prepared for U.S. Environmental Protection Agency. Woodlot Alternatives, Inc., Topsham, ME.

Weston Solutions, Inc. 2004a. *Ecological Risk Assessment for General Electric (GE)/Housatonic River Site, Rest of River,* vol. 5. Environmental Remediation Contract DACW33-00-D-0006, Task Order 0003, DCN GE-100504-ACJS. GE/Housatonic River Project, Pittsfield, Massachusetts.

Weston Solutions, Inc. 2004b. A stochastic population model incorporating PCB effects for wood frogs (*Rana sylvatica*) breeding in vernal pools associated with the Housatonic River, Pittsfield to Lenoxdale, Massachusetts. Attachment E.4 in *Ecological Risk Assessment for General Electric (GE)/Housatonic River Site, Rest of River,* vol. 5. Environmental Remediation Contract DACW33-00-D-0006, Task Order 0003, DCN GE-100504-ACJS. GE/Housatonic River Project, Pittsfield, Massachusetts.

8

Potential Effects of Freshwater and Estuarine Contaminant Exposure on Lower Columbia River Chinook Salmon (*Oncorhynchus tshawytscha*) Populations

JULANN A. SPROMBERG
LYNDAL L. JOHNSON

The Columbia River provides critical habitat for 13 stocks of threatened and endangered salmon species in the Pacific Northwest. For at least some period of time, all of these stocks use the Columbia River estuary for a migration corridor, and ocean-type stocks that outmigrate as subyearlings, such as the lower Columbia River (LCR) fall chinook, use it for a more extended period of rearing. Hydroelectric dams, overharvest, hatchery production, and habitat degradation are commonly cited as causes of the long-term decline in Columbia River salmon stocks (National Research Council 1996). However, one component of that habitat degradation—toxic contamination—has not been well explored, although it is increasingly recognized as a potential limiting factor for salmon recovery in watersheds with significant agricultural, urban, or industrial land use coverage (National Marine Fisheries Service 1998). Toxic contaminants may pose a particular threat to the LCR chinook salmon evolutionarily significant unit (ESU), which was listed as threatened under the Endangered Species Act (ESA) in 1999 (National Marine Fisheries Service 1999).

Contaminant sources

Contaminants such as metals, current-use pesticides, pharmaceuticals, and persistent organic pollutants have been found in water, sediment, and biota samples in the LCR (Fuhrer et al. 1996; TetraTech Inc. 1996; Johnson et al. 2007b; LCREP 2007). Salmon in the Columbia River generally come into contact with contaminants within the water column, through their diet, and from interaction with contaminated bed and suspended solids. Resultant exposures can impair the health and performance of individual salmon (e.g., growth, survival) and ultimately reduce population productivity.

The LCR from the Bonneville Dam to the mouth, where fish from the LCR ESU reside, is the most urbanized section of the river, including the major population centers of Portland, Oregon, and Vancouver, Washington. Contaminants enter the LCR from urban runoff and from a variety of municipal and industrial dischargers, including aluminum smelters, pulp and paper plants, wood product facilities, chemical manufacturers, and wastewater treatment plants (TetraTech Inc. 1996). Agricultural runoff is another major source of contamination, containing both legacy organochlorine pesticides, such as DDT, and current-use pesticides, such as atrazine, carbaryl, and diazinon (Wagner et al. 2006). Tributaries of the Columbia River may also contribute to the contaminant load depending on their discharge (or volume) and land use in the watershed; they are a particularly likely source for inputs from agricultural and mining operations (Wentz et al.1998; Majewski et al. 2003; Fuhrer et al. 1996).

Contaminant effects

Life-history models of chinook salmon have shown that individual survival through the first year of life is an important determinant of the overall population growth rate (Ratner et al. 1997; Kareiva et al. 2000; Spromberg and Meador 2005). Therefore, an effective strategy for increasing the population growth rate of wild runs may be to reduce anthropogenic mortality in these early life stages. Chinook salmon undergo tremendous physiological and anatomical transformations during their first-year freshwater residence, including development, somatic growth, imprinting on their natal stream, and smoltification. These processes are just a few that can be affected by contaminant exposure and can lead to reductions in survival and reproductive rates. Toxic stressors, as well as other factors, may directly and indirectly influence these vital rates.

Sublethal toxic effects can influence the age/stage-specific vital rates of individuals and potentially produce population-level impacts. While immune suppression directly relates to individual health, it can also alter survival rates via delayed mortality. Reproductive toxicity effects can alter the reproductive contribution of individuals and age groups within a population well after exposure. Inhibition of somatic growth potentially decreases both survival and reproductive contribution. Through these pathways, the toxic impacts alter life-history parameters. Direct sublethal effects on juvenile chinook salmon resulting from contaminant exposure during freshwater or estuarine residence may include immune suppression, behavior changes, growth inhibition, and impaired reproductive function.

Arkoosh et al. (1998, 2001) observed compromised immune function in juvenile chinook collected from urban estuaries in Puget Sound. Significant differences in mortality two months after removal from the contaminated areas suggest that individuals exposed to chemical contaminants may have a higher predisposition to infection. Salmon exposed to anticholinesterase insecticides have altered swimming behavior (Sandahl et al. 2005; Brewer et al. 2001; Beauvais et al. 2000) and antipredator behaviors (Scholz et al. 2000) that can result in increased mortality.

Many contaminants, including metals and organics, inhibit somatic growth in fish (Roch and McCarter 1984; Finlayson and Verrue 1985; Glubokov 1990; Stein et al. 1995; Hansen et al. 2002). Current-use pesticides, and particularly anticholinesterase insecticides, have been shown to interfere with salmon swimming behavior (Sandahl et al. 2005; Brewer et al. 2001; Beauvais et al. 2000), feeding behavior (Sandahl et al. 2005), and foraging behavior (Morgan and Kiceniuk 1990), potentially culminating in reduced somatic growth.

Contaminant exposures during early life stages have been shown to alter reproductive functions such as fecundity, egg survival, spawning behavior, and time to reproductive maturity in affected fish (Milnes et al. 2006; Ankley and Johnson 2004). There is particular concern about environmental pollutants that exhibit hormonal activity and may cause endocrine disruption or directly affect gamete viability (Kime 1999; Milnes et al. 2006). Contaminant-associated impacts on growth and condition may also affect reproduction. Fish with a lower condition factor (i.e., lower weight:length ratio) have increased oocyte atresia and decreased fecundity (Kurita et al. 2003; Laine and Rajasilta 1999; Trippel et al. 1997). In addition, a direct relationship between reduced fecundity and female body length has been observed over time in chinook and coho salmon at the University of Washington hatchery (Quinn et al. 2000). Heintz et al. (2000) suggested that fish surviving a juvenile growth inhibition could share an average length distribution with unexposed fish upon returning to their natal stream, but could exhibit reduced weights, which may result in lower reproductive success (Roni and Quinn 1995). Because the onset of sexual maturation is also influenced by size, slow growth could increase the age at reproduction.

Delayed effects, or impacts that extend beyond the life stage in which the exposure occurs, have been observed in salmon via several different mechanisms. Chinook exposed to contaminants early in their life exhibited immunosuppression well after leaving the contaminated areas (Arkoosh et al. 1998; Loge et al. 2005). Juvenile chinook salmon exposed to polycyclic aromatic hydrocarbons (PAHs) experienced both somatic growth reduction and increased mortality during the overwinter starvation that followed (Meador et al. 2006). In addition, Heintz et al. (2000) exposed pink salmon (*O. gorbuscha*) to crude oil during embryonic development and, after releasing "healthy looking" smolts that survived the exposure, observed a significant reduction in their survival to maturity compared to unexposed fish.

Juvenile growth is a critical determinant of freshwater and marine survival for chinook salmon (Higgs et al. 1995), and individual salmon size has been linked to age-specific survival rates, age at reproductive maturity, fecundity, and spawning success (Healey 1991; Beamish and Mahnken 2001; Zabel and Achord 2004). Reductions in the somatic growth rate of salmon fry and smolts are believed to result in increased size-dependent mortality (West and Larkin 1987; Healey 1982) primarily because they must reach a critical or threshold size to successfully transition from fresh water to seawater (smoltification) (Beamish and Mahnken 2001). A second threshold size has been hypothesized for survival over the first winter in the open ocean (Beamish and Mahnken 2001). A recent study on size-selective mortality in chinook salmon from the Snake River (Zabel and Williams 2002) found that naturally reared wild fish did not return to spawn if they were below a size threshold when they migrated to the ocean. Although Zabel and Williams studied stream-type chinook, similar trends in size-selective mortality have been observed in juvenile ocean-type chinook (Connor et al. 2004). Mortality is likely higher among smaller and slower growing salmonids because they are more susceptible to predation during their first year (Healey 1982; Holtby et al. 1990). These studies suggest that the primary influence of reductions in somatic growth caused by contaminant exposure will result in increased first-year mortality.

LCR fall chinook salmon hatch and rear in fresh water and the LCR estuary and, as discussed above, are potentially exposed to toxic compounds during this period as evidenced by the contaminant levels measured in juvenile chinook body burdens and stomach contents (Johnson et al. 2007a, 2007b; LCREP 2007). In this study, we investigated a potential range of population-level effects of reduced first-year survival, delayed survival impacts, and reproductive inhibition resulting from contaminant exposure in the LCR and estuary on 22 populations of fall chinook salmon. Specific questions include whether the effects of toxicant exposure manifest differently at the population and metapopulation level, and how these differences change when the intensity of exposure varies between populations.

Methods

Study species and area

This study focuses on the potential effects of toxic contaminant exposure during freshwater and estuarine residence of 22 populations of fall chinook salmon within the LCR chinook ESU (figure 8.1). Chinook salmon are anadromous, living in both freshwater and marine environments, and semelparous, reproducing only once. The general life history consists of hatching and rearing in fresh water, smoltification (physiological changes to transition from fresh water to saltwater), migration to the ocean, maturation at sea, and return migration to the natal freshwater system for spawning followed shortly by death. The anadromous life-history strategy may have evolved to take advantage of freshwater rearing habitat as well as abundant nutrient availability in the ocean to maximize somatic growth and reproductive capability (Taylor 1990; Brannon et al. 2004). A single delayed reproductive effort allows chinook salmon greater somatic growth in the ocean, and the resulting increased size leads to increased fecundity and offspring survival to emergence (Healey 1991).

The 22 populations of chinook salmon composing the fall run of the LCR ESU (figure 8.1) exhibit primarily an ocean-type life-history strategy. Subyearling chinook begin their migration to sea as newly emerged fry or as fingerlings, and return to spawn in the tributaries of the Columbia River at 3–5 years of age (Morrow 1980; Beauchamp et al. 1983), with a low percentage of precocious males returning earlier (<4% of second-year fish). Since subyearlings begin their downstream migration at a small size, it is important that they feed and grow during their migration in order to reach an appropriate size to undergo smoltification. Feeding during migration through known contaminated habitats, such as in the LCR mainstem and estuary, increases the potential for contaminant exposure in the water column and through their diet. Subyearling migrants spend several months rearing in the Columbia River estuary before entering the ocean and can be found in the estuary from March to August (Fresh et al. 2005; Myers et al. 2006). Dispersal in the form of straying of adult spawners occurs among the 22 populations of fall chinook in the LCR, and the populations taken as a whole can be assessed as a metapopulation. Our study excludes the spring chinook in the LCR chinook ESU because they exhibit a stream-type life-history strategy and spend little time in the mainstem and estuary (Myers et al. 2006). While they may be exposed to contaminants in their respective tributaries, they have limited

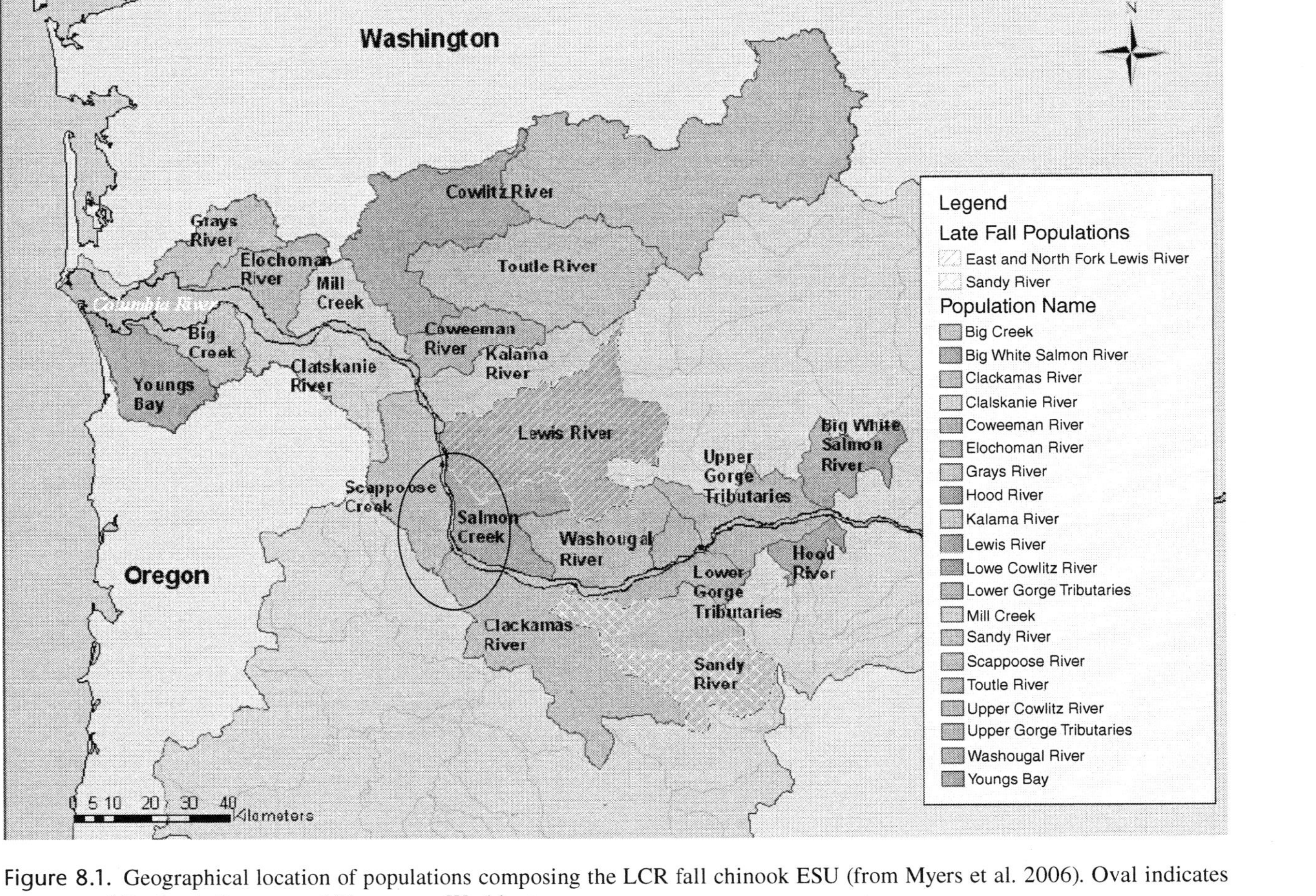

Figure 8.1. Geographical location of populations composing the LCR fall chinook ESU (from Myers et al. 2006). Oval indicates area around Portland, Oregon, and Vancouver, Washington.

potential for exposure to contaminants in the LCR mainstem and estuary, which is the focus of this study.

The study area encompasses the LCR mainstem and estuary from the mouth upstream to the confluences of the Big White Salmon or Hood River. The land use in the LCR drainage can be categorized as approximately 74% forest, 17% agriculture, and 5% urban (TetraTech Inc. 1996; Maher et al. 2005; Sheer and Steel 2006). While the urban area makes up a small percentage of overall acreage, it has a disproportionate effect on river water use and quality (Fuhrer et al. 1996). Major tributaries entering the LCR include the Grays, Cowlitz, Kalama, Lewis, Washougal, and Big White Salmon rivers in Washington and the Youngs, Clatskanie, Willamette, Sandy, and Hood rivers in Oregon. These contribute more than 25% of the total runoff that enters the mainstem of the Columbia River (TetraTech Inc. 1996).

The primary sources of contaminant load in the LCR originate from human activities. More than 100 point sources are known to discharge directly into the LCR (Fuhrer et al. 1996), including municipal wastewater treatment plants, fish hatcheries, and industrial dischargers. In addition, nonpoint sources of contamination (e.g., surface water runoff, combined sewer overflows, atmospheric deposition, accidental spills, and transport from upriver sources) and uncontrolled discharges from septic tanks, marinas, and moorage contribute to pollution loads. The approximately 55 hazardous waste sites and landfills located within a mile of the river also pose a threat when contaminants such as metals, polychlorinated biphenyls (PCBs), and a host of other organics leach into groundwater or nearby surface water and are transported to the river. Most of the landfills, hazardous waste sites, and point sources of contaminants are located around Portland, Oregon, and Vancouver and Longview, Washington (figure 8.2). However, past operation of the hydroelectric facilities has also led to PCB contamination of the area around the Bonneville Dam complex (URS Corporation 2004). As noted above, tributaries of the Columbia River may also contribute to the contaminant load. Monitoring data from U.S. Geological Survey suggest that the greatest contaminant contributions are probably coming from the Willamette River (Wentz et al. 1998; Fuhrer et al. 1996).

Population model

Using RAMAS Metapop, we constructed a five-age matrix model with reproductive contributions from salmon three, four, and five years of age for the LCR fall chinook to assess impacts of low-level contamination on dynamics and trends in abundances. We estimated survival rates from Greene and Beechie (2004) and Fast et al. (1988) to represent the ocean-type life history with a five-month redd residence, two-month stream residence, two-month estuary residence, and three-month nearshore residence for the first year. Ocean survival rates, age distribution, sex ratio, and fecundity of spawning adults were derived for the LCR fall chinook and are listed in table 8.1. All 22 populations within the metapopulation use an identical transition matrix. The population models assume no hatchery augmentation, but demographic parameters integrate both wild and naturally produced descendents of hatchery fish. The parameters produce a population growth rate of 1.0598, which is increasing. The vital rates used in constructing the control transition matrix are derived from various populations under different conditions, and therefore the models are not an attempt to estimate

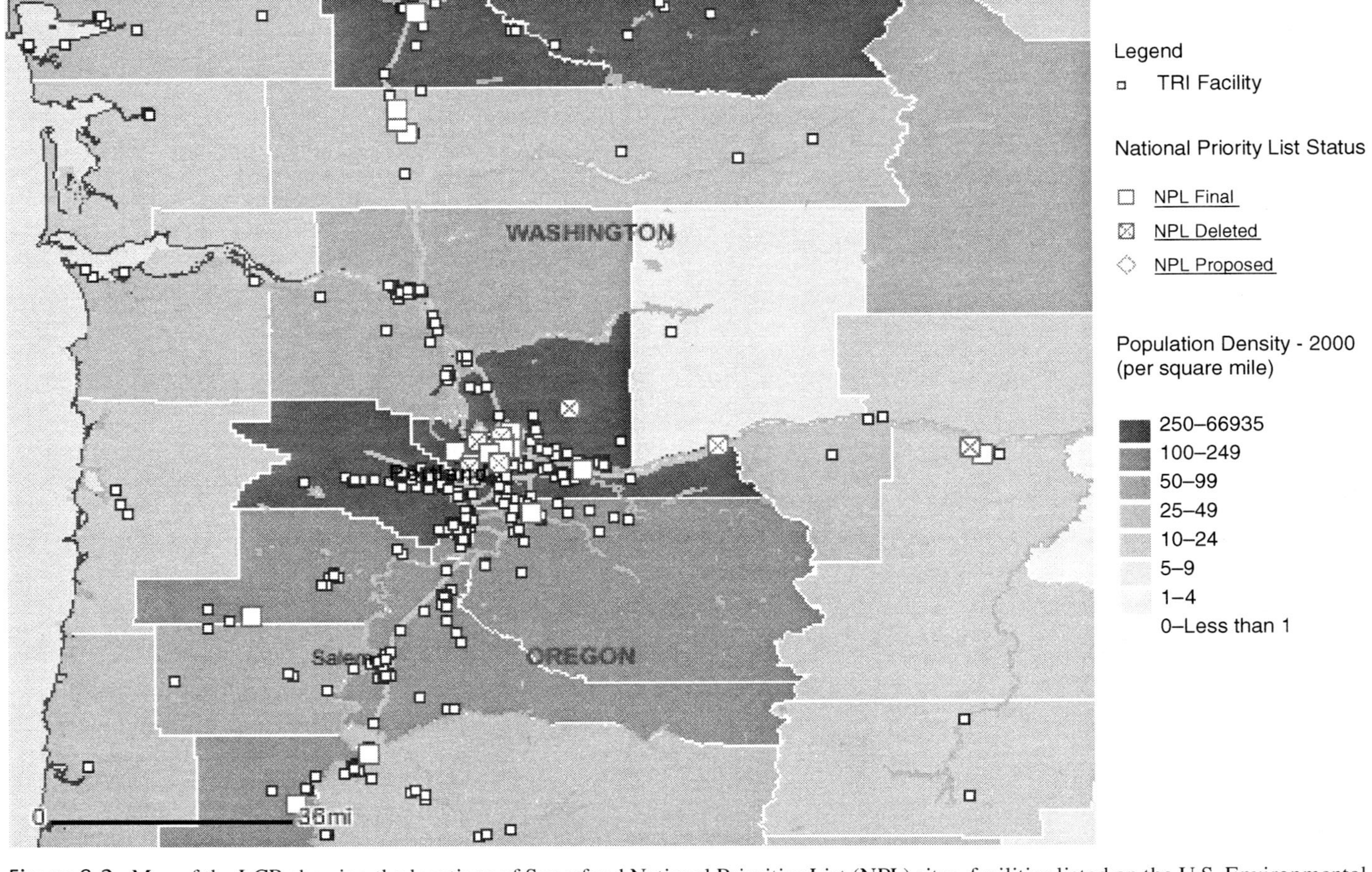

Figure 8.2. Map of the LCR showing the locations of Superfund National Priorities List (NPL) sites, facilities listed on the U.S. Environmental Protection Agency's Toxic Release Inventory (TRI) program, and human population density (from toxnet.nlm.nih.gov/ U.S. National Library of Medicine 2007).

Table 8.1. Demographic parameters used in developing the transition matrix for lower Columbia ocean-type chinook salmon.

	Annual survival rate	Proportion sexually mature[a]	Female sex ratio of spawners[a]	Fecundity[a]	Upstream survival
Age 1	0.00581[b]	0			
Age 2	0.5[c]	0.04	0		
Age 3	0.6[c]	0.15	0.2	3167	0.37
Age 4	0.8[c]	0.70	0.6	4716	0.37
Age 5	0.9[c]	1.0	0.6	5827	0.37

[a] Howell et al. (1985).
[b] Greene and Beechie (2004), Fast et al. (1988) control value before density-dependent factors included.
[c] PSCCTC (2002).

actual population abundance for a particular location or time. In fact, these parameters may be overly optimistic. Many of the LCR fall chinook populations are not currently considered naturally self-sustaining and persist only because of large-scale hatchery production (Good et al. 2005).

Resource or habitat limitations normally influence the first-year survival rate in density-dependent fish populations (DeAngelis et al. 1980). It has been suggested that salmon listed under the ESA may have too low an abundance to trigger most density-dependent dynamics (Kareiva et al. 2000). For this reason, we assumed a ceiling density dependence that will not provide compensatory growth rates for low abundances. The density dependence was based upon age 1 abundance with a carrying capacity of 1,250,000 and low standard deviation of 1,250. This produces a spawning population of approximately 1,900 which falls within the historic range of spawner abundances for many of the populations of fall chinook in the LCR ESU (Myers et al. 2006).

Stochasticity was introduced with survival coefficient of variation at 20% and fecundity coefficient of variation at 12% using the RAMAS Standard Deviation matrix set to Autofill. We believe this to be a conservative estimate of the highly variable annual conditions relating to resource availability but reflective of values reported in the literature (Greene and Beechie 2004; Howell et al.1985; McElhany et al. 2003, 2006, Pacific Salmon Commission Chinook Technical Committee 2002).

The fall runs of the LCR chinook ESU include 22 separate populations, based upon the definition that there is less than 10% exchange of individuals between any two distinct populations (Myers et al. 2006). To simplify calculations and analysis and since data are not available for straying between all the populations, the arrangement and location of the generic identical model populations were defined (figure 8.3) and differ from the map in figure 8.1. We numbered the populations 1–22 from downstream to upstream on alternating sides of the Columbia River mainstem (figure 8.3). The distance between two adjacent populations was defined to be two units, and the distance across the river was one unit. Distance-dependent dispersal among all 22 populations (462 pathways) was calculated using the dispersal equation

$$m = ae^{[(d^c)/b]},$$

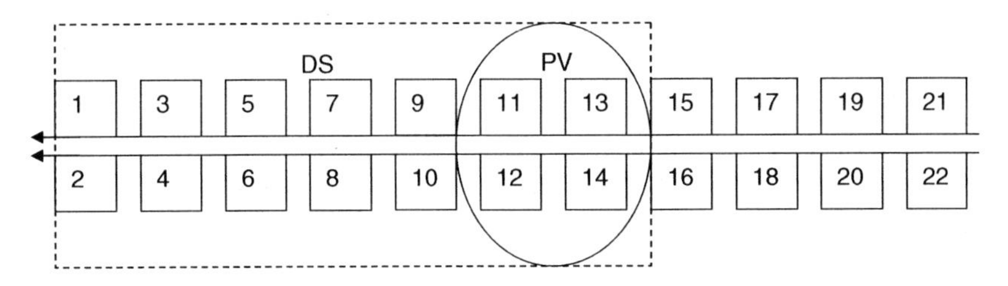

Figure 8.3. Locations of the 22 modeled populations along the river. The oval indicates populations experiencing higher exposure in the Portland/Vancouver (PV) exposure scenario. The rectangle indicates populations experiencing higher exposure in the downstream (DS) exposure scenario. Arrowheads indicate direction of river flow.

where a = 0.18, b = 0.68, c =0.52, and d is distance (Akçakaya 2005). The equation allows some dispersal among populations 1 and 22 while keeping the exchange between any two populations less than 10%, the defined level of distinguishing separate populations (McElhany et al. 2006). Applying the above equation to the distance matrix produced a maximum one-way dispersal between two populations of 4.1% and total cumulative dispersal from a single population to all 21 others ranged from 12.0% to 19.0%. Dispersal rates between two populations were assumed to be symmetrical. Straying rates were assumed to be the same for all adult ages, which is consistent with current theories of straying (Quinn 2005).

Scenarios

Previous studies have shown a variety of water column and sediment contaminants as well as salmon prey body burden in the LCR (Buck 2004). We know from previous laboratory experiments that exposure to low levels of some of these compounds causes somatic growth reduction, immune suppression, and reproductive dysfunction, as discussed above. In ongoing monitoring studies, contaminants that have not been well studied, including brominated fire retardants and pharmaceuticals, have been found in the LCR (LCREP 2007). Since the potential impacts of the mixture of contaminants present in the LCR and estuary are unknown, we assumed conservative impacts on growth, immune function, and reproduction that may lead to delayed mortality or reduced reproductive output. The conservative impacts were applied by reducing survival and/or reproductive rates at a level of 10% or 20%, levels that would be difficult to measure in field experiments and often considered not statistically significant, although they may be biologically significant (Spromberg and Meador 2005).

Since the ocean-type chinook populations in the LCR spend some time rearing in the estuary, exposure to contaminants in the estuary more likely affects these populations. The basic impact scenario represents direct effects leading to mortality, such as a reduction in growth rate, resulting in a 20% reduction in young of the year survival (YOYS scenario). The YOYS scenario was used for comparison to the other impact scenarios. The additional impact scenarios are based upon the presence in water, sediment, prey, and fish of other contaminants known or suspected of causing growth, behavior, immune, or reproductive effects (Johnson et al. 2007b; Fresh et al. 2005).

Delayed cumulative mortality due to immune suppression or growth reduction resulting in effects over several years was modeled as a 20% reduction in YOYS (S1) and a 10% reduction in each of the remaining survival parameters (table 8.1). This scenario combines young of the year effects with delayed mortality, and is designated YD scenario. Reproductive impacts were modeled with cumulative impacts of a 20% reduction in YOYS, 10% reduction in the remaining survival rates, and a 10% reduction in reproductive output for all mature age groups, designated YDR scenario. The analysis was run first on one solitary population not connected to a metapopulation and then on 22 generic populations representing the LCR fall chinook connected as a metapopulation.

Current and previous monitoring studies on contaminants such as metals, PCBs, PAHs, pesticides, and pharmaceuticals in the LCR water, sediment, and biota have shown higher concentrations around the Portland, Oregon, and Vancouver, Washington, regions, as well as downstream areas and the estuary (TetraTech Inc. 1996; Roy F. Weston Inc. 1998; National Oceanic and Atmospheric Administration 1999; McCarthy and Gale 2001; Sethajintanin et al. 2004; Hinck et al. 2004; Buck et al. 2005; Johnson and Norton 2005; Johnson et al. 2007b). For this reason, we applied the differential exposure scenarios to three subsets of the metapopulation and assessed impacts to the whole metapopulation, including potential for localized extinctions. First, each scenario was applied to all of the populations, simulating conditions where all populations were equally affected by contaminant exposure. Next, we assumed higher exposure levels (YD and YDR scenarios) for the subsets of populations occupying the more contaminated areas, with exposure limited to YOYS for the other populations. In one case, all the populations downstream of the Portland area (DS, populations 1–14, figure 8.3) were more heavily impacted. Finally, only the four populations representative of the Portland/Vancouver (PV) area were more heavily impacted (populations 11–14, figure 8.3).

Model scenarios were run for 100 years, with 500 replications of each scenario. The initial condition for each population was 1,250,000 individuals allocated to each age based upon the stable age distribution of the control transition matrix, calculated as the right-normalized eigenvector corresponding to the dominant eigenvalue (Caswell 2001). These initial values placed the age 1 abundance slightly below the carrying capacity. Assessment endpoints were abundance and distribution at end of run, as well as the probability of quasi extinction or time to extinction. Quasi extinction was defined as fewer than 100 individuals at age 3, 4, and 5 for a single population and fewer than 2,200 individuals at age 3, 4, and 5 in the metapopulation. One hundred individuals age 3, 4, and 5 in the entire population results in fewer than 20 mature adults returning as spawners per year. Sensitivity and elasticity analyses were conducted using RAMAS Eigenanalysis.

Results

Our investigation focused on the population-level effects of sublethal contaminant exposure on populations of fall chinook salmon in the LCR. This was done by applying defined impact scenarios to spatially explicit transition matrices connected by dispersal patterns using RAMAS Metapop software. The elasticity analysis of the baseline matrix revealed that changes to the first- and second-year survival rates will produce larger per unit effects on population growth rate than will changes in other matrix elements (data not shown). The sensitivity analysis (data not shown) predicts the first-year (young of the year) survival to be more important to population growth, demonstrating that first-year survival has the greatest influence on population growth rate. The combined sensitivity and elasticity analysis determining the relative importance of the three reproductive pathways (ages 3, 4, and 5) confirms that age 4 reproductive output contributes the greatest to population growth rate. These analyses concur with previous models of ocean-type chinook salmon (Kareiva et al. 2000; Ratner et al. 1997; Spromberg and Meador 2005).

We first assessed the three impact scenarios on a single isolated population. The population growth rates and time to quasi extinction for those scenarios that resulted in quasi extinction are shown in table 8.2. Although the matrix population growth rate was slightly increasing for the YOYS scenario ($\lambda = 1.006$), the modeled stochasticity in survival and reproductive rates resulted in quasi extinction for 13% of the single-population runs. The YOYS scenario also resulted in a decline in abundance over time, with an adult abundance of 1,490 (standard deviation = 1,440) at 100 years. The two scenarios incorporating delayed effects on survival and reproduction (YD and YDR) reached quasi extinction in every case (100%), with mean times to extinction of 52 and 39 years, respectively.

The metapopulation model output predicted partial protection, or buffering, from impacts over all of the scenarios assessed with regard to the magnitude of effect resulting from the same insult. When all 22 subpopulations experienced the same impact scenarios, the time to quasi extinction increased and the probability of quasi extinction decreased compared with the solitary population output (table 8.2). The total metapopulation abundances (ages 3, 4, and 5 summed over all 22 populations) over time for the three impact scenarios are depicted in figure 8.4. The average size of a population within the metapopulation at 100 years under the YOYS scenario was 2,780 (standard deviation = 1,140), compared to control at 5,400 (standard deviation = 1,320).

When the exposure scenarios were applied to subsets of the metapopulation, different patterns and trends emerged (figures 8.5 and 8.6). When the 14 downstream

Table 8.2. Population growth rates and time to quasi extinction or probability of quasi extinction for a solitary population and the whole metapopulation exposed to the impact scenarios described in the text.

Population/ impact scenario	% Reduction in vital rates			Population growth rate	Probability of quasi extinction at 100 years	Time to quasi extinction
	YOY survival	Annual ocean survival	Reproductive output			
Solitary population						
YOYS	20	0	0	1.006	0.13	NA
YD	20	10	0	0.951	1.0	52 years
YDR	20	10	10	0.928	1.0	39 years
All 22 populations						
YOYS	20	0	0	1.006	0.0	NA
YD	20	10	0	0.951	1.0	74 years
YDR	20	10	10	0.928	1.0	50 years

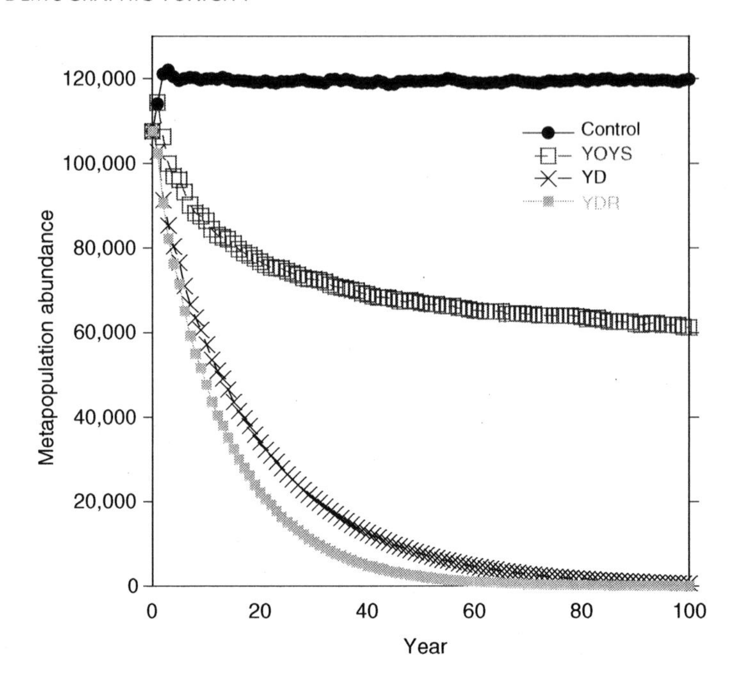

Figure 8.4. Mean total metapopulation abundance for all 22 populations exposed to the young of the year survival (YOYS), YOYS + delayed mortality (YD), and YOYS + YD + reproductive inhibition (YDR) scenarios. Abundance includes ages 3, 4, and 5. Mean (standard deviation) abundances at year 100: control, 120,000 (8,000); YOYS, 61,200 (9,300); YD, 620 (250); and YDR 56 (30).

populations were exposed, abundance declined gradually until populations 1 and 2, at the mouth of the river, were greater than 95% below the mean of the YOYS scenario outputs, as well as nearing the level of quasi extinction (figure 8.5). The upstream populations, whose exposure level did not change, experienced a 40–60% reduction. The assumption that the four populations near the Portland/Vancouver region experienced the YD scenario resulted in a greater than 50% reduction in local population abundance of populations 11–14 below the average projected when all 22 populations exposed to the YOYS scenario. In addition, an 11–33% decrease in abundance was observed for the other populations, whose exposure did not change from the basic YOYS scenario (figure 8.6). The additional reproductive impacts produced the same pattern, with a decrease in relative abundance (figures 8.5 and 8.6).

Discussion and Conclusion

Several interesting points result from the fall chinook models. If fall chinook show straying levels similar to those we modeled, this low-level straying may be augmenting some impacted and declining sink populations that cannot maintain themselves without strays from other populations while depleting populations that otherwise experience lesser or no impacts. This augmentation could mask the effects of contaminant

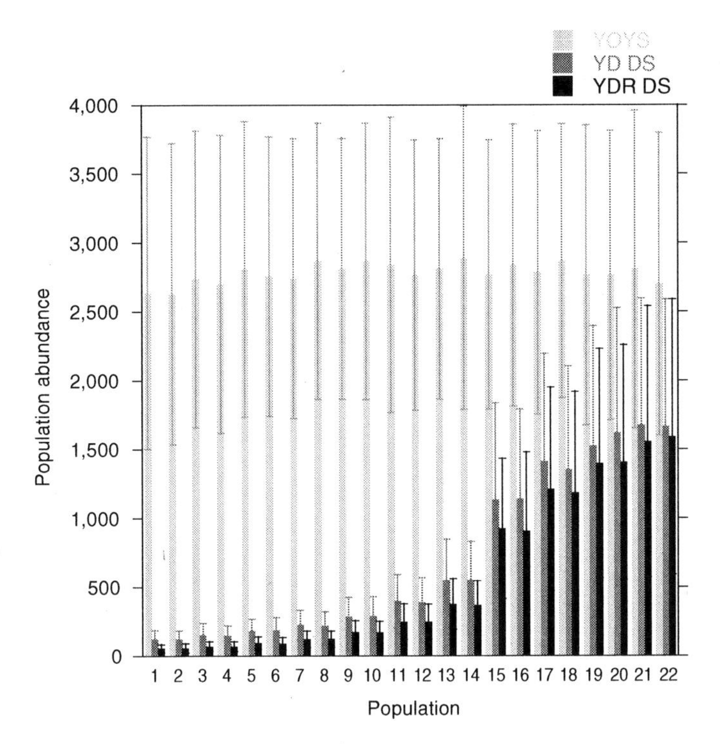

Figure 8.5. Mean abundance in each of the 22 runs after 100-year projection (includes age 3, 4, and 5 individuals) for all populations exposed (YOYS) and the populations at and downstream (DS) of the Portland/Vancouver area exposed equally to the YD and YDR scenarios.

exposure if the magnitude of the stray rate into a population is comparable to the level of impact. The protective effects can be seen in the model output when impacted populations in the metapopulation do not respond to the same extent, when measured as abundance or time to quasi extinction, as a solitary population modeled with the same impact scenario (table 8.2). This outcome, even when all the populations experience the same effects, can be partially attributed to differences in annual variability of vital rates between populations. Populations having good returns one year may supplement populations experiencing lower returns that year through straying. Reducing the modeled dispersal rates resulted in similar abundance trends with a lower magnitude of response. When the PV populations were subjected to the highest exposure scenario, rather than experiencing quasi extinction within 40 years, they are reduced only to approximately 31% of the YOYS abundance (figure 8.6). However, the protection of a single population may come at a cost for other populations within the ESU. The indirect impact on other unexposed populations due to a toxic insult in an exposed population that reduces the net immigration to the unexposed populations has been termed "action at a distance" (Spromberg et al. 1998). This change in net immigration can result from either creating a sink population or reducing a source population that otherwise would be self-sustaining and produce excess individuals that disperse to other areas. In our model, action at a distance appears as a reduction in the abundance

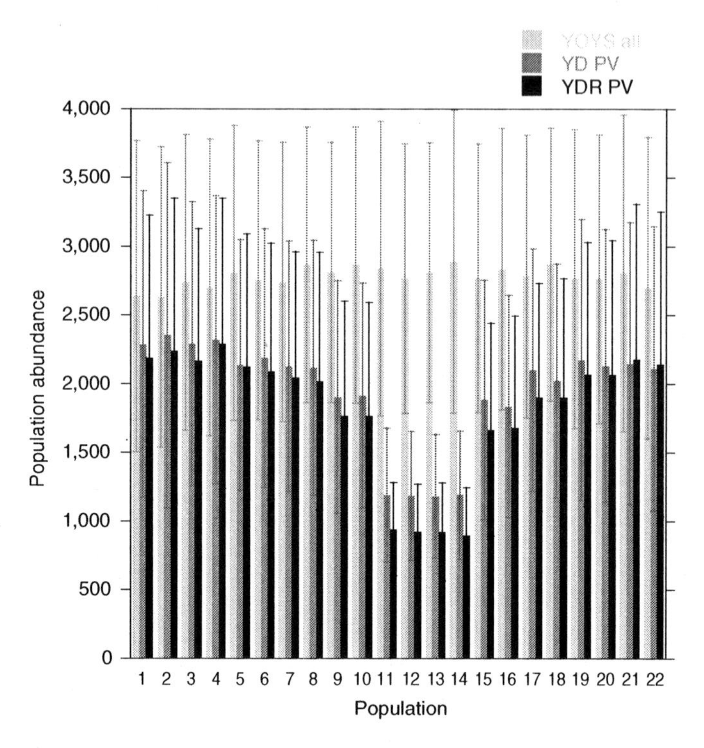

Figure 8.6. Mean abundance in each of the 22 runs after 100-year projection (includes age 3, 4, and 5 individuals) for all populations exposed (YOYS) and the Portland/Vancouver (PV) populations exposed to YD and YDR scenarios.

of populations not exposed in the PV and DS scenarios (figures 8.5 and 8.6). So, even without direct contaminant exposure, the perturbations experienced by chinook salmon populations at contaminant hotspots may be influencing the abundance and dynamics of a population at the farthest extent of the ESU. In addition, the total metapopulation abundance declined for both the PV and DS scenarios.

While it may be difficult to observe low-level impacts in LCR fall chinook, sublethal effects from toxic exposures have been implicated in population declines through field studies of other species. Declines in estuarine and coastal fish populations were noted by Matthiessen and Law (2002) while reviewing impacts of estuary and near-shore contamination. Kroglund and Finstad (2003) observed lower marine survival in Atlantic salmon (*Salmo salar*) after presmolt exposure to low concentrations of aluminum in acidic fresh water. Baumann et al. (1990) found decreased survival in brown bullhead with PAH-associated liver tumors. In addition, there is a growing body of evidence demonstrating that sublethal effects on individual organisms are occurring after exposure to environmentally relevant concentrations. Field studies by Arkoosh et al. (1998, 2001) documented significant and long-term immunosuppression in chinook salmon with concentrations of total PAHs in stomach contents as low as 8 ppm wet weight, lower than the 10–25 ppm documented in stomach contents from fish in urban estuaries and 10 ppm measured in juvenile chinook stomach contents in the

LCR (Johnson et al. 2007a, 2007b; LCREP 2007). Heintz et al. (2000) simulated environmental exposure to weathered crude oil in a laboratory setting by exposing pink salmon eggs and larvae to extracts of contaminated sediments and observed decreased growth and marine survival when exposed to 5.4 ppb PAHs in water during development. Observed effects on vital rates resulting from environmentally realistic contaminant concentrations tested in these studies support our simplified assumption of low-level effects in our exposure scenarios. These examples also suggest that, although toxic impacts on the LCR salmon populations have not been directly observed, they may be occurring but are masked by stray supplementation among populations.

The YOYS scenario was used for comparison to the other impact scenarios for several reasons. Available data for constructing the control transition matrix, without any contaminant-related reductions, produced a population growth rate of 1.05, which resulted in populations increasing to the carrying capacity and remaining there. This does not fit the available data for most of these depressed and declining populations (Myers et al. 2006). In addition, a study by Loge et al. (2005) suggests that fish migrating through the LCR are experiencing toxic effects, mediated through immune suppression, that potentially could decrease first-year survival by approximately 18%. Because all of the ocean-type populations in the LCR spend time rearing in the estuary, contaminants present are more likely to affect these populations. The additional impact scenarios were based upon the presence of other contaminants in water, sediment, prey, and fish known or suspected of causing growth, behavior, immune, or reproductive effects (Fresh et al. 2005; Johnson et al. 2007b).

Despite the simplification applied in the current analysis, it is valuable as a heuristic evaluation. The demographic model used in this study is a simplification of the actual complexity of chinook salmon population dynamics and probably does not reflect actual extinction risks. Fall chinook populations in the LCR are influenced by the pressures of freshwater habitat resource fluctuations, hatchery production, harvest policies, and interaction with native and exotic predators. In addition, the autocorrelated temporal dynamics of marine production have a major influence on population extinction risks. The study indicates that contaminants can have a major influence on chinook populations. If we imagine applying a 10–20% additional harvest rate to these populations, there is little doubt that extinction risk would be affected. The effects of contaminant exposure may not be as directly apparent, but likely produce similar results. This is particularly significant since contaminant exposure in the LCR most likely affects first-year survival rates, which elasticity and sensitivity analyses have shown to be the strongest individual determining factor for the population growth rates.

Exposure of chinook populations to mixtures of contaminants has been ongoing for generations (Fuhrer et al. 1996), and our results suggest that it may be limiting productivity of the ESU in regions outside of known contaminant hotspots. Straying and connectivity within the ESU may therefore be masking impacts in some areas while undermining conservation efforts in others. For example, a habitat restoration project in an area where contaminant exposure is minimal may not result in the expected increased productivity due to effects of contaminants on other areas of the ESU acting as sink populations. Conversely, cleanup of a contaminant hotspot affecting only one or two populations might lead to unexpected increases in productivity in populations from sites not directly affected by the contaminants by converting a sink population to a source population. Increasing the number and extent of source populations could provide a larger abundance of strays to augment all of the populations within the ESU and increase population and total ESU productivity. While contaminants may not directly affect the majority of the populations within the ESU to a large degree, it is apparent that their direct, delayed, and indirect effects as well as the metapopulation dynamics should be considered in ESU-wide restoration and management decisions.

RAMAS Metapop provided an efficient tool for simulating the potential toxic impacts to metapopulations and their subpopulations. Once location and dispersal patterns are established, manipulating each population for impact scenarios can be done with relative ease. The salmon life-history strategy of ocean migration and semelparity places management focus on the abundance of returning spawners rather than the total abundance of age classes. RAMAS did not allow for spawner abundance to be calculated as an output within the confines of the program. Back-calculations were required to determine the appropriate levels of quasi extinction and other values. In addition, the only time that straying is measurable is during the freshwater spawning migration. Therefore, it would be more accurate to include only the spawners in dispersal calculations. This difference could produce greater connection among model populations than would otherwise be observed. We believe the trends seen in this study are accurate, but the magnitude of potential connection may be overestimated by this difference in calculation of dispersal.

Acknowledgments We thank Reşit Akçakaya and John Stark for the invitation to contribute this chapter. Bernadita Anulacion assisted greatly with the maps. We also thank Michelle McClure, Paul McElhany, Kate Macneale, and Nat Scholz for their helpful comments for improving the manuscript.

References

Akçakaya HR. 2005. *RAMAS GIS: Linking Spatial Data with Population Viability Analysis* (version 5.0). Applied Biomathematics, Setauket, New York. USA.

Ankley GT, Johnson RD. 2004. Small fish models for identifying and assessing the effects of endocrine-disrupting chemicals. *ILAR J* 45(4):469–483.

Arkoosh MR, Casillas E, Huffman P, Clemons E, Evered J, Stein JE, Varanasi U. 1998. Increased susceptibility of juvenile chinook salmon from a contaminated estuary to *Vibrio anguillarum*. *Trans Am Fish Soc* 127:360–374.

Arkoosh MR, Clemons E, Huffman P, Kagley AN, Casillas E, Adams N, Sanborn HR, Collier TK, Stein JE. 2001. Increased susceptibility of juvenile chinook salmon to vibriosis after exposure to chlorinated and aromatic compounds found in contaminated urban estuaries. *J Aquat Animal Health* 13:257–268.

Baumann PC, Harshbarger JC, Hartman KJ. 1990. Relationship between liver tumors and age in brown bullhead populations from two Lake Erie tributaries. *Sci Total Environ* 94:71–87.

Beamish RJ, Mahnken C. 2001. A critical size and period hypothesis to explain natural regulation of salmon abundance and the linkage to climate and climate change. *Prog Oceanogr* 49:423–437.

Beauchamp DA, Shepard MF, Pauley GB. 1983. Species profiles: Life-histories and environmental requirements of coastal fishes and invertebrates (Pacific Northwest): Chinook salmon. FWS/OBS-82/11.6. U.S. Fish and Wildlife Service Division of Biological Services, Slidell, LA.

Beauvais SL, Jones SB, Brewer SK, Little EE. 2000. Physiological measures of neurotoxicity of diazinon and malathion to larval rainbow trout (*Oncorhynchus mykiss*) and their correlation with behavioral measures. *Environ Toxicol Chem* 19:1875–1880.

Brannon EL, Powell MS, Quinn TP, Talbot A. 2004. Population structure of Columbia River basin chinook salmon and steelhead trout. *Rev Fish Sci* 12:99–232.

Brewer SK, Little EE, DeLonay AJ, Beauvais SL, Jones SB, Ellersieck MR. 2001. Behavioral dysfunctions correlate to altered physiology in rainbow trout (*Oncorynchus mykiss*) exposed to cholinesterase-inhibiting chemicals. *Arch Environ Contam Toxicol* 40:70–76.

Buck J. 2004. Environmental contaminants in aquatic resources from the Columbia River. Final Report 1130–1F02 and 1261–1N04. U.S. Fish and Wildlife Service, Oregon Fish and Wildlife Office, Portland, OR.

Buck JA, Anthony RG, Schuler CA, Isaacs FB, Tillitt DE. 2005. Changes in productivity and contaminants in bald eagles nesting along the lower Columbia River, USA. *Environ Toxicol Chem* 24:1779–1792.

Caswell H. 2001. *Matrix population models: Construction, analysis, and interpretation.* Sunderland, MA: Sinauer.

Connor WP, Smith SG, Andersen T, Bradury SM, Burum DC, Hockersmith EE, Schuck ML, Mendel GW, Bugert RM. 2004. Postrelease performance of hatchery yearling and subyearling fall chinook salmon released into the Snake River. *N Am J Fish Manag* 24:545–560.

DeAngelis DL, Svoboda LJ, Christensen SW, Vaughan DS. 1980. Stability and return times of Leslie matrices with density-dependent survival: Applications to fish populations. *Ecol Model* 8:149–163.

Fast DE, Hubble JD, Kohn MS. 1988. Yakima River Spring Chinook Enhancement Study, Annual Report FY 1988. Project 82–16. U.S. Department of Energy, Bonneville Power Administration, Division of Fish and Wildlife.

Finlayson BJ, Verrue KM. 1985. Toxicities of butoxyethanol ester and propylene glychol butyl ether ester formulations of 2,4-dichlorophenoxy acetic acid (2,4-D) to juvenile salmonids. *Arch Environ Contam Toxicol* 14:153–160.

Fresh KL, Casillas E, Johnson LL, Bottom DL. 2005. Role of the estuary in the recovery of Columbia River basin salmon and steelhead: An evaluation of the effects of selected factors on population viability. NOAA Technical Memorandum. Northwest Fisheries Science Center. National Oceanographic and Atmospheric Administration, U.S. Department of Commerce, Seattle, WA.

Fuhrer GJ, Tanner DQ, Morace JL, McKenzie SW, Skach KA. 1996. Water quality of the lower Columbia River basin—analysis of current and historical water quality data through 1994. U.S. Geological Survey Water Resources Investigations Report 95–4294.

Glubokov AI. 1990. Growth of three species of fish during early ontogeny under normal and toxic conditions. *Voprosy ikhtiologii* 30:137–143.

Good TP, Waples RS, Adams P (eds). 2005. Updated status of federally listed ESUs of West Coast salmon and steelhead. NOAA Technical Memorandum NMFS-WFSC-66. U.S. Department of Commerce.

Greene CM, Beechie TJ. 2004. Consequences of potential density-dependent mechanisms on recovery of ocean-type chinook salmon (*Oncorhynchus tshawytscha*). *Can J Fish Aquat Sci* 61:590–602.

Hansen JA, Lipton J, Welsh PG, Morris J, Cacela D, Suedkamp MJ. 2002. Relationship between exposure duration, tissue residues, growth, and mortality in rainbow trout (*Oncorhynchus mykiss*) juveniles sub-chronically exposed to copper. *Aquat Toxicol* 58:175–188.

Healey MC. 1982. Timing and relative intensity of size-selective mortality of juvenile chum salmon (*Oncorhynchus keta*) during early sea life. *Can J Fish Aquat Sci* 39:952–957.

Healey MC. 1991. Life history of chinook salmon (*Oncorhynchus tshawytscha*). Pages 311–394 in C. Groot and L. Margolis (eds). *Pacific salmon life histories.* University of British Columbia Press, Vancouver, BC.

Heintz RA, Rice SD, Wertheimer AC, Bradshaw RF, Thrower FP, Joyce JE, Short JW. 2000. Delayed effects on growth and marine survival of pink salmon *Oncorhynchus gorbuscha* after exposure to crude oil during embryonic development. *Mar Ecol Prog Series* 208:205–216.

Higgs DA, MacDonald JS, Levings CD, Dosanjh BS. 1995. Nutrition and feeding habits in relation to life history stage. Pages 161–315 in Groot C, Margolis L, Clarke WC (eds). *Physiological ecology of Pacific salmon*. University of British Columbia Press, Vancouver, BC.

Hinck JE, Schmitt CJ, Bartish TM, Denslow ND, Blazer VS, Anderson PJ, Coyle JJ, Dethloff GM, Tillitt DE. 2004. Biomonitoring of environmental status and trends (BEST) program: Environmental contaminants and their effects on fish in the Columbia River basin. U.S. Geological Survey, Scientific Investigations Report 2004–5154.

Holtby LB, Andersen BC, Kadowak RK. 1990. Importance of smolt size and early ocean growth to interannual variability in marine survival of coho salmon (*Oncorhynchus kisutch*). *Can J Fish Aquat Sci* 47:2181–2194.

Howell P, Jones K, Scarnecchia D, LaVoy L, Kendra W, Ortmann D. 1985. Final Report: Stock assessments of Columbia River anadromous salmonids. Vol. 1: Chinook, coho, chum, and sockeye stock summaries. Project 83–335. U.S. Department of Energy, Bonneville Power Administration, Division of Fish and Wildlife.

Johnson A, Norton D. 2005. Concentrations of 303(d) listed pesticides, PCBs and PAHs measured with passive samplers deployed in the lower Columbia River. Washington State Department of Ecology publication 05–03–06. Environmental Assessment Program, Olympia, WA.

Johnson LL, Ylitalo GM, Arkoosh MR, Kagley AN, Stafford CL, Bolton JL, Buzitis J, Anulacion BF, Collier TK. 2007a. Contaminant exposure in outmigrant juvenile salmon from Pacific Northwest estuaries of the United States. *Environ Monitor Assess* 124:167–194.

Johnson LL, Ylitalo GM, Sloan CA, Anulacion BF, Kagley AN, Arkoosh MR, Lundrigan T, Larson K, Siipola M, Collier TK. 2007b. Persistent organic pollutants in outmigrant juvenile chinook salmon from the Lower Columbia Estuary, USA. *Sci Total Environ* 374:342–366.

Kareiva P, Marvier M, McClure M. 2000. Recovery and management options for spring/summer chinook salmon in the Columbia River basin. *Science* 290:977–979.

Kime DE. 1999. A strategy for assessing the effects of xenobiotics on fish reproduction. *Sci Total Environ* 225:3–11.

Kroglund F, Finstad B. 2003. Low concentrations of inorganic monomeric aluminum impair physiological status and marine survival of Atlantic salmon. *Aquaculture* 222:119–133.

Kurita Y, Meier S, Kjesbu OS. 2003. Oocyte growth and fecundity regulation by atresia of Atlantic herring (*Clupea harengus*) in relation to body condition throughout the maturation cycle. *J Sea Res* 49:203–219.

Laine P, Rajasilta M. 1999. The hatching success of Baltic herring eggs and its relation to female condition. *J Exp Mar Biol Ecol* 237:61–73.

Loge F, Arkoosh MR, Ginn TR, Johnson LL, Collier TK. 2005. Impact of environmental stressors on the dynamics of disease transmission. *Environ Sci Technol* 39:7329–7336.

Lower Columbia Estuary Partnership (LCREP). 2007. Lower Columbia River and Estuary Ecosystem Monitoring: Water Quality and Salmon Sampling Report. Available: www.lcrp.org.

Maher M, Sheer MB, Steel EA, McElhany P. 2005. Atlas of salmon and steelhead habitat in the Oregon lower Columbia and Willamette basins. Report for the Willamette-Lower Columbia Technical Recovery Team. Produced by the NOAA Northwest Fisheries Science Center. Available: www.nwfsc.noaa.gov/trt/atlas_salmon.htm.

Majewski MS, Kahle SC, Ebbert JC, Josberger EG. 2003. Concentrations and distribution of slag-related trace elements and mercury in fine-grained beach and bed sediments of Lake Roosevelt, Washington, April–May 2001. U.S. Geological Survey Water Resources Investigations Report 03–4170.

Matthiessen P, Law RJ. 2002. Contaminants and their effects on estuarine and coastal organisms in the United Kingdom in the late twentieth century. *Environ Poll* 120:739–757.

McCarthy KA, Gale RW. 2001. Evaluation of persistent hydrophobic organic compounds in the Columbia River Basin using semipermeable membrane devices. *Hydrol Process* 15:1271–1283.

McElhany P, Backman T, Busack C, Heppell S, Kolmes S, Maule A, Myers J, Rawding D, Shively D, Steel A, Steward C, Whitesel T. 2003. Interim report on viability criteria for Willamette and lower Columbia basin Pacific salmonids. Willamette/Lower Columbia Technical Recovery Team. Available: www.nwfsc.noaa.gov/trt/wlc_viabrpt/body.pdf.

McElhany P, Busack C, Chilcote M, Kolmes S, McIntosh B, Myers J, Rawding D, Steel A, Steward C, Ward D, Whitesel T, Willis C. 2006. Revised viability criteria for salmon and steelhead in the Willamette and lower Columbia basins. Draft report. Willamette/Lower Columbia Technical Recovery Team and Oregon Department of Fish and Wildlife.

Meador JP, Sommers FC, Ylitalo GM, Sloan CA. 2006. Altered growth and related physiological responses in juvenile chinook salmon (*Oncorhynchus tshawytscha*) from dietary exposure to polycyclic aromatic hydrocarbons (PAHs). *Can J Fish Aquat Sci* 63:2364–2376.

Milnes MR, Bermudez DS, Bryan TA, Edwards TM, Gunderson MP, Larkin ILV, Moore BC, Guillette LJ Jr. 2006. Contaminant-induced feminization and demasculinization of nonmammalian vertebrate males in aquatic environments. *Environ Res* 100:3–17.

Morgan MJ, Kiceniuk JW. 1990. Effect of fenitrothion on the foraging behavior of juvenile Atlantic salmon. *Environ Toxicol Chem* 9:489–495.

Morrow JE. 1980. *The freshwater fishes of Alaska*. Alaska Northwest Publishing, Anchorage, AK.

Myers J, Busack C, Rawding D, Marshall A, Teel D, Van Doornik DM, Maher MT. 2006. Historical population structure of Pacific salmonids in the Willamette River and lower Columbia River basins. NOAA Technical Memorandum NMFS-NWFSC-73. U.S. Department of Commerce.

National Marine Fisheries Service. 1998. Factors contributing to the decline of chinook salmon: An addendum to the 1996 West Coast steelhead factors for decline report. Available: Protected Resources Branch, NMFS, 525 NE Oregon St., Suite 500, Portland, OR 97232.

National Marine Fisheries Service. 1999. Endangered and threatened species; threatened status for three chinook salmon evolutionarily significant units (ESUs) in Washington and Oregon, and endangered status for one chinook salmon ESU in Washington [990303060-9071-02; I.D. 022398C, March 24, 1999]. *Fed Reg* 64(57):14308–14328.

National Oceanic and Atmospheric Administration. 1999. *Preliminary natural resource survey: Lower Willamette River, Portland, Oregon*. Seattle, WA.

National Research Council. 1996. *Upstream: Salmon and society in the Pacific Northwest*. National Academy Press, Washington, DC.

Pacific Salmon Commission Chinook Technical Committee. 2002. Pacific Salmon Commission Joint Chinook Technical Committee report: Annual exploitation rate analysis and model calibration. Report TCCHINOOK (02)-3. Vancouver, BC.

Quinn TP. 2005. *The behavior and ecology of Pacific salmon and trout*. University of Washington Press, Seattle, WA.

Quinn TP, Peterson J, Gallucci V. 2000. Trends in adult coho and chinook salmon life history at the University of Washington's hatchery. Draft annual report to Metro King County, Seattle, WA.

Ratner S, Lande R, Roper BB. 1997. Population viability analysis of spring chinook salmon in the South Umpqua River, Oregon. *Conserv Biol* 11:879–889.

Roch M, McCarter JA. 1984. Metallotionein induction, growth, and survival of chinook salmon exposed to zinc, copper, and cadmium. *Bull Environ Contam Toxicol* 32:478–485.

Roni P, Quinn TP. 1995. Geographic variation in size and age of North American chinook salmon. *N Am J Fish Manag* 15:325–345.

Roy F. Weston Inc. 1998. Portland Harbor sediment investigation report Multnomah County, Oregon. EPA910/R-98-006. Prepared for U.S. Environmental Protection Agency. Roy F. Weston Inc., Seattle, WA.

Sandahl JF, Baldwin DH, Jenkins JJ, Scholz NL. 2005. Comparative thresholds for acetylcholinesterase inhibition and behavioral impairment in coho salmon exposed to chlorpyrifos. *Environ Toxicol Chem* 24:136–145.

Scholz NL, Truelove NK, French BL, Berejikian BA, Quinn TP, Casillas, E, Collier TK. 2000. Diazinon disrupts antipredator and homing behaviors in chinook salmon (*Oncorhynchus tsawytscha*). *Can J Fish Aquat Sci* 57:1911–1918.

Sethajintanin D, Johnson ER, Loper BR, Anderson KA. 2004. Bioaccumulation profiles of chemical contaminants in fish from the lower Willamette River, Portland Harbor, Oregon. *Arch Environ Contam Toxicol* 46(1):114–23.

Sheer MB, Steel EA. 2006. Lost watersheds: barriers, aquatic habitat connectivity and salmon persistence in the Willamette and Lower Columbia basins.*Trans Amer Fish Soc* 135:1654–1669.

Spromberg JA, Meador JP. 2005. Population-level Effects on chinook salmon from chronic toxicity test measurement endpoints. *Integr Environ Assess Manag* 1:9–21.

Spromberg JA, John BM, Landis WG. 1998. Metapopulation dynamics: Indirect effects and multiple distinct outcomes in ecological risk assessment. *Environ Toxicol Chem* 17:1640–1649.

Stein JE, Hom T, Collier TK, Brown DW, Varanasi U. 1995. Contaminant exposure and biochemical effects in outmigrant juvenile chinook salmon from urban and nonurban estuaries of Puget Sound, Washington. *Environ Toxicol Chem* 14:1019–1029.

Taylor EB. 1990 Environmental correlates of life-history variation in juvenile chinook salmon *Oncorhynchus tshawytscha* (Walbaum). *J Fish Biol* 37:1–17.

TetraTech Inc. 1996. Lower Columbia River bi-state program—the health of the river, 1990–1996. Integrated Technical Report 0253-01, prepared for Oregon Department of Environmental Quality and Washington Department of Ecology. Tetra Tech, Inc., Redmond, WA.

Trippel EA, Morgan MJ, Frechet A, Rollet C, Sinclair A, Annand C, Beanlands D, Brown L. 1997. Changes in age and length at sexual maturity of Northwest Atlantic cod, haddock and pollock stocks, 1972–1995. Canadian Technical Report of Fisheries and Aquatic Sciences.

URS Corporation. 2004. Site characterization report. Bradford Island Landfull Bonneville Lock and Dam Project, Cascade Locks, OR. Prepared for the U.S. Army Corps of Engineers, Portland District. URS Corporation, Portland, OR.

Wagner RJ, Frans LM, Huffman RL. 2006. Occurrence, distribution, and transport of pesticides in agricultural irrigation-return flow from four drainage basins in the Columbia Basin Project, Washington, 2002–04, and comparison with historical data. U.S. Geological Survey Scientific Investigations Report 2006–5005.

Wentz DA, Bonn BA, Carpenter KD, Hinkle SR, Janet ML, Rinella FA, Ulrich MA, Waite IR, Laenen A, Bencala KE. 1998. Water quality in the Willamette Basin 1991–95. U.S. Geological Survey Circular 1161.

West CJ, Larkin PA. 1987. Evidence of size-selective mortality of juvenile sockeye salmon (*Oncorhynchus nerka*) in Babine Lake, British Columbia. *Can J Fish Aquat Sci* 44:712–721.

Zabel RW, Achord S. 2004. Relating size of juveniles to survival within and among populations of chinook salmon. *Ecology* 85:795–806.

Zabel RW, Williams JG. 2002. Selective mortality in chinook salmon: What is the role of human disturbance? *Ecol Appl* 12:173–183.

9

Water Flea *Daphnia pulex*:

Population Recovery after Pesticide Exposure

JOHN D. STARK

Cladocerans such as *Daphnia* species are an important component of aquatic ecosystems because they are primary consumers feeding on algae and bacteria and serve as a food source for other aquatic organisms, including fish and invertebrates. Although *Daphnia* species are not threatened or endangered, they are often used as indicator species for estimating the effects of pollutants on aquatic ecosystems.

The most commonly generated toxicity data for *Daphnia* and other aquatic organisms are acute mortality estimates (LC_{50}, the concentration of a chemical that kills 50% of a population). The advantage of short-term toxicity data such as LC_{50} is that it costs much less to produce than do longer term data. However, short-term toxicity data may not provide enough information about potential effects on populations because chronic exposures can sometimes result in much higher mortality levels than predicted by acute exposures and because sublethal effects can occur that may affect populations at several levels most notably by decreasing fecundity (Forbes and Calow 1999; Stark and Banks 2003).

Even though it has been four decades since Rachel Carson's seminal work on the effects of pesticides on biological communities (Carson 1962), pesticides are found almost universally throughout the surface water systems of the United States (Larson et al. 1999). A large body of literature has been developed on the toxicity of pesticides to aquatic life (Scholz et al. 2000; Munn et al. 2006). However, studies on the effects of pesticides on population viability are less common (Kammenga and Laskowski 2000).

Pesticides are a major industry accounting for $32 billion in sales worldwide, with production in the United States accounting for $11 billion of the market (Kiely et al. 2004). Although pesticide use declined through the 1980s, it is once again on the upswing. For example, herbicide and insecticide use per acre on U.S. corn has increased by an order of magnitude from the early 1960s (National Research Council 2000). These statistics indicate that pesticides are a major ecological issue that needs to be investigated further.

The pesticide industry has responded to the negative image of pesticides by developing a new generation of purportedly safer products, including genetically modified crops and selective pesticides that appear to be much less damaging to nontarget organisms than to pest species (Stark and Banks 2001). One such product is spinosad, a neonicotinoid insecticide that has been shown to be selectively more toxic to pest than to beneficial species (Williams et al. 2003).

Traditional evaluation of pesticide effects on many organisms involves the development of acute mortality data, often presenting results as the acute LC_{50}. These data have been shown to be inadequate as predictors of longer term effects on populations because they do not take into account sublethal effects or population processes (Stark 2005, 2006). The development of life table data and ensuing demographic parameters have been shown to provide much better information with regard to the actual effects of pesticides and other toxicants on populations (Forbes and Calow 1999).

Daphnia often have extremely high reproductive rates and short generation times. For example, population growth rates (λ) of 1.3/day are common in the laboratory. One of the great difficulties associated with modeling *Daphnia pulex* population data is that this group of organisms goes through five nonreproductive life stages in the first five days of life and then reproduces almost every day for the remainder of the life span, which can be 60 days. For example, *D. pulex* produces live young and has six basic life stages: juvenile 1 (0–1 days old), juvenile 2 (1–2 days old), juvenile 3 (3–4 days old), juvenile 4 (4–5 days old), adolescent stage (5–6 days old), and the adult stage (≥6 days old) (Stark and Vargas 2003). *Daphnia pulex* molts every day for five days until reaching the adult stage. This makes use of stage-based models awkward for studying this species, so age-based models are often used (Stark et al. 2004).

In this chapter, I describe the effects of the insecticide spinosad on the water flea *Daphnia pulex* evaluated using demographic data developed in the laboratory, followed by modeling with RAMAS Metapop.

Methods

Pesticides evaluated

Spinosad consists of a mixture of spinosyn A and D, which are fermentation products of the soil bacterium *Saccharopolyspora spinosa* (actinomycetes) (Crouse et al. 2001). Spinosad is a neurotoxin and acts as a contact and stomach poison (DowElanco 1996; Salgado et al. 1998). This insecticide is a neonicotinoid and thus acts in a similar manner to nicotine in the nervous systems of animals.

Daphnia exposure and development of data

Daphnia pulex were exposed to formulated spinosad (Success; 240 g active ingredient per liter [ai/l]; Dow AgroSciences, Indianapolis, IN, USA). Control populations were unexposed and compared to the populations exposed to spinosad. The 48-hr LC_{50} for spinosad and *D. pulex* was reported to be 129 μg ai/l (Stark and Banks 2001). In a previous demographic study on the effects of spinosad on *D. pulex*, Stark and Vargas (2003) found that population growth rate of *D. pulex* declined in a concentration-dependent manner between 2 and 10 μg ai/l, with extinction occurring after exposure to 10 μg ai/l. Therefore, in this study, the following spinosad concentrations were evaluated: 2, 4, 6, 8, and 10 μg ai/l. These concentrations were equivalent to the 48-hr acute LC_4, LC_7, LC_9, LC_{11}, and LC_{12}, respectively (Stark and Banks 2001; Stark and Vargas 2003).

For each nominal concentration tested, 25 ml of test solution was transferred into a 30-ml plastic cup, and one neonate was transferred into the test container using a disposable glass pipette. *Daphnia* used in this study were obtained from cultures at or beyond the third filial (F_3) generation. Thirty individuals were tested for each concentration. Test containers were held in an environmental chamber at 25°C. *Daphnia* were moved to newly made pesticide solutions, or fresh dilution water in the case of the controls, every other day (Walthall and Stark 1997). Survival and reproduction were measured weekly until all animals had died. The following demographic parameters were determined in this study: L_x, the proportion of females surviving at the start of the start of the age interval; M_x, the average number of female offspring produced per female by age x; and λ, the population growth rate, the factor by which a population increases in size from time t to time $t + 1$ (Carey 1993). For a complete description of the calculation of these parameters, see Carey (1993).

Model development

Nine weekly age classes were used in the models developed in this study (table 9.1). RAMAS Metapop (Akçakaya 2005) was used to develop an age-structured model with one-week age classes. All models were run using ceiling density dependence with 1,000,000 individuals set as the maximum population size (K). With ceiling density dependence, the population grows exponentially until it reaches a prespecified ceiling (number of individuals) and then remains at that level (Akçakaya 2005). The starting population vector consisted of 100 individuals in the (deterministic) stable age distribution. Simulations had one replication and 12 time steps (weeks). Population recovery was determined by comparing the time it took each spinosad-exposed population

Table 9.1. Age classes used in the RAMAS matrix models.

Class	Age range
Age class 1	0–6 days old
Age class 2	7–13 days old
Age class 3	14–20 days old
Age class 4	21–27 days old
Age class 5	28–34 days old
Age class 6	35–41 days old
Age class 7	42–48 days old
Age class 8	49–55 days old
Age class 9	56–62 days old

to reach the ceiling (1,000,000 individuals) to the time it took the control to reach 1,000,000 individuals. Stark et al. (2004) have advocated the use of recovery time as an endpoint when modeling the effects of stressors on populations. Effects of toxicants can be compared by comparing recovery times of toxicant-exposed populations to control populations.

Results

The effects of spinosad on survival (L_x) of *D. pulex* were concentration dependent, indicating that survival declined as spinosad concentration increased (figure 9.1). Little mortality occurred in any of the concentrations tested two days after exposure, which corresponded to results of the 48-hr acute mortality data presented by Stark and Banks (2001). However, mortalities increased greatly thereafter and were 50%, 63%, and 97% in the 6, 8, and 10 μg ai/l concentrations, respectively, two weeks after the start of the study (figure 9.1).

The number of female offspring per surviving female (M_x) and population growth rate declined as concentrations of spinosad increased (figures 9.2 and 9.3).

Modeling results showed that λ values were reduced 34%, 61%, 75%, 78%, and 97% compared to the control population after exposure to 2, 4, 6, 8, and 10 μg ai/l spinosad (table 9.2). Exposure to 10 μg ai/l spinosad resulted in $\lambda < 1$, indicating that this population was headed toward extinction.

Results of RAMAS Metapop models indicated that it took control populations four weeks to grow from 100 to 1,000,000 individuals (figure 9.4). It took *D. pulex* populations 5, 11, 14, and 15 weeks to reach 1,000,000 individuals after exposure to 2, 4, 6, and 8 μg/l spinosad, respectively indicating that populations would be delayed from 1 to 11 weeks. The *D. pulex* population exposed to 10 μg ai/l spinosad became extinct.

There was virtually no difference between control population projections and those of a population with a simulated 12% mortality (figure 9.5).

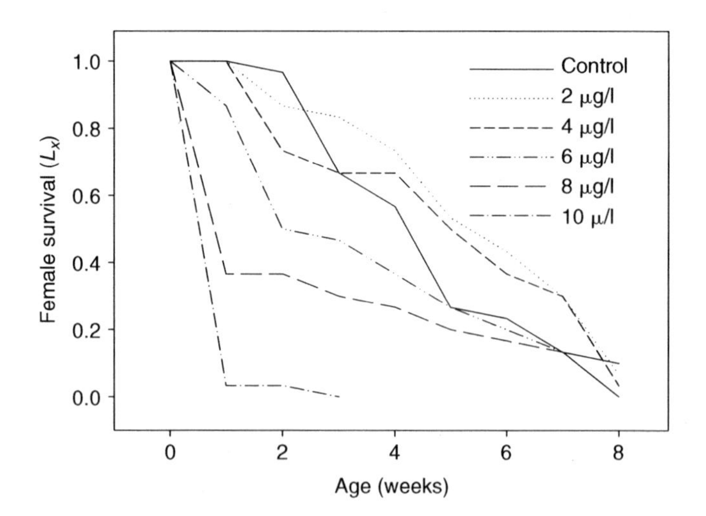

Figure 9.1. Female survival (L_x) over time after exposure to a range of spinosad concentrations.

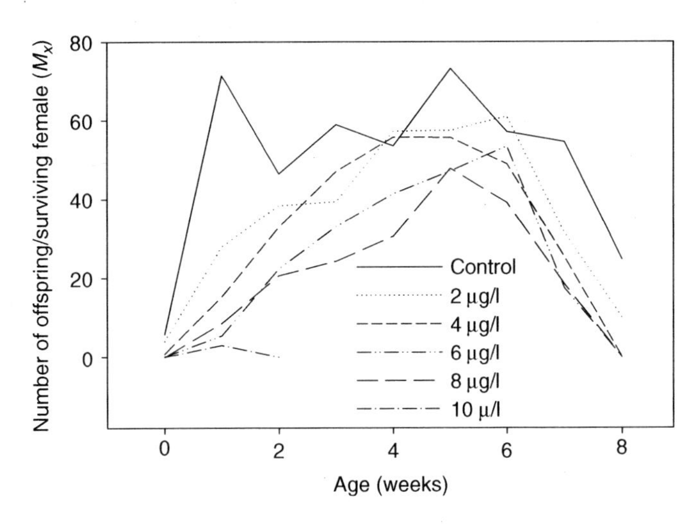

Figure 9.2. Number of offspring per surviving female over time after exposure to a range of spinosad concentrations.

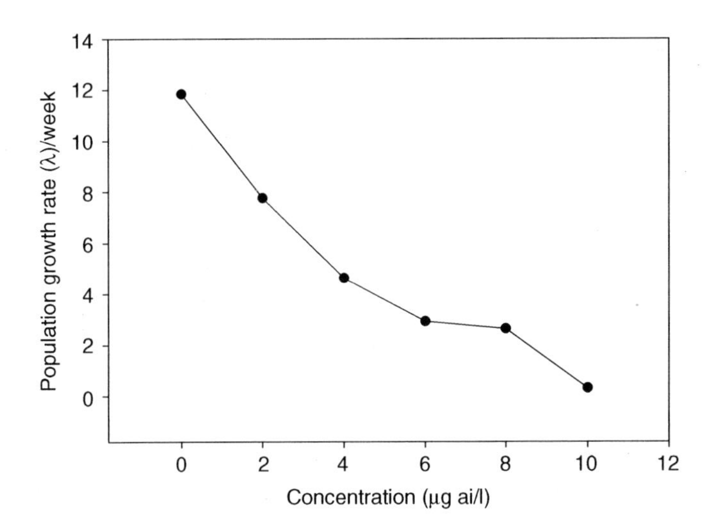

Figure 9.3. Population growth rate (λ) per week after exposure to a range of spinosad concentrations.

Table 9.2. Relationship among spinosad concentrations and various endpoints developed in this study.

	Concentration (μg ai/l)				
Endpoint	2	4	6	8	10
48-hr acute LC estimate	4	7	9	11	12
Predicted acute mortality	4	7	9	11	12
Reduction in λ compared to control (%)	34	61	75	78	97
Population recovery time (weeks)	1	7	10	11	Extinct

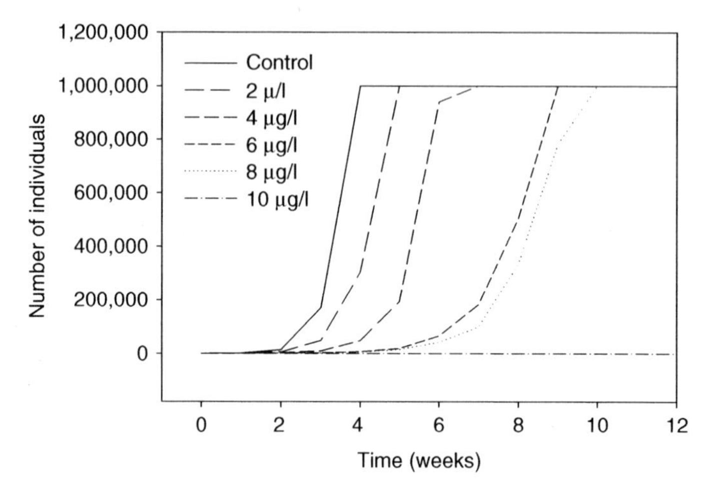

Figure 9.4. Population projections of *D. pulex* with ceiling density dependence after exposure to a range of spinosad concentrations.

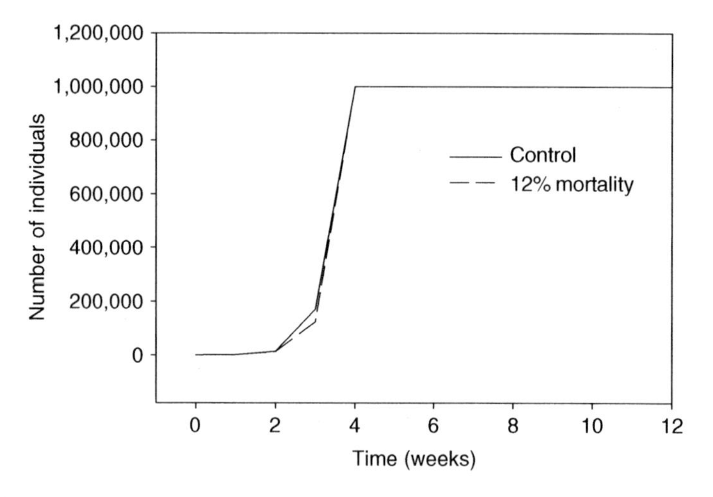

Figure 9.5. Population projection of *D. pulex* with ceiling density dependence after survival was reduced 12%.

Discussion

It is interesting to note that although the concentrations evaluated were equivalent to 48-hr LC estimates ranging from 4% to 12% mortality, mortalities two weeks after the start of the study were 50%, 63%, and 97% in the 6, 8, and 10 µg ai/l concentrations, respectively. This shows that mortality increased substantially in *D. pulex* over time after exposure to spinosad, a result that would not be evident from 48-hr mortality. Additionally, spinosad significantly reduced maternity in *D. pulex* at the

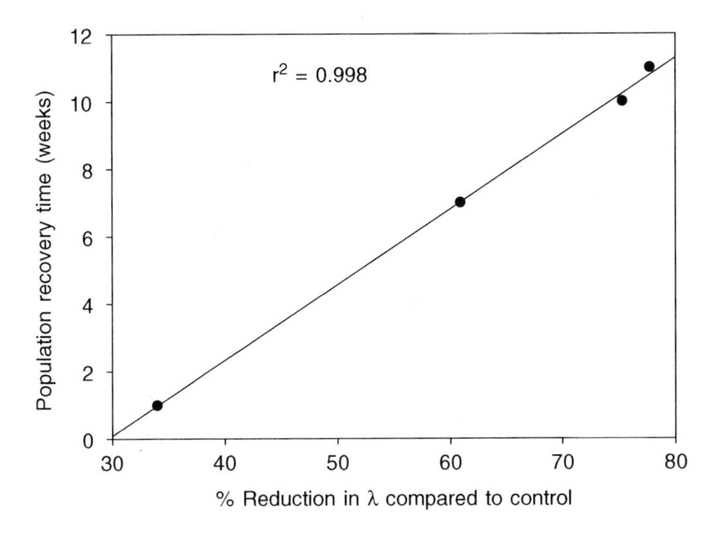

Figure 9.6. Regression of recovery time (weeks) and percent reduction in population growth rate (λ) compared to the control for *D. pulex* populations exposed to a range of spinosad concentrations.

concentrations evaluated. Thus, this insecticide caused high levels of both mortality and reductions in the number of offspring at concentrations predicted to cause ≤12% acute mortality.

In many demographic toxicity studies, reductions in population growth rate are observed (for a list of some of these studies, see Stark and Banks 2003). However, unless exposure to a pollutant causes negative growth rate or extinction, it is difficult to determine what reductions in λ mean to a population. In this study, λ was reduced 34%, 61%, 75%, and 78% compared to the control population after exposure to increasing spinosad concentrations. These reductions corresponded to population recovery times of 1, 7, 10, and 11 weeks (table 9.2). A regression analysis of the relationship between population recovery time and percent reduction in λ from the control revealed a strong statistically significant relationship (r^2 = 0.998) between these two variables (figure 9.6). However, the linearity assumption is clearly limited to a small range, as evidenced by the fact that the highest concentration (10 μg/l) caused extinction (i.e., infinite recovery time). Although this relationship has been elucidated only for one pesticide, one species, and in a limited range of concentrations, it may hold for other species and pollutants. If this proves to be true, then recovery times at moderate concentrations could be predicted from reductions in λ in future studies.

Another way to interpret recovery time is to compare them to generation time, the time it takes a newborn female to replace herself R_0-fold (Carey 1993). Generation time for *D. pulex* is approximately three weeks (Stark and Vargas 2003). Therefore, exposure to spinosad at 8 μg/l would result in a recovery time period during which an unexposed population could complete almost four generations.

Results of this study indicate that exposure of *D. pulex* to spinosad at the concentrations tested would result in significant recovery times. What is most interesting is that *D. pulex* populations were so negatively affected after exposure to spinosad at concentrations that were predicted to cause ≤12% mortality by acute mortality estimates. The large delays seen in *D. pulex* populations after exposure to low concentrations of spinosad were due to increased mortality after 48-hr exposure and substantial declines in the production of offspring, a sublethal effect. Traditional methods of estimating toxicity that only take into account mortality, such as the LC_{50}, would not be able to predict these population outcomes.

References

Akçakaya, H.R. 2005. RAMAS Metapop: Viability Analysis for Stage-Structured Metapopulations (version 5.0). Applied Biomathematics, Setauket, New York.

Carey, J.R. 1993. Applied Demography for Biologists with Special Emphasis on Insects. Oxford University Press, New York.

Carson, R. 1962. Silent Spring. Houghton-Mifflin, Boston.

Crouse, G.D., Sparks, T.C., Schoonover, J., Gifford, J., Dripps, J., Bruce, T., Larson, L., Garlich, L., Hatton, C., Hill, R.L., Worden, T.V., and Martynow, J.G. 2001. Recent advances in the chemistry of spinosyns. Pest Management Science 57: 177–185.

DowElanco. 1996. Spinosad Technical Guide. Form No. 200–03–001. Indianapolis, IN.

Forbes, V.E., and Calow, P. 1999. Is the per capita rate of increase a good measure of population-level effects in ecotoxicology? Environmental Toxicology and Chemistry 18: 1544–1556.

Kammenga, J., and Laskowski, R. 2000. Demography in Ecotoxicology. John Wiley and Sons, West Sussex, UK.

Kiely, T., Donaldson, D., and Grube, A. 2004. Pesticide Industry Sales and Usage: 2000 and 2001 Market Estimates. U.S. Environmental Protection Agency, Washington, DC. Available: www.epa.gov/oppbead1/pestsales/01pestsales/market_estimates2001.pdf.

Larson, S.J., Gilliom, R.J., and Capel, P.D. 1999. Pesticides in Streams of the United States—Initial Results from the National Water-Quality Assessment Program. Water-Resources Investigations Report 98–4222. U.S. Geological Survey, Sacramento, CA.

Munn, M.D., Gilliom, R.J., Morgan, P.W., and Nowell, L.H. 2006. Pesticide Toxicity Index for Freshwater Aquatic Organisms, 2nd ed. Investigations Report 2006–5148. U.S. Geological Survey, Reston, Virginia.

National Research Council. 2000. The Future Role of Pesticides in US Agriculture. National Academy Press, Washington, DC.

Salgado, V.L., Sheets, J.L., Watson, G.B., and Schmidt, A.L. 1998. Studies on the mode of action of spinosad: The internal effective concentration and the concentration dependence of neural excitation. Pesticide Biochemistry and Physiology 60: 103–110.

Scholz, N., Truelove, N.K., French, B.L., Berejikian, B.A., Quinn, T.P., Casillas, E., and Collier, T.K. 2000. Diazinon disrupts antipredator and homing behaviors in chinook salmon (*Oncorhynchus tshawytscha*). Canadian Journal of Fisheries and Aquatic Sciences 57: 1911–1918.

Stark, J.D. 2005. How closely do acute lethal concentration estimates predict effects of toxicants on populations? Integrated Environmental Assessment and Management 1: 109–113.

Stark, J.D. 2006. Toxicity endpoints used in risk assessment: What do they really mean? SETAC Globe 7(2): 29–30.

Stark, J.D., and Banks, J.E. 2001. "Selective pesticides": Are they less hazardous to the environment? BioScience 51: 980–982.

Stark, J.D., and Banks, J.E. 2003. Population-level effects of pesticides and other toxicants on arthropods. Annual Review of Entomology 48: 505–519.

Stark, J.D., and Vargas, R.I. 2003. Demographic changes in *Daphnia pulex* (Leydig) after exposure to the insecticides spinosad and diazinon. Ecotoxicology and Environmental Safety 56: 334–338.

Stark J.D., Banks, J.E., and Vargas, R.I. 2004. How risky is risk assessment? The role that life history strategies play in susceptibility of species to stress. Proceedings of the National Academy of Sciences of the USA 101: 732–736.

Walthall, W.K., and Stark, J.D. 1997. Comparison of two population-level ecotoxicological endpoints: the intrinsic (rm) and instantaneous (ri) rates of increase. Environmental Toxicology and Chemistry 16: 1068–1073.

Williams, T., J. Valle, J., and Viñuela, E. 2003. Is the naturally derived insecticide spinosad compatible with insect natural enemies? Biocontrol Science and Technology 13: 459–475.

10

Lymnaea stagnalis

The Effects of Experimental Demographic Reduction on Population Dynamics

MARIE-AGNÈS COUTELLEC
THIERRY CAQUET
LAURENT LAGADIC

A population-level estimation of toxicant effects is of primary interest for ecotoxicological risk assessment, because populations often represent elementary units for both ecological and evolutionary processes. Population-level parameters such as growth rate, *r*, have proven to be more relevant to reflect ecotoxicological impact of toxicants than are life-history traits measured at the individual level, over a wide range of species and substances (Forbes and Calow, 1999), and the higher relevance of population-level approaches is acknowledged for different taxonomic groups (Sibly et al., 2005). To study the impact of a presumably toxic substance on a population, demographic approaches such as life table response experiments can be used (Mauri et al., 2003), under fixed, random, or regression designs, depending on the degree of control of the factor tested, and whether or not it is continuous (Caswell, 2000, 2001). Such designs are relevant because they provide answers at the desired level (the whole life table is the dependent variable) and allow one to include interactions with factors other than the toxicant, such as local ecological components or genetic characteristics of the populations in the statistical test model. Whereas demographic methodology can be applied to laboratory-raised as well as natural populations, its efficacy will strongly depend on the accuracy of vital rate estimates.

In the context of risk assessment, small populations cannot be treated in the same way as large populations, because some factors of minor importance in the latter become prominent in the former. This applies to demographic and genetic stochasticities, which expose small populations to increased risks of extinction (Bürger and Lynch, 1997). Empirical studies of toxicant effects on populations cannot strictly distinguish such potential effects from more specific sublethal ones.

In the present study, the effect of a toxicant on population size was experimentally simulated in outdoor pond mesocosms where populations of a freshwater snail developed from various initial numbers of individuals and levels of genetic relatedness. The aim was to estimate the consequences of demographic bottlenecks due to toxicants on further performances, in terms of population dynamics. This experimental approach was preferred to a direct study of toxicant effects on populations, because it allowed us (1) to focus exclusively on potential effects on numbers without impairment in survivors and (2) to control for the level of demographic reduction, instead of depending upon an undetermined effect. Moreover, genetic risks of fitness reduction due to the accumulation of deleterious mutations are enhanced in small and inbred populations (see Keller and Waller, 2002, and references therein). Such indirect consequences of a toxicant effect on demography were taken into account in the experimental design by crossing the level of genetic relatedness with the number of founders used to create the populations. Results are expected to provide an empirical calibration to model the potential demogenetic effects of xenobiotics, with a particular emphasis on population long-term responses and extinction risks. Demographic predictions and extinction risks were estimated using RAMAS GIS (Akçakaya, 2005). A joint study of population genetics parameters in the set of experimental populations is concurrently performed using microsatellite markers.

Methods

Study species and area

Lymnaea stagnalis (Linné, 1758; Mollusca, Pulmonata, Basommatophora) is a palearctic and nearctic freshwater gastropod inhabiting ponds and lakes. It is a simultaneous hermaphroditic snail, which can reproduce by self-fertilization and cross-fertilization. Eggs are contained in gelatinous egg masses made of acid mucopolysaccharides (Geraerts and Joosse, 1984), laid on various supports, usually near the water surface. Eggs develop within the egg mass and start to hatch after 10 days of incubation at 20°C. Time between first and last hatchings in a clutch is also around 10 days (Coutellec, unpublished observation; Geraerts and Joosse, 1984).

In this species, reproduction is annual and occurs in summer, usually over a single period (birth-pulse reproduction). Most individuals are semelparous and usually die after reproduction (Noland and Carriker, 1946; Berrie, 1965; Brown, 1979). From an ecotoxicological point of view, *L. stagnalis* is handy as a model invertebrate for freshwater risk assessment, especially in water bodies surrounding areas of pesticide application such as agricultural zones. Moreover, it is a suitable model to investigate the impact of pollutants on individual reproductive performances and juvenile and adult life-history traits (Gomot, 1998; Jumel et al., 2002; Coeurdassier et al., 2004; Coutellec and Lagadic, 2006; Lagadic et al., 2007).

The *L. stagnalis* strain used in the present study originated from natural ponds (Le Rheu, France) and is maintained at the INRA (Institut National de la Recherche Agronomique) Experimental Unit of Aquatic Ecology and Ecotoxicology (Rennes, France). Snails are mass-reared under laboratory conditions (14/10-hr light/dark cycle, 20 ± 1°C) in 30-L aquaria containing dechlorinated tap water and siliceous sand supplemented with maerl as a calcium source. They are fed twice a week with organic lettuce, and a third of the water volume is renewed weekly.

Families used to create experimental populations were initiated from 105 adults isolated from the stock. Individual size was inferred from the shell height from the external rim of the aperture to the apex, measured with an ocular micrometer (nearest 0.025 mm) or a caliper (nearest 0.1 mm). The snails were surveyed for oviposition over a nine-day period (January 15–24, 2002). One egg mass per snail was individually incubated in the laboratory, and around 30 hatchlings per clutch were further reared as groups until May 20, 2002. After this period, eight sibs per lineage were reared in isolation until transplantation to outdoor conditions (July 23, 2002). At that time, one-quarter of the snails were sexually mature (self-fertilization), and this proportion was stable across the four predefined population sizes (chi-square [df = 3] = 0.193, $p = 0.979$). Mean size of these mature snails was 25.65 mm (SD = 1.59), and their mean age was 178.9 days (SD = 3.43). Populations did not differ significantly for these two traits ($F_{(14;160)} = 0.737$, $p = 0.734$; $F_{(14;160)} = 0.921$, $p = 0.538$, respectively).

Snails were introduced into 16 circular 9-m^3 outdoor experimental ponds (3.2 m diameter × 0.90 m depth) located at the Rennes site of the Experimental Unit of Aquatic Ecology and Ecotoxicology (Caquet et al., 2001). Each pond was lined with a Polyane film (Celloplast SAS, Ballee, France). Sediment consisted of 400 L of a mixture of sand (60%), clay (10%), and natural uncontaminated sediment (30%). Sediments were introduced, and the ponds were filled with 7 m^3 of tap water in March–April 2002. The ponds were then allowed to stabilize during four months before introducing the snails. During the maturation period, plants (*Glyceria maxima*) and inocula of field-collected phytoplankton and zooplankton were repeatedly introduced. Spontaneous development of plants (*Zannichellia palustris*, *Myriophyllum* sp.) and colonization by insects (*Chironomidae*, *Chaoboridae*, *Notonectidae*, *Coenagrionidae*, *Aeschnidae*, *Baetidae*) further increased the ecological realism of the ponds (Hanson et al., 2007; Caquet et al., 2007).

Experimental methods

Experimental demographic reduction

Lineages were used to create 16 experimental populations under two conditions of demographic bottleneck (80 and 8 individuals, respectively) and four conditions of genetic bottleneck (80, 20, 8, and 2 lineages, respectively; table 10.1). For size treatments 1 and 3 (no relatedness between individuals), replicate populations were made from the same lineages, whereas (for technical reasons) different lineages were used to create replicate populations under treatments 2 and 4.

Toxicant effects were simulated in a way that only a few adults and subadults survived in bottlenecked populations. This corresponds to a case where all preadult stages have been killed while adult survivors were not impaired.

Life-history traits

Populations have been surveyed for individual size structure for an 18-month period assumed to cover two reproduction events (April 2003 to October 2004). During this

Table 10.1. Conditions of demographic and genetic reductions applied to experimentally founded populations of *Lymnaea stagnalis*.

Population size treatment	Number of founders	Number of lineages involved	Effective population size (N_e)[a]	Number and code of replicate populations
1: large, no relatedness	80	80	80	4 (B1 to B4)
2: large, relatedness	80	20	52.67	4 (B5 to B8)
3: small, no relatedness	8	8	8	4 (A1 to A4)
4: small, relatedness	8	2	4.67	4 (A5 to A8)

[a] N_e calculated as ½(probability of full-sib mating × coefficient of coancestry) under the hypothesis of full-sib families and zero initial inbreeding coefficient.

period, size structure was recorded every 21 days. Snails were sampled on the inner walls of the mesocosms, from the water surface to 40 cm depth (pulmonate snails need frequent contact with the atmosphere to breath), or just below the water surface, where vascular plants and filamentous algae provided floating substrates for animals. Sampled individuals were reintroduced into the ponds after enumeration and shell length measurement. This method limited disturbance of the ponds due to repeated sampling. Snails were collected on one-quarter to one-half of the wall surface band (40-cm depth). When abundance was very low, they were collected on the total perimeter of the mesocosm inner wall. Low abundances (<35 individuals) observed at more than two successive sampling dates were interpreted as a sign of population reduction or low size.

Several life-history traits were inferred from the temporal change in size structure. Whenever possible, cohorts were separated by a decomposition of size-frequency distributions using NORMSEP, a maximum likelihood separation of normally distributed components of the size-frequency sample (Hasselblad, 1966; Pauly and Caddy, 1985) provided in the FiSAT 2 freeware (version 1.1.3; Gayanilo and Pauly, 1997). Modal progression was followed in order to identify size/age at next recruitment and at death in each population. These estimates of life-history traits in populations were compared to those obtained in the laboratory from sampled individuals (see below).

Reproduction

In the summer of 2004, five adult snails per population were caged in each pond for 15 days (July 27 to August 12). Sample size was deliberately small, to limit population disturbance.

Adult snails sampled in July 2004 and June 2005 were brought to the laboratory (14/10-hr light/dark cycle, 20 ± 1°C) in order to estimate and compare individual reproductive output among populations. Ten snails per population were sampled in 2004, and an average of 12 snails in 2005. Animals were placed into plastic boxes and fed with organic lettuce. The water volume was 150 ml per individual. Water was renewed every week. Egg masses were collected twice a week, and the number of eggs was counted to estimate fecundity.

In summer 2004, because no oviposition was observed after the first week of individual isolation, snails from the same population were temporarily paired until copulation could be observed and a first clutch was laid; they were then returned to isolation for a period of 63 days.

In summer 2005, snails were kept isolated from the sampling date to death. Fecundity (selfing or outcrossing, owing to the fact that allosperm storage can last several weeks in freshwater snails) was measured over 65 days, after which oviposition started to decrease dramatically. This period fits well with observations on reproduction in outdoor populations. Size at death was measured.

Fertility

Clutches laid by the snails in the laboratory or in the ponds (caged individuals) were incubated under laboratory conditions and used to estimate hatching rates. Some clutches were also collected directly in the ponds (2004 and 2005). Hatching rates were multiplied by individual fecundities to estimate adult fertility. The estimated fertility was used to calculate age-class fecundity (F_i) in RAMAS models and simulations.

Survival

From hatching date, juveniles were reared as sibships of 15 snails (2004 data set), and survival was measured after 50, 100, and 150 days. In 2005, survival at 50 days was measured on groups of 10 sibs (data not used in the present study). After that period, snails were reared in isolation until sexual maturity under self-fertilization (starting on average from 160 days under laboratory conditions). Survival data were used as vital rates in RAMAS models and simulations.

Population Model

Matrix projection and deterministic predictions

Parameters described above as vital rates were used to construct Leslie matrices for the 16 populations. The time step of the model was 100 days, as was age class length (four classes). Although the life cycle could have been reduced to two age classes (juvenile and adult stages), this decomposition was preferred because we searched for age-specific sensibilities, which could affect population dynamics. Survival of age classes 1 and 2 (100-day length) was inferred from our laboratory estimations up to 200 days, whereas later survival could not be used because it was too high to be realistic (almost 100% under our rearing conditions). Thus, for class 3 (200–300 days), we used data from the literature on survivorship to sexual maturity, as estimated in natural populations of *L. stagnalis* showing a life cycle otherwise similar to ours ($l_{sm} = 0.016$; Brown, 1979). We assumed reproduction to occur in this class, under the hypothesis of postbreeding census (Akçakaya et al., 1999; Caswell, 2001). Survivors entering the last class were assumed not to reproduce. The reproductive output of snails that were caged *in situ* during 15 days was adjusted to the duration of an age class (100 days), whereas laboratory measurements were performed only until mortality started to be observed among sampled snails (63 and 65 days in 2004 and 2005, respectively). Demographic models were based on reproduction averaged between these two conditions.

Fertility (m_i) was estimated by multiplying the number of eggs laid per individual by the mean hatching rate. Age-class fecundity (F_i) was calculated as $P_i \times m_i$ (post-breeding census).

Leslie matrices were used to calculate population intrinsic growth rate (λ, population multiplication rate) and stable age structure, and to perform perturbation analysis (sensitivity and elasticity of λ to the model parameters). Predicted population size is presented for step 15, which roughly corresponds to four years.

Standard deviation and stochastic simulations

Survival

Standard deviations (SDs) for survival values were calculated on an individual basis (parental snail). This was used to reflect interindividual variation within populations, which may be related to the degree of inbreeding in the population. This parameter was thought to be particularly useful for populations of identical number of founders but different levels of genetic relatedness.

Fecundity

As fecundity was inferred from the average reproduction measured under external and laboratory conditions, the SD between the two conditions was used, because it was assumed to reflect variation between environmental conditions.

Simulations were performed separately for each population (i.e., the metapopulation structure was not taken into account), which appeared as the most realistic approach for the set of studied populations. Considering the site configuration, snails escaping from one pond (which in itself is unlikely) could hardly survive to enter another one. Population trajectories and extinction risks were compared among treatments (= initial population demographic size). Simulation duration was set to 40 steps, that is, about 10 years.

Results

Size-structure distributions and their temporal evolution

A summary of observed population structure over the period of 18 months is provided in figure 10.1. Cohorts were identified and the evolution of individual size was estimated for each of them. Size and age at death were inferred from the examination of cohort first and last date of observation, respectively. Depending on the population, two to four distinct cohorts could be identified. This number was not related to initial population size.

In some populations, adult founders from July 2002 were still present in spring 2003. The age of these snails could be calculated since their date of birth was known (laboratory-raised families). This cohort (C0) was observed in all large populations (B1 to B8) and fewer than half of the small ones (A6 to A8). Mean size of these snails at the last date of observation was 33.58 mm (SD = 2.19) in large populations and 31.42 mm (SD = 4.72) in small populations. Their mean age at the last date of observation

Population	Cohort	2003 Spring	2003 Summer	2003 Fall	2003 Winter	2004 Spring	2004 Summer	2004 Fall
B1	C0	C0						
	C1	C1	C1	C1	C1	C1	C1	?
	C2		C2	C2	C2	C2	C2	C2
	C3						C3	C3
B2	C0	C0						
	C1	C1	C1	C1	C1	C1	C1	
	C2		C2	C2	C2	C2	C2	C2
	C3							
B3	C0	C0						
	C1	C1	C1	C1	C1	C1	C1	C1
	C2			C2	C2	C2	?	?
	C3						C3	C3
B4	C0	C0						
	C1	C1	C1	C1	C1	C1	C1	C1
	C2			C2	C2	C2	C2	C2
	C3							
B5	C0	C0						
	C1	C1	C1	C1	C1	C1	C1	C1
	C2		C2	C2	C2	C2	?	?
	C3					C3	C3	C3
B6	C0	C0						
	C1	C1	C1	C1	C1	C1	C1	C1
	C2							
	C3					C3	C3	C3
B7	C0	C0						
	C1	C1	C1	C1	C1	C1	?	?
	C2		C2	C2	C2	C2	C2	C2
	C3						C3	C3
B8	C0	C0						
	C1	C1	C1	C1	C1	C1	C1	C1
	C2		C2	C2	C2	C2	C2	C2
	C3						C3	C3
A1	C1	C1	C1					
	C2		C2	C2	C2	C2	C2	C2
	C3						C3	C3
A2	C1	C1	C1	C1	C1			
	C2		C2					
	C2'			C2'	C2'	C2'	C2'	C2'
	C3					C3	C3	C3
A3	C1	C1	C1	C1				
	C2	C2	C2	C2	C2	C2	C2	C2
	C3					C3	C3	C3
A4	C1	C1	C1					
	C2	C2	C2	C2	C2	C2	C2	C2
	C2'			C2'	C2'	C2'	C2'	C2'
	C3						C3	C3
A5	C1	C1	C1	C1	C1	C1	C1	C1
	C2		C2	C2	C2	C2	C2	?
	C3					C3	C3	C3
A6	C1	C1	C1	C1	C1	C1	C1	C1
	C2			C2	C2	C2	C2	C2
	C3					C3	C3	C3
A7	C0	C0						
	C1	C1	C1	C1	C1	C1	C1	C1
	C2		C2	C2	C2	C2	C2	?
	C3					C3	C3	C3
A8	C0	C0						
	C1	C1	C1	C1	C1	C1	C1	C1
	C2			C2	C2	C2	?	?
	C3						C3	C3

Figure 10.1. Summary of the temporal evolution of the different cohorts identified in the 16 populations during the 18-month survey of size-structure distributions.

was 508.62 days (SD = 31.66) in large populations and 574 days (SD = 96.9) in the small ones, suggesting that C0 snails grew faster and died earlier in initially large populations than in the small ones. However, none of the observed differences was significant (Kruskal-Wallis test, df = 1, $p = 0.413$ for size, and $p = 0.214$ for age).

Cohort C1 either was present at the first sampling date (April 4, 2003) or appeared during spring 2003. In large and small populations, C1 disappeared in four and three of eight replicate populations, respectively. In all cases, C1 was already present on the first date of sampling, and the age at death was estimated under the hypothesis that these snails stemmed from the reproduction of cohort C0 during the summer of 2002. Mean size at last observation (interpreted as size at death) was 25.69 mm (SD = 4.11) in large populations (B1, B2, B3, B7) and 25.92 mm (SD = 2.77) in small populations (A1, A2, A4). The estimation of age at death in C1 was 615 days (SD = 157.6) on average for large populations, whereas it was only 421.67 days (SD = 43.88) in small populations. However, size and age at death did not differ significantly among population treatments (Kruskal-Wallis test, df = 1, $p = 0.724$, and $p = 0.157$, respectively).

Adult snails sampled in early summer of 2005 died massively after intensive reproduction during two months (65 days) under laboratory conditions. Size at death was not different between large (mean = 28.15 mm, SD = 2.41, $n = 99$ snails) and small populations (mean = 28.72, SD = 2.37, $n = 103$; analysis of variance, $p = 0.095$).

Cohort C2 appeared in 2003, during summer/fall, except in A3 and A4 populations, where C2 was observed earlier (spring 2003), and in the B6 population, where no recruitment was observed until spring 2004. Populations A2 and A4 presented a particular scheme with two successive and distinct recruitments in 2004 (C2 and C2′).

In spring/summer 2004, a new recruitment event was observed (cohort C3) in all populations except B2 and B4. In population B4, the absence of the cohort C3 may be related to population size, which was very small during winter (the mean sample size was 23.5 snails, from October 6, 2003, to March 2, 2004). In the summer of 2005, no new recruitment could be observed in B2 and B4 (a single sample was collected on June 28), and this pattern is likely to result from the missing C3 cohort of 2004.

Size and age at sexual maturity were inferred from the temporal evolution of the cohorts. When compared between large and small initial population abundances, the observed differences in age and size at sexual maturity were not significant (Kruskal-Wallis test, df = 1, $p = 0.092$ and $p = 0.248$, respectively), although snails from “A” populations appeared to be mature about 100 days before snails from “B” populations (A populations: 341.75 days, SD = 127.01; B populations: 463.12 days, SD = 135.14) and tended to be smaller (A populations: 18.49 mm, SD = 5.46; B populations: 20.66 mm, SD = 4.55). Under laboratory conditions, a different relationship was observed, the mature snails from A populations being significantly larger (mean = 25.49 mm, SD = 1.43) than the B mature snails (mean = 24.88 mm, SD = 1.62; $F_{(1;137)} = 5.544$, $p = 0.020$). “A” snails also tended to mature later (mean age = 182.53 days, SD = 15.73) than “B” snails (mean = 179.07 days, SD = 15.65). However, the difference was not statistically significant ($F_{(1;137)} = 1.686$, $p = 0.196$).

Demographic models

Due to the general lack of statistical significance, differences observed in population parameters were not taken into account in the demographic models. A single matrix “structure” was used for all the populations, based on a 100-day time step, four age

Table 10.2. Parameters [mean values (standard deviation)] of the Leslie matrices used to simulate dynamic evolution of 16 experimental populations of *L. stagnalis,* as well as λ values and stable age distributions.

Population	Class 1 (100 days)	Class 2 (200 days)	Class 3 (300 days)	Class 4 (400 days)	λ	Stable age vector
B1	0	0	1.663(1.008)	0	1.064	0.413
	0.815(0.221)	0	0	0		0.317
	0	0.888(0.096)	0	0		0.264
	0	0	0.022	0		0.005
B2	0	0	2.157(1.246)	0	1.022	0.481
	0.602(0.300)	0	0	0		0.283
	0	0.822(0.218)	0	0		0.228
	0	0	0.032	0		0.007
B3	0	0	2.152(0.632)	0	1.052	0.472
	0.651(0.279)	0	0	0		0.291
	0	0.833(0.119)	0	0		0.231
	0	0	0.032	0		0.006
B4	0	0	2.116(0.621)	0	1.131	0.436
	0.844(0.127)	0	0	0		0.326
	0	0.809(0.099)	0	0		0.233
	0	0	0.023	0		0.005
B5	0	0	1.660(0.206)	0	1.078	0.420
	0.776(0.188)	0	0	0		0.302
	0	0.973(0.059)	0	0		0.273
	0	0	0.021	0		0.005
B6	0	0	2.147(1.470)	0	1.082	0.444
	0.796(0.170)	0	0	0		0.326
	0	0.742(0.198)	0	0		0.224
	0	0	0.027	0		0.006
B7	0	0	3.552(0.723)	0	1.292	0.539
	0.624(0.107)	0	0	0		0.261
	0	0.973(0.038)	0	0		0.196
	0	0	0.026	0		0.004
B8	0	0	3.300(0.359)	0	1.123	0.544
	0.547(0.171)	0	0	0		0.265
	0	0.786(0.158)	0	0		0.185
	0	0	0.037	0		0.006
A1	0	0	3.370(0.460)	0	1.213	0.525
	0.651(0.263)	0	0	0		0.282
	0	0.813(0.111)	0	0		0.189
	0	0	0.030	0		0.005
A2	0	0	3.400(0.738)	0	1.223	0.535
	0.611(0.265)	0	0	0		0.267
	0	0.882(0.080)	0	0		0.193
	0	0	0.030	0		0.005
A3	0	0	2.290(1.038)	0	1.128	0.458
	0.765(0.228)	0	0	0		0.311
	0	0.819(0.193)	0	0		0.226
	0	0	0.026	0		0.005

continued

Table 10.2. *Continued.*

Population	Class 1 (100 days)	Class 2 (200 days)	Class 3 (300 days)	Class 4 (400 days)	λ	Stable age vector
A4	0	0	2.164(1.842)	0	1.062	0.460
	0.711(0.282)	0	0	0		0.308
	0	0.779(0.226)	0	0		0.226
	0	0	0.029	0		0.006
A5	0	0	2.632(0.844)	0	1.172	0.495
	0.662(0.258)	0	0	0		0.280
	0	0.925(0.065)	0	0		0.221
	0	0	0.026	0		0.005
A6	0	0	4.237(1.012)	0	1.347	0.548
	0.676(0.151)	0	0	0		0.275
	0	0.853(0.104)	0	0		0.174
	0	0	0.028	0		0.004
A7	0	0	2.169(1.715)	0	1.081	0.469
	0.673(0.152)	0	0	0		0.292
	0	0.866(0.084)	0	0		0.234
	0	0	0.027	0		0.006
A8	0	0	2.922(0.153)	0	1.146	0.496
	0.701(0.275)	0	0	0		0.304
	0	0.734(0.179)	0	0		0.195
	0	0	0.031	0		0.005

classes, and only one reproductive age class (class 3), under a life cycle with annual reproduction followed by massive death in the adult reproductive and postreproductive age classes.

Leslie matrix parameters with SDs are given in table 10.2, which also presents matrix characteristics (finite rate of increase, λ, and stable age distribution vector) for each population. Population growth rates were similar among small (λ mean value = 1.172, SD = 0.091) and large populations (λ mean value = 1.106, SD = 0.083) (Kruskal-Wallis test, df = 1, $p = 0.093$). No effect of the initial effective population size (N_e) between population growth rates (type 1, B1 to B4: λ = 1.067, SD = 0.046; type 2, B5 to B8: λ = 1.144, SD = 0.101; type 3, A1 to A4, λ = 1.157, SD = 0.076; type 4, A5 to A8, λ = 1.187, SD = 0.113) could be detected statistically (Kruskal-Wallis test, df = 3, $p = 0.186$).

Sensitivity of λ to fecundity and survival parameters is summarized on figure 10.2. Sensitivity to fecundity was generally weaker than sensitivity to survival. Considering the initial population size, a trend of large populations to show higher sensitivity of λ to fecundity and less sensitivity to survival was suggested, but without statistical significance (see SD values in figure 10.2). Elasticities are not presented since they all equal one-fourth, probably due to the occurrence of only one nonzero fecundity age class, and two prereproductive age classes to which λ is equally sensitive (Caswell, 2001).

Simulations were performed using SDs for the fecundity of class 3 and survival of the two first age classes, 1,000 replicates, and a duration set to 40 steps of 100 days (around 10 years). Initial abundances were set to reflect the proportion of mature snails among founders (0.25). Thus, for large (small) populations, initial abundances were 60 (6) submature snails plus 20 (2) mature snails.

Results of the simulations are summarized in table 10.3 and figure 10.3, A and B. Average predicted population sizes (and SD) are given at step number 15 (4 years, i.e., around the age of the populations at the time of analyses). Mean size ranged from 128 to 3,830 individuals among initially large populations (type B) and from 20 to 690 snails among initially small populations (type A). In most cases, SDs were higher than mean values. Expected minimum abundances (figure 10.3A) reflect the minimum size that could be reached at least once during the course of simulations (interval extinction risk, 40 steps). Not surprisingly, minimum abundances were positively related to initial abundances, and thus much smaller in initially small populations (1.1–12.3 individuals) than in large populations (11.1–127.1 individuals) (Kruskal-Wallis test, df = 1, $p = 0.001$). Terminal extinction risk (figure 10.3B) indicates the probability that at least one population would go extinct during the course of simulations. In the present simulations, no significant difference could be detected between large (mean probability = 0.158, SD = 0.193) and small populations (mean probability = 0.309, SD = 0.238; Kruskal-Wallis test, df = 1, $p = 0.127$). Considering the important influence of initial abundance on the results, we examined the behavior of the 16 populations, starting from other numbers than the real ones (N_0 values of 5, 8, 54, 80, and 160 individuals) in order to compare population dynamics on a same basis. Results of these simulations are summarized by population type (A or B) on figure 10.3, C and D. For large N_0 values (54 and above), type A populations tended to yield larger minimum sizes than did type B populations. Similarly, under small N_0 (five and eight individuals), B populations tended to show higher risks of extinction than A populations. However, SDs were very high and prevent drawing any clear conclusion.

Discussion

Model predictions

From deterministic predictions, all the populations showed positive and similar dynamics ($\lambda > 1$). No clear difference could be detected between initially large and small populations, or between the four types of initial effective population size (N_e), despite a slight (but nonsignificant) trend for initially larger populations to exhibit lower growth rates. However, the application of transition matrices to the number of individuals at the time of population foundation led to sharp differences between type A and B populations, as shown in table 10.3, with an example of population size predicted after about four years.

Sensitivity of λ to vital rates was of the same level for the 16 populations, that is, lower sensitivity to fecundity than to survival for classes 1 and 2. Similar sensitivities were obtained in the prosobranch gastropod *Potamopyrgus antipodarum* under various conditions (Jensen et al., 2001). The present results are consistent with patterns described in other basommatophorans (e.g., *Biomphalaria glabrata*, Salice and Miller, 2003) and more generally in semelparous life cycles, for which population growth rate is most sensitive to juvenile survival (Calow et al., 1997; see also Forbes et al.,

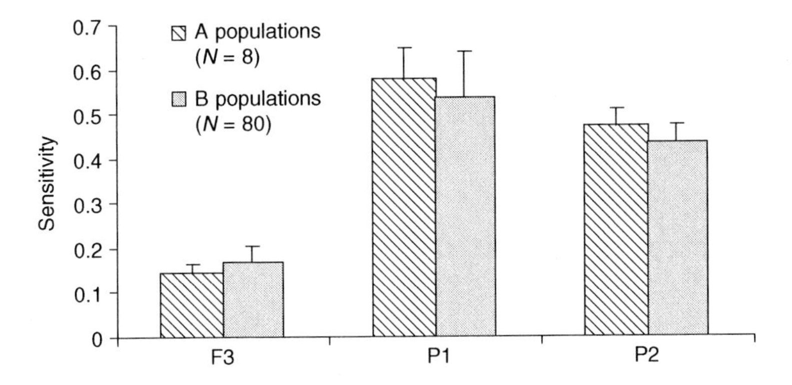

Figure 10.2. Mean sensitivities (SD) of λ to the Leslie matrix parameters: initially small populations (type A) and initially large populations (type B).

Table 10.3. Predicted average population size under RAMAS simulations: Mean values and SDs at step 15 (about four years after population foundation).

Population	Average abundance (SD) with stochasticity
B1	201.6 (473)
B2	127.6 (283.6)
B3	187.1 (237.4)
B4	522.5 (457)
B5	254.9 (157.3)
B6	297.9 (629.7)
B7	3830.0 (2332)
B8	457.4 (430.3)
A1	147.0 (160.8)
A2	175.0 (199.4)
A3	49.0 (82.1)
A4	20.4 (69.9)
A5	90.6 (113.8)
A6	690.5 (675.5)
A7	25.5 (57.9)
A8	66.8 (76.7)

2001, for a benthic macroinvertebrate semelparous life cycle). Slight differences were observed between large and small populations, with lower mean sensitivity to fecundity and higher mean sensitivity to survival in small populations. However, these indices varied widely within population types, preventing any firm conclusion.

The use of literature data on survivorship to sexual maturity may have introduced some bias in the modeling of the studied populations. However, such data appeared much more realistic than our estimations of survival beyond 200 days under laboratory conditions, which led to demographic explosion for all the populations. Prereproductive survival (P1 and P2) was probably also overestimated under laboratory conditions,

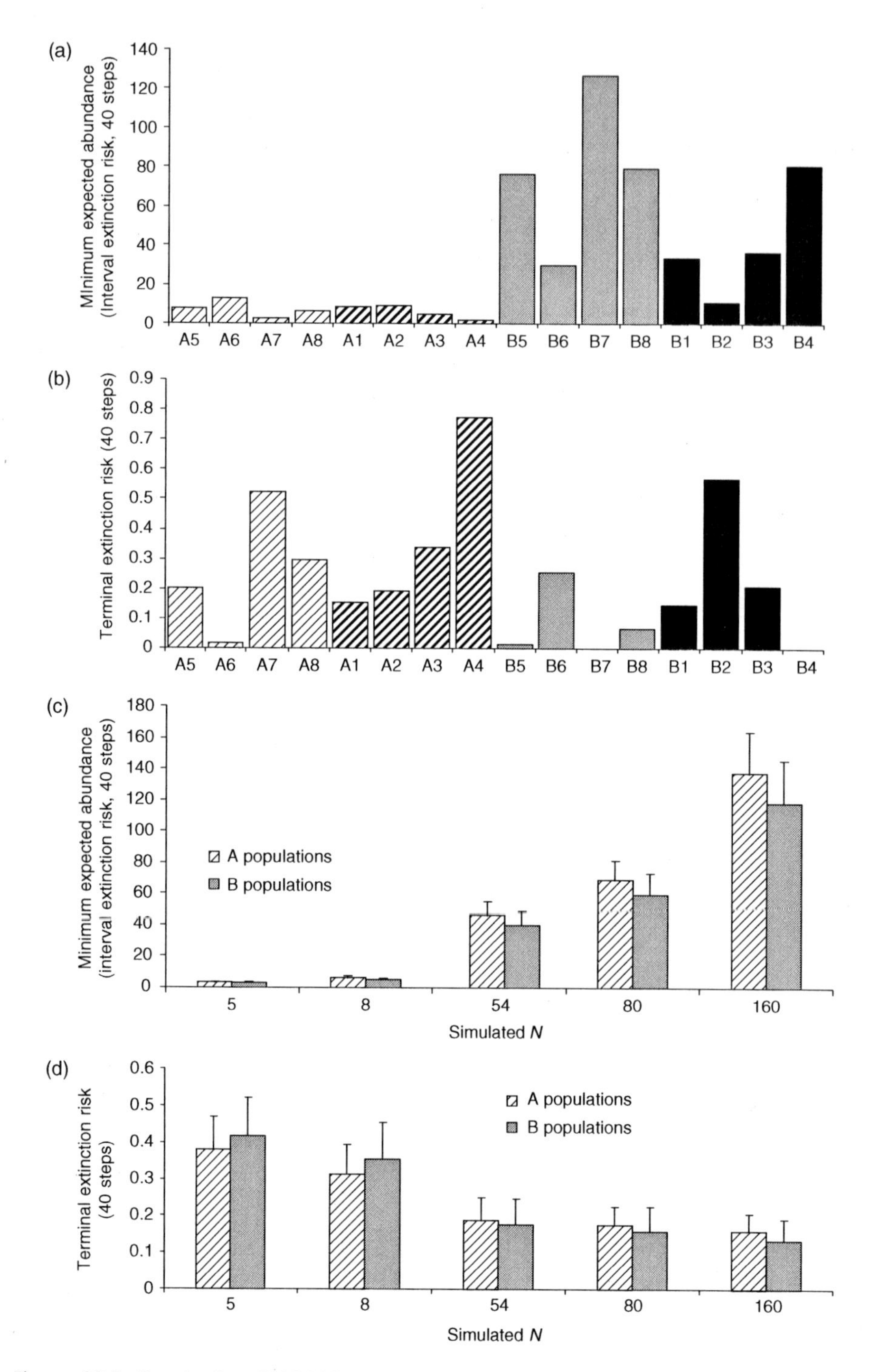

Figure 10.3. Results from RAMAS simulations (40 steps ~ 10 years) under demographic stochasticity and SD for fertility and survival to 200 days. (A) Expected minimum abundance (interval extinction risk). (B) Terminal extinction risk. Bars denote initial N_e: black, 80; gray, 52.67; hatched black, 8; hatched white, 4.67. (C and D) Mean expected minimum abundance (SD) and mean terminal extinction risk (SD) for each population type (A or B), using varied N_0 values (including initial N_e values).

because (1) density dependence could not be taken into account (Stark and Banks, 2000), and (2) in the ponds, juveniles were subjected to environmental fluctuations (e.g., water temperature and level) and interspecific interactions such as predation or parasitism, which did not occur in the laboratory. Comparison of survival between groups of various sizes and isolation levels would be a way to account for density dependence, which might be important for many traits in *L. stagnalis* (Seugé and Bluzat, 1984; van Duivenboden et al., 1985). Nevertheless, our estimations are consistent with the fact that survival is likely to increase with age (before the adult stage) in *L. stagnalis*, as suggested elsewhere under toxic conditions (Coeurdassier et al., 2004).

The introduction of SDs in vital rates and stochasticities in population dynamics allowed the estimation of extinction risks in the course of 10 years, even under population increase ($\lambda > 1$). This highlights the importance of variation in vital rates and argues for the use of demographic stochasticity (Laskowski, 2000). The present study emphasizes the primary role of initial population size on extinction indices. Thus, the minimum abundances expected in the 40 steps run were significantly lower in A populations than in B populations, but this was also the case for all the 16 populations when reduced abundances were used as starting points. More interesting, the whole set of tested N_0 values showed a trend of A populations (real $N_0 = 8$) to generate higher minimum expected abundances than B populations (real $N_0 = 80$), which, on their own, are more likely to go to extinction for small N_0 than are A populations. However, populations behaved very differently within each type (see SD values on figure 10.3, C and D). This suggests the need to increase the number of replicate populations and points to the importance of power analysis in experimental designs such as the one used in the present study.

Another explanation for the failure of the study to detect clear long-term consequences of the experimental demographic (and genetic) reductions may be related to a low level of genetic variation within the original populations used to create founder lineages. *L. stagnalis* was found to be quite insensitive to inbreeding depression, as measured through fitness under self-fertilization relative to that under outbreeding (Coutellec and Lagadic, 2006). Such a pattern may involve a mating system with historically high selfing rates, resulting in increasing homozygosity, decreasing genetic variability and a purge process of the deleterious genetic load (see references in Coutellec and Lagadic, 2006). The hypothesis of weak genetic variability in *L. stagnalis* is supported by the generally low levels of molecular variation observed in these experimental populations (microsatellite markers; unpublished results). We are also currently performing a study of genetic load accumulation since the populations were created, by crossing individuals from populations of the lowest initial effective size ($N_e = 4.67$ individuals) in order to see if genetic stochasticity due to small size and isolation has been significant in these experimental populations.

Implications for ecotoxicological risk assessment

In the present study, the use of variation (either between individuals or between the environmental conditions of trait measurements) and demographic stochasticity allowed models to clearly reflect the increased risk of extinction due to population reduction. The results show that a single event of demographic reduction can be prominent on population fate, compared to the influence of vital rates used in projection models. More precisely, the results reflect a particular case where long-term effects of population reduction have not been clearly detected on life-history variation and vital rates, whereas only direct demographic reduction (simulated N_0) was significant in increasing the risk of extinction. However, some differences in the pattern of responses were detected between large and small populations, suggesting higher sensitivity to demographic reduction in large populations than in the small ones (lower average growth rate, higher extinction risk when N_0 is artificially reduced, and lower minimum expected abundances when N_0 is artificially increased). Although these differences were not significant under our design, they should deserve further attention, because they can provide information about the effect of past demography on population response to new sources of reduction (e.g., those induced by the exposure to a toxicant). In the present study, the effect of controlled initial conditions on population demography may have been masked or modulated by additional factors or events since population foundation. The resulting global demographic effect is reflected through the vital rates (and their variation) and population growth rates, which can be used for extinction risk estimation. Our experimental study shows that it may be difficult to link observed population dynamics to particular factors or events, such as the exposure to a toxicant. However, chronic exposure conditions may lead to a different conclusion.

Acknowledgments This study was part of a program granted by the "Interface Recherche-Expertise pour l'Evaluation du Risque Ecotoxicologique," coordinated by the INRA-DGAL Structure Scientifique Mixte. We thank the U3E technical staff for mesocosm settling and survey. We are also grateful to M. Roucaute for temporal sampling, and to J. Moreau for measuring life-history traits in Summer 2004.

References

Akçakaya, H.R. 2005. *RAMAS GIS: Linking Spatial Data with Population Viability Analysis* (version 5.0). Applied Biomathematics, Setauket, New York.

Akçakaya, H.R., Burgman, M.A., and Ginzburg, L.R. 1999. *Applied Population Ecology: Principles and Computer Exercises using RAMAS Ecolab 2.0*, 2nd ed. Sinauer, Sunderland, Massachusetts.

Berrie, A.D. 1965. On the life cycle of *Lymnaea stagnalis* (L.) in the west of Scotland. *Proc. Malacol. Soc. London*, 36: 283–295.

Brown, K. 1979. The adaptive demography of four freshwater pulmonate snails. *Evolution*, 33: 417–432.

Bürger, R., and Lynch, M. 1997. Adaptation and extinction in changing environments. Pages 210–239 in R. Bijlsma and V. Loeschcke (eds.), *Environmental Stress, Adaptation and Evolution*. Birkhäuser Verlag, Basel.

Calow, P., Sibly, R.M., and Forbes, V. 1997. Risk assessment on the basis of simplified life-history scenarios. *Environ. Toxicol. Chem.*, 16: 1983–1989.

Caquet, T., Lagadic, L., Monod, G., Lacaze, J.-C., and Couté, A. 2001. Variability of physico-chemical and biological parameters between replicated outdoor freshwater lentic mesocosms. *Ecotoxicology*, 10: 51–66.

Caquet, T., Hanson, M.L., Roucaute, M., Graham, D.W., and Lagadic, L. 2007. Influence of isolation on the recovery of pond mesocosms to the application of an insecticide. II. Benthic macroinvertebrate responses. *Environ. Toxicol. Chem.*, 26: 1280–1290.

Caswell, H. 2000. Life table response experiments in ecotoxicology. Pages 43–55 in J. Kammenga and R. Laskowski (eds.), *Demography in Ecotoxicology*. John Wiley and Sons, Chichester, UK.

Caswell, H. 2001. *Matrix population models*, 2nd ed. Sinauer, Sunderland, Massachusetts.

Coeurdassier, M., de Vaufleury A., Scheifler, R., Morhain, E., and Badot, P.-M. 2004. Effects of cadmium on the survival of three life-stages of the freshwater pulmonate *Lymnaea stagnalis* (Mollusca: Gastropoda). *Bull. Environ. Contam. Toxicol.*, 72: 1083–1090.

Coutellec, M.-A., and Lagadic, L. 2006. Effects of self-fertilization, environmental stress and exposure to xenobiotics on fitness-related traits of the freshwater snail *Lymnaea stagnalis*. *Ecotoxicology*, 15: 199–213.

Forbes, V., and Calow, P. 1999. Is the per capita rate of increase a good measure of population-level effects in ecotoxicology? *Environ. Toxicol. Chem.*, 18: 1544–1556.

Forbes, V., Calow, P., and Sibly, R.M. 2001. Are current species extrapolation models a good basis for ecotoxicological risk assessment? *Environ. Toxicol. Chem.*, 20: 442–447.

Gayanilo, F.C., Jr., and Pauly D. (eds.). 1997. *FAO-ICLARM Stock Assessment Tools (FiSAT). Reference Manual.* FAO Computerized Information Series (Fisheries) No. 8. Food and Agriculture Organization of the United Nations, Rome.

Geraerts, W.P.M., and Joosse, J. 1984. Freshwater Snails (Basommatophora). Pages 141–207 in A.S. Tompa (ed.), *The Mollusca*, Vol. 7. *Reproduction*. Academic Press, London.

Gomot, A. 1998. Toxic effects of cadmium on reproduction, development, and hatching in freshwater snail *Lymnaea stagnalis* for water quality monitoring. *Ecotoxicol. Environ. Saf.*, 41: 288–297.

Hanson, M.L., Lagadic, L., Babin, E., Azam, D., Coutellec, M.-A., Knapp, C.W., Graham, D.W., and Caquet, T. 2007. Influence of isolation on the recovery of pond mesocosms to the application of an insecticide. I. Study design and planktonic community responses. *Environ. Toxicol. Chem.*, 26: 1265–1279.

Hasselblad, V. 1966. Estimation of parameters for a mixture of normal distributions. *Technometrics*, 8: 431–444.

Jensen, A., Forbes, V., and Parker, E.D., Jr. 2001. Variation in cadmium uptake, feeding rate, and life-history effects in the gastropod *Potamopyrgus antipodarum*: Linking toxicant effects on individuals to the population level. *Environ. Toxicol. Chem.*, 20: 2503–2513.

Jumel, A., Coutellec M.-A., Cravedi J.-P., and Lagadic, L. 2002. Nonylphenol polyethoxylate adjuvant mitigates the reproductive toxicity of fomesafen to the freshwater snail *Lymnaea stagnalis* in outdoor experimental ponds. *Environ. Toxicol. Chem.*, 21: 1876–1888.

Keller, L.F., and Waller, D. 2002. Inbreeding effects in wild populations. *T.R.E.E.*, 17: 230–241.

Lagadic, L., Coutellec, M.-A., and Caquet, T. 2007. Endocrine disruption in aquatic pulmonate molluscs: Few evidences, many challenges. *Ecotoxicology* 16: 45–59.

Laskowski, R. 2000. Stochastic and density-dependent models in ecotoxicology. Pages 57–71 in J. Kammenga and R. Laskowski (eds.), *Demography in Ecotoxicology*. John Wiley and Sons, Chichester, UK.

Mauri, M., Baraldi, E., and Simonini, R. 2003. Effects of zinc exposure on the polychaete *Dinophilus gyrociliatus*: A life-table response experiment. *Aquat. Toxicol.*, 65: 93–100.

Noland, L. E., and Carriker, M.R. 1946. Observations on the biology of the snail *Lymnaea stagnalis appressa* during twenty generations in laboratory culture. *Am. Midl. Nat.*, 36: 467–493.

Pauly, D., and Caddy, J.F. 1985. A modification of Bhattacharya's method for the analysis of mixtures of normal distributions. *FAO Fish. Circ.* 781: 1–16.

Salice, C.J., and Miller, T.J. 2003. Population-level responses to long-term cadmium exposure in two strains of the freshwater gastropod *Biomphalaria glabrata*: Results from a life-table response experiment. *Environ. Toxicol. Chem.*, 22: 678–688.

Seugé, J., and Bluzat, R. 1984. The effects of population density on growth and fecundity of the pond snail *Lymnaea stagnalis* L. *Zool. Jb. Physiol.*, 88: 505–514.

Sibly, R.M., Akçakaya, H.R., Topping, C.J., and O'Connor, R.J. 2005. Population-level assessment of risks of pesticides to birds and mammals in the UK. *Ecotoxicology*, 14: 863–876.

Stark, J., and Banks, J. 2000. The toxicologists' and ecologists' point of view—unification through a demographic approach. Pages 9–23 in J. Kammenga and R. Laskowski (eds.), *Demography in Ecotoxicology*. John Wiley and Sons, Chichester, UK.

van Duivenboden, Y.A., Pieneman, A.W., and ter Maat, A. 1985. Multiple mating suppresses fecundity in the hermaphrodite freshwater snail *Lymnaea stagnalis*: A laboratory study. *Anim. Behav.*, 33: 1184–1191.

11

Pollution, Stochasticity, and Spatial Heterogeneity in the Dynamics of an Age-Structured Population of Brown Trout Living in a River Network

ARNAUD CHAUMOT
SANDRINE CHARLES

Present assessment of potential impacts of pollution on natural populations mainly calls upon the extrapolation of exposure experiments performed on small groups of individuals (bioassays). But toxicological perturbations of the environment occur in complex landscapes, so the development of pertinent ecotoxicological decision-making tools requires taking into account the spatial dimension. That is to say, both the spatial heterogeneity of the environmental contamination and the spatial dynamics of exposed populations must be taken into account. Such a change of scale is a necessary step because refuge or "action at distance" processes in spatially fragmented populations can emerge (Sherratt and Jepson, 1993; Spromberg et al., 1998), because some areas or patches of the habitat could be key locations for the endangered population (Chaumot et al., 2003a) or because migratory and spatial behaviors could be disturbed by the presence of a toxin (Woodward et al., 1995; Hansen et al., 1999). That is why Caswell (1996) underlines the importance of integrating the spatial dimension into ecotoxicological population models. Therefore, focus on a larger spatial scale is required in order to study toxic effects on population dynamics. Spatial modeling in a metapopulation approach is a way to address this issue. From an ecotoxicological point of view, it is also necessary to use age-structured models because fish, like many other species, exhibit age-dependent sensitivity to toxicants (Sorensen, 1984; Buhl and Hamilton, 1991).

We previously developed multiregion matrix population models to explore how the demography of a hypothetical brown trout population living in a river network varies in response to different spatial scenarios of cadmium contamination (Chaumot et al., 2002, 2003a, 2003b). Cadmium is highly toxic, and this compound is a common pollutant of freshwater systems, with different possible sources of emission (industrial effluents, phosphate fertilizers, sewage sludge treatment). We present here how we used RAMAS Metapop to go further in analyzing the pertinence of ecotoxicological population endpoints defined in our previously deterministic modeling attempts. In the case of brown trout populations living in western France, we do not really deal with metapopulations *sensu stricto* (Levin, 1970; Hanski and Simberloff, 1997)—a population of local populations with migratory links and where local dynamics are influenced by migratory flows only at a long time scale (allowing recolonization after local extinction). The studied resident trout population is effectively dispatched in a river network, which constitutes a discrete habitat. But subpopulations are highly connected, because spatial structure mainly corresponds to an age population structure: Upstream areas constitute nurseries, and downstream areas are occupied by older stages, where trout feed and grow (Maisse and Baglinière, 1991). Two massive migration events take place during the year: an upstream migration of adult breeders late in fall, and symmetrically, a downstream migration of one-year-old trout in spring. We will demonstrate here how such patterns of dispersal can be translated using RAMAS Metapop.

First, this study aims to show how we can handle spatially explicit and age-structured models in an ecotoxicological issue using RAMAS Metapop. We consider chronic cadmium contamination in one downstream location of the river network (high density of breeders). The tested concentrations are supposed to influence only the reproductive process and not survival. We previously demonstrated that fecundity is the most relevant to understand population level impacts of cadmium pollution (Chaumot et al., 2003a). We report here the influence of increasing pollutant concentrations on population features. Second, we take advantage of the flexibility of RAMAS Metapop to incorporate demographic stochasticity and to compare the use of average abundances (supplied by deterministic methodology) with the use of extinction risk concepts (stochastic approach) as population endpoints for decision making. Third, in a previous contribution (Chaumot et al., 2003b), we stressed the need to take into account toxic-induced migratory disruptions in the assessment of pollution impacts on populations. This seemed to be highly important when density-dependent mechanisms are implied in the dynamics of the exposed population. In a stochastic approach, we analyze here how such migratory effects of a contaminant affect extinction risk of the trout population.

Methods

This study follows the development of a model relative to a hypothetical brown trout population exposed to cadmium pollution in a river network (Chaumot et al., 2003a); all details for biological parameterization are fully explained in this previous contribution. Here, we stress the biological aspects, which lead to specific choices in the modeling strategy.

Study species and area

Brown trout from western French populations (Bretagne, Normandie) present two types of life history. All of them spawn in river systems close to the sea, but some of these fish (called "resident trout") spend all their life cycle in river systems and never migrate to the sea. The other migratory individuals (called "sea trout") breed and live in fresh water during the first year before smolting; the older juvenile and adult stages take place in the sea (Baglinière et al., 1987, 1989, 1994; Gouraud et al., 1997, 1998; Gouraud, 1999; Maisse and Baglinière, 1991). We focus here on resident trout. The breeding takes place in winter, and the first reproduction mostly occurs during the two-year-old stage. Thus, since censusing is performed in spring or in fall, trout populations are classically described by three age classes (Maisse and Baglinière, 1991; Elliott, 1994): young of the year (YOY 0+), one-year-old juveniles (Juv 1+), and adults. For this study, we refer to a life cycle with a census late in spring, in the end of May (figure 11.1). "Resident trout" does not mean that fish do not move inside river networks. In fact, Maisse and Baglinière (1991) report spatial segregation between young stages and older stages, where the former are present in upstream areas (close to their hatching location) and the latter establish territories in more downstream stretches of river. Supporting this spatial segregation, two migration events occur during the year (figure 11.1). Late in autumn, adults migrate upstream for spawning; one-year-old juveniles symmetrically migrate downstream in spring and colonize the downstream

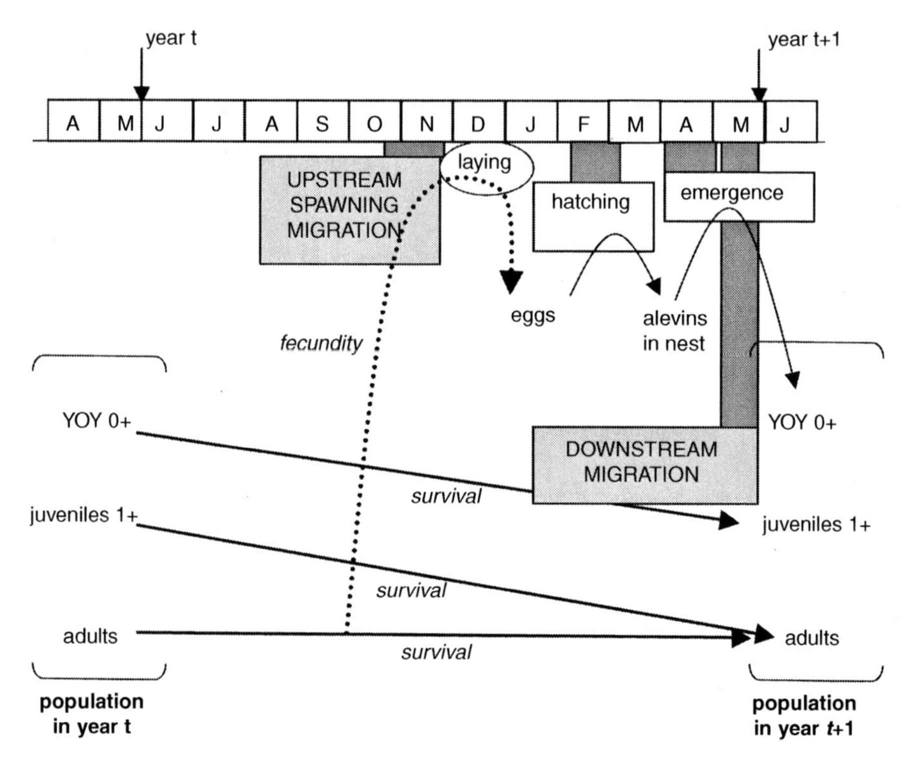

Figure 11.1. Life cycle of resident brown trout *Salmo trutta* in river networks of western France (Bretagne, Normandie). The time step of the model is one year, and the census happens late in spring (end of May). Arrows indicate demographic transitions between the three age classes considered (young of the year [YOY 0+], one-year-old juveniles [1+] and breeders [adults]). Gray boxes stand for the two major spatial shifts occurring in the population: an upstream migration of breeders in fall and a "symmetric" downstream migration of one-year-old juveniles in spring. Adults go back to downstream streams just after spawning (not shown on the figure).

levels (Gouraud et al., 1997, 1998; Gouraud, 1999). Adult trout survive after spawning, and they go back to their initial location in the river network after the breeding event (Schuck, 1945; Solomon and Templeton, 1976; Ovidio, 1999). We can assume that they are absent from their living area one or two months during the year.

In this theoretical approach, we consider a fish population that is spatially distributed in an arborescent river network, composed of 15 stretches of river (also called "patches") organized in four levels from up to downstream: eight patches in level 1, four in level 2, two in level 3, and one patch in the most downstream level 4 (figure 11.2, left). The surface of each of the upstream stretches of river (patches 5–8 and 12–15) is 1,000 m^2. It is assumed to be 2,000 m^2 in the second level (due to confluence), 4,000 m^2 in the third level (patch 2 and 9), and 8,000 m^2 for patch 1. Spatial heterogeneity affects demographic rates where survival and fecundity depend on the level in the river network. Moreover, density-dependent regulation occurs during the recruitment (Maisse and Baglinière, 1991; Héland, 1991; Elliott, 1994): Young of the year establish territories after the emergence, and so the production of juveniles in each stretch of river is limited by local carrying capacity.

Parameter estimation

Demographic and migratory rates

The different values for the survival rates were estimated by Charles et al. (1998) based on data from Maisse and Baglinière (1991). First, the survival rates during the first year (from YOY 0+ to Juv 1+ stages) are maximal in the upstream level (estimated at 0.07); it is half as high in the second level, and so on, until the most downstream level (patch 1). This decreasing gradient is observed because the habitat conditions in the first levels of the river networks (e.g., oxygen, bed [gravel], temperature) result in higher survival rates of young stages. Inversely, adult survival rates increase from up to downstream: The maximum (of 0.4) is reached in patch 1 and is divided by 2 for each transition to the next upstream level (0.2 for patches 2 and 9, 0.1 for patches 3, 4,

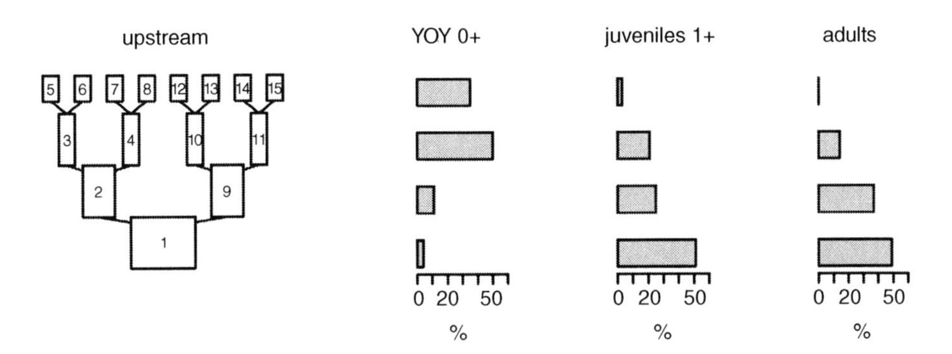

Figure 11.2. Spatial structure of the brown trout population. (Left) Arborescent river network considered for the study: 15 streams organized in four levels from upstream to downstream. Pollution takes place in patch 1 (downstream). (Right) Simulated spatial repartition of the population (end of May) in the unpolluted scenario (average terminal frequencies computed from 100 simulations during 50 years).

etc.). Annual survival rates of Juv 1+ individuals are maximal in the third level (0.4 in patches 2 and 9) and are divided by 2 for the following levels (0.2 in levels 4 and 2, and 0.1 in level 1 upstream). Such patterns of heterogeneity in survival rates between river stretches are explained by the fact that feeding and growth areas for the older age classes are situated downstream in river networks. Fecundity rates (number of eggs per female) are size dependent (Maisse and Baglinière, 1991). We use the relationship between the size and the fecundity established by Gouraud et al. (1998) for a population in the river Oir in Normandie:

$$\text{fecundity} = 5.2719 \text{ size (mm)} - 689.12 \tag{11.1}$$

Moreover, the size of breeders depends on the spatial location in the river network of each individual: Trout living in downstream levels are bigger and so produce more eggs than do trout living in more upstream levels (Baglinière et al., 1987). Based on data from the river Scorff (Bretagne) (Baglinière et al., 1987, 1994), from the river Oir (Normandie) (Gouraud et al., 1998), and from other French rivers (Maisse and Baglinière, 1991), we estimated average sizes for the females: 200 mm in the upstream level, 230 mm and 270 mm in the second and third levels, and 300 mm in the more downstream location (patch 1). The fertilization rate is assumed to be 0.9 (Maisse and Baglinière, 1991), and the survival rate in nests (from fertilized egg stage to newly emerged fry stage) to be 0.5 (J.L. Baglinière, personal communication; Liebig, 1998).

The migratory patterns are the same as in Chaumot et al. (2003a). Spawning migration occurs only in the upstream direction. Breeders living in patch 1 mainly (68%) migrate to the third level for spawning, 14% to the second, and 12% to the first level. Adults from patches 2 and 9 (level 3) mostly (80%) spawn in level 1 (upstream) and 14% in level 2; 83% of breeders from level 2 migrate to the level 1, while trout from the upstream level stay in their location. Symmetrically, 50% of the juveniles born in levels 1, 2, and 3 migrate to the patch 1 in spring. Twenty-five percent of the young trout produced in levels 1 and 2 move to the third level (patches 2 and 9), and 20% of the juveniles born in the upstream level swim down to the second level.

Ecotoxicological assumptions

Only chronic pollution is considered in this study (low concentrations of cadmium throughout the year). Cadmium effects on demographic parameters are expressed through dose–response curves. We refer to chronic bioassays from the literature performed on a long-term scale (one year of exposure; Benoit et al., 1976; Brown et al., 1994). We process nonlinear regressions with bioassays results in order to estimate dose–response curves. For that, we used the function nls (nonlinear least squares) from the Stat package in R Software (R Development Core Team, 2004). Brown et al. (1994) for brown trout *Salmo trutta*, and Benoit et al. (1976) for brook trout *Salvenilus fontinalis*, described the insensitivity of young stages to chronic cadmium concentrations (at least up to 30 µg/L for brown trout). Hatchability and annual survival rate during the first year (YOY 0+ survival rate) are thus not affected by chronic exposure. Fecundity and annual survival rate of older stages are reduced by such a chronic exposure. Dose–response curves are the same as in Chaumot et al. (2002). The lack of data for brown trout led us to estimate the curve of adult fecundity reduction from data collected for brook trout (Benoit et al., 1976). Thus, we refer to the dose response for relative fecundity, rf:

$$\text{rf}(\text{concentration}) = \exp[0.237 - 1.52 \log(\text{concentration})] \div (1 + \exp[0.237 - 1.52 \log(\text{concentration})]), \quad (11.2)$$

where "log" is the natural log. This equation supplies a half-maximal effective concentration (EC_{50}) of 1.3 μg/L, and the fecundity is reduced to almost zero for concentrations greater than 10 μg/L [rf(10) = 4%], which means fecundity at this concentration is about 4% of fecundity at zero concentration. According to dose–response curves in Chaumot et al. (2002), we assume that the survival of Juv 1+ and adults is not affected by cadmium concentrations less than 10 μg/L. For instance, Brown et al. (1994) report an EC_{50} value of 30 μg/L for the reduction of annual survival of brown trout weighing approximately 100 g. Here, we consider chronic pollution occurring in the most downstream stretch of river (patch 1) with cadmium concentrations from 0 to 10 μg/L. Thus, the only effect of contamination on demography is the reduction in fecundity of trout living in patch 1 (though they can spawn anywhere in the river network due to upstream migration).

To include such dose–response curves, we built the model making certain restrictive ecotoxicological assumptions. First, we assumed that the contamination of patch 1 is homogeneous and that the reduction of fecundity (the deficit in eggs) due to the pollutant is deterministically calculated using the dose–response function. Moreover, we assumed that the pollution is constant throughout the year and every year. Second, we assumed that the exposure occurred only in the stretch of river where the trout spend most of the year; in other words, we did not take into account the break of exposure during the spawning event. We accept this simplifying assumption by considering that breeders are absent from downstream areas at most two months in the year. Third, we assumed that the exposure of an individual during a given year does not influence demographic parameters relative to this individual during the next year. In other words, the exposure to toxicant gives rise to damages during the year of exposure, and the history of exposure of individuals is reinitialized in the beginning of each time step of the demographic process.

Lastly, the dispersal pattern could be disturbed by the presence of the metal (Woodward et al., 1995; Hansen et al., 1999). Unfortunately, very little is known about these potential toxic-induced behaviors (avoidance or attraction). We globally tested the possible impact of these migratory perturbations by comparing two versions of the model: One considers a deterministic model for the upstream spawning migration (the coefficient of variation for dispersal [CV] is null in the dialog box relative to stochasticity), and the other performs the simulations according to a stochastic model for dispersal (CV = 1).

Population model

Demographic process

Because of spatial heterogeneity and because demographic processes take place within each of the 15 stretches of river, we define 15 "populations" and 15 associated stage matrices. Each of these matrices describes transitions among the three age classes from year to year in order to translate the life cycle (figure 11.1). These matrices are Leslie matrices (Caswell, 2000): Survival rates between age classes are on the sub-diagonal, adult survival rate on the last row and last column (because adults survive after breeding), and one fecundity term on the first row and third column (only adults produce young trout each year). Our model includes only females. While survival rates directly correspond to the values discussed above, the calculation of the fecundity term is more complex. This term represents the number of YOY 0+ females in spring produced by each adult female present in the patch one year earlier. The census happens at the end of May, and the spawning event at the end of November (figure 11.1). Thus, the fecundity element in the stage matrix is the product of adult survival during six months (estimated by the square root of the annual survival rate of adult in the considered patch), the fecundity rate (calculated from the mean size of females in the patch; see equation 11.1), the fertilization rate (0.9), the survival rate in nests (0.5), and the sex ratio (0.5). In patch 1, where the contaminant is present in polluted scenarios, this fecundity term is multiplied by the dose–response curve (equation 11.2).

Density dependence

According to the observations of Héland (1991), the density-dependent regulation of recruitment is characterized by a carrying capacity in each stretch of river. We used a ceiling function to model this process. The carrying capacity is defined in each of the 15 stretches of river as a maximum number of Juv 1+ individuals. We approximated these thresholds according to field estimates of densities (Gouraud et al., 1997, 1998), which yielded 50 one-year-old females in level 1 (upstream), 100 in level 2, 200 in level 3, and 400 in level 4 (downstream).

Migratory process

The dispersal matrix is applied after the demography in RAMAS Metapop. Adults migrate upstream for spawning and come back to their resident area in the river network. Since the fecundity rates depend on the resident and not on the spawning locations, we simply translated the upstream breeding migration as a movement of young trout. Thus, the dispersal matrix concerns only the first age class in the model (this is specified in the "Relative dispersal" parameter in the Stages dialog). During one step of simulation, the stage matrices are first applied to the population vector of the previous year (in spring), which supplies the number of YOY 0+, juveniles, and adults present in spring for the next year. Afterward, the dispersal matrix "dispatches" the young trout in the network according to the upstream pattern of migration of breeders. Then we encountered difficulty in modeling the downstream migration of juveniles. The dispersal modeling is not totally stage specific in the software: It offers only the possibility of homothetic patterns of migration for the different stages (with stage-specific coefficients). This method is not suitable here for modeling the upstream spawning migration and the downstream movements of juveniles, which are totally disconnected. To solve this problem, we artificially integrated translocations of juveniles using "population management actions." This has to be done cautiously because the coefficients of these translocations are not the direct migration rates, because the different population managements are taken into account successively. For instance,

50% of the juvenile trout in patch 5 migrate to patch 1 and 25% to patch 2. To translate these rates, we introduced in the model a first translocation of 50% of the juveniles of patch 5 to patch 1, and a second translocation of 50% from patch 5 to patch 2, because the proportion is then calculated on the remaining individuals after the first translocation.

Stochasticity and statistical aspects

We have no data to quantify the stochasticity associated with demographic and migratory rates. Nevertheless, we were interested in evaluating the bias introduced by the use of deterministic models, such as those we previously developed (Chaumot et al., 2002, 2003a, 2003b). In order to examine this influence on endpoints employed for decision making in ecotoxicology, we compared the outputs from a totally deterministic release of the model (demographic stochasticity is not used), with the same outputs if we consider binomial sampling errors for demographic transitions. Alternatively, we tried to analyze the potential impacts of toxic-induced perturbations of migratory patterns by looking at the impact of pollution whether or not stochasticity for dispersal rates is taken into account (parameter CV is 0 or 1). For this comparison, we performed Kolmogorov-Smirnov tests under "Comparison of results" in RAMAS. The determination of the different dose–response curves reported in the results is completed by nonlinear regression using the Stat package in R Software (R Development Core Team, 2004).

Results

Following the unpolluted scenario, the model (considering only demographic stochasticity) yields realistic population features compared to observed populations in the field (Maisse and Baglinière, 1991; Gouraud et al., 1998). Abundances reach equilibrium values with a good biological likelihood (top curve in figure 11.3). The spatial distribution of trout in the river network is stage specific (figure 11.2): YOY gather in the upstream levels; Juv 1+ and adult trout live mainly in the two downstream levels. This is totally consistent with the ecological requirements of the three different stages of trout: Upstream stretches of river in the network are nursery areas (gravel, oxygenized waters), while more downstream stretches offer good habitat conditions for feeding and growing. Note that if we remove density-dependent regulations (ceiling), all local subpopulations are growing, although each of them presents a stage matrix with a growth rate (dominant eigenvalue) less than 1. This is explained by the stage- and direction-specific dispersal, which allows the different age classes living in levels of the river network with good age-specific conditions of habitat. For instance, the weak values of YOY 0+ survival in downstream patches does not influence the demography of these patches, because trout living there (mostly Juv 1+ and adults) have spent their first year in upstream levels (where the survival of young stages is high). The isolated analysis of the stage matrix is thus inefficient to translate the dynamics of each local subpopulation. To resume the analysis under unpolluted conditions, outputs of the stochastic model draw a likely description of a brown trout population living in a river network. Therefore, our work aims at analyzing patterns of perturbation of these population features under different pollution scenarios.

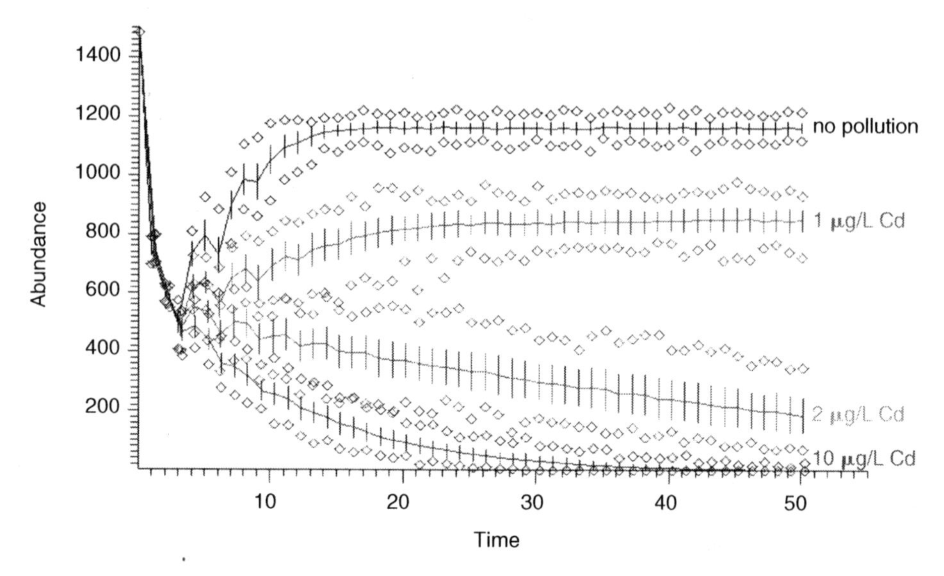

Figure 11.3. Trajectory summary: Evolution of the abundance for different intensities of cadmium pollution. Abundance is the total number of one-year-old juveniles (Juv 1+) and adults in the river network. The curves are completed using 100 replications. Only demographic stochasticity is considered (dispersal processes are deterministic). This figure is a snapshot from the Comparison of Results package in RAMAS (see Results | Trajectory summary): Lines report mean abundances, bars stand for standard deviation, and points are minimum and maximum values.

First, we present two examples of these modifications. Density-dependent models supply abundance values, which can constitute good ecotoxicological endpoints for environmental management. Here, we see that increasing concentrations of pollution badly affect the abundance of trout in the network, and contamination can lead to extinction (figure 11.3). We built a "population" dose–response curve in order to underline this decreasing pattern. To do this, we linked the mean of the total terminal abundance (juveniles and adults) of simulation during 50 years with the logarithm of the emitted concentration of cadmium (figure 11.4). Then, classical ecotoxicological parameters such as EC_{50} (concentration leading to half of the maximum effect caused by the pollutant) can be determined using this dose–response curve and defined on a population level. Note that such a dose–response function has to be analyzed cautiously because it depends on the duration of the simulations; for example, the curve obtained in the year 20 (figure 11.3) would be shifted to the right compared to the one for year 50 (figure 11.4).

Pollution also affects the age structure of the population. Based on the final abundances of the different age classes (in the RAMAS Metapop program, see Results → Final Stage Abundances), we observed aging in the population (figure 11.5). This can be understood as a consequence of the toxicant-induced reduction in fecundity of the breeders in patch 1. That underlines the key role played by the adults of the downstream stretch of river in the renewal of the global population in the network.

Second, we examined how demographic stochasticity can affect ecotoxicological endpoints. We consider a pollution amount of 3 µg/L Cd in patch 1. Simulations

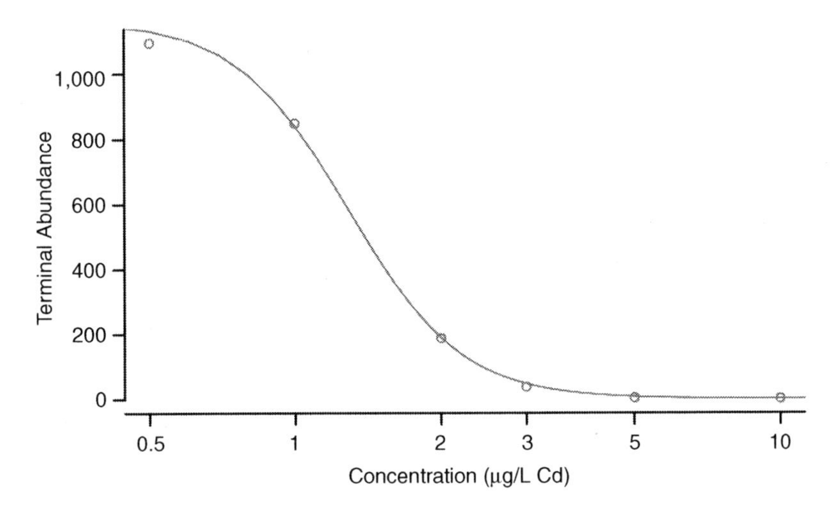

Figure 11.4. "Population" dose–response curve for cadmium. Terminal average abundances (total number of one-year-old juveniles (Juv 1+) and adults) are estimated for different intensities of pollution (logarithmic scale). These abundances correspond to final values reported in figure 11.3 (time step = 50; demographic stochasticity only).

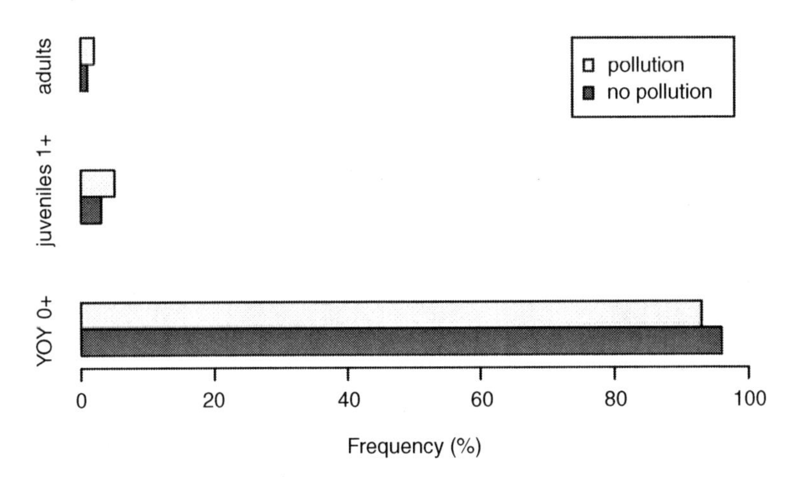

Figure 11.5. Age structure (in spring) simulated in a clean river network and in a cadmium-polluted one (3 µg/L Cd in patch 1). Frequencies are calculated from the final-stage abundances obtained in year 50 with the deterministic version of the model (stochasticity is not used in demographic and dispersal processes).

based on deterministic assumptions (demographic stochasticity is not used) end with a population abundance of 98 trout (juveniles and adults) in the network. This value is reached before the 50th year (data not shown), so we can consider this value an equilibrium. Running simulations with demographic stochasticity supplies average terminal abundance but also information on the potential future trajectory of the population (see Results → Interval Extinction Risk and Results → Terminal Extinction Risk in RAMAS Metapop). The Interval Extinction Risk section shows that the probability of abundance below 98 trout during a new run of 50 years is 0.99, and that the expected minimum abundance is 30.7 fish (data not shown). The "terminal extinction risk" curve tells us that the probability to obtain a final abundance below 98 trout 50 years later is 0.98 (figure 11.6). It appears that endpoints based on deterministic assumptions underestimate the endangering effect of pollution on population level.

Third, we aimed at disclosing the effect of migratory disruptions induced by the pollutant. For this, we compared the interval extinction risk curves built for increasing concentrations of pollutant in two distinct cases: one in which the dispersal model is deterministic (CV = 0), and the other in which stochasticity is taken into account in the migratory processes (CV = 1). We made this comparison using two dose–response curves, where response is the expected minimum abundance (provided by the program in Results → Interval Extinction Risk) for increasing cadmium concentrations (figure 11.7). Random variations of migratory rates during the upstream migration of breeders amplify the negative effects of contamination (shift of the dose–response curve down). We report the interval extinction risk curves for a concentration of 3 μg/L considering two intensities of stochasticity in dispersal rates (CV = 0 and CV = 1; figure 11.8). When we integrated random variations in migratory rates, the expected minimum abundance was reduced from 30.7 to 5.7 trout, and the extinction risk (intercept of the curve with the vertical axis) increased from 0.01 to 0.32. The difference between the two curves is highly significant according to the test performed in RAMAS Metapop under Comparison of Results (Kolmogorov-Smirnov test statistic: $D = 0.66$; p-value $< 10^{-4}$).

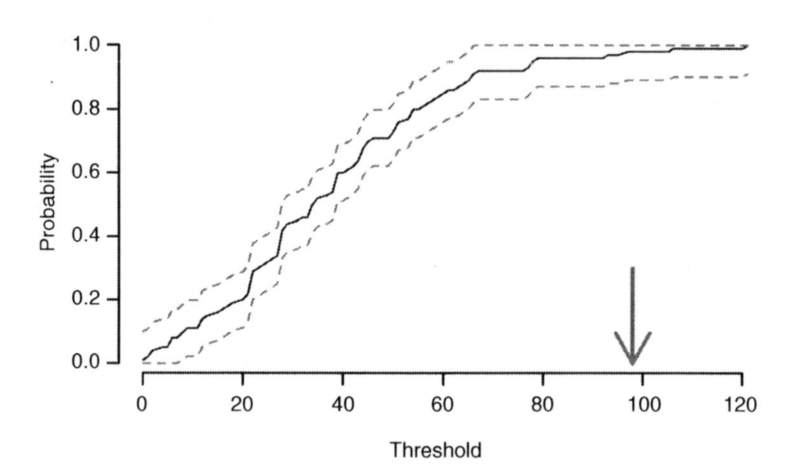

Figure 11.6. Terminal extinction risk: Output from the model considering demographic stochasticity in a polluted scenario (3 μg/L Cd in the patch 1) based on 100 simulations during 50 years (see Results | Terminal Extinction Risk). The figure reports the probability (and the associated 95% confidence interval, dashed lines) that the total number of one-year-old juveniles (Juv 1+) and adults will be less than a given threshold after 50 years. The arrow indicates the "equilibrium" value of abundance (value at year 50) supplied by the deterministic model for the same scenario of pollution.

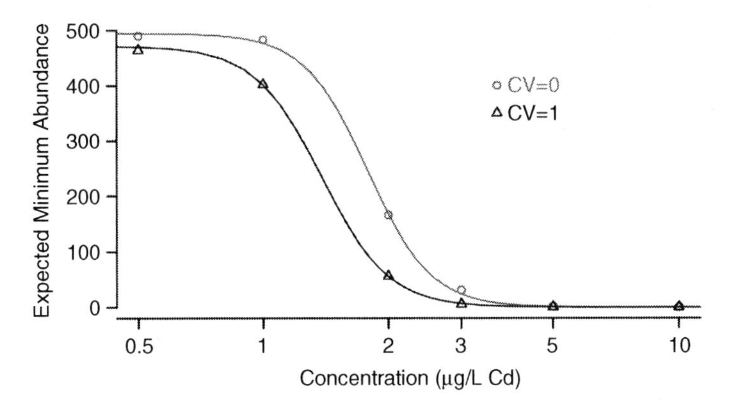

Figure 11.7. Effect of stochasticity in dispersal rates for increasing cadmium pollution scenarios: CV = 0 (circles), the model considers only demographic stochasticity; CV = 1 (triangles), the model takes into account both demographic and dispersal stochasticity (CV = 1 for dispersal rates of the upstream spawning migration). The expected minimum abundance (i.e., expected minimum total number of one-year-old juveniles (Juv 1+) and adults during the next 50 years) is supplied under Results | Interval Extinction Risk in RAMAS Metapop. All values are calculated using 100 replications of 50 years.

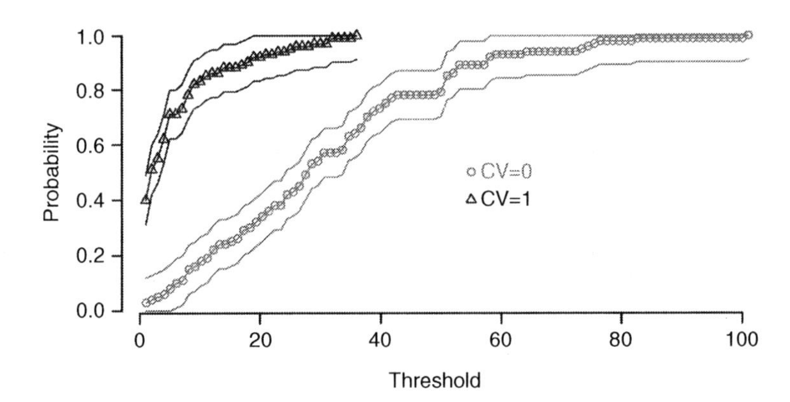

Figure 11.8. Interval extinction risk: Effect of stochasticity in dispersal rates considering a pollution scenario of 3 µg/L Cd in patch 1. CV = 0 (circles), demographic stochasticity only; CV = 1 (triangles), both demographic and dispersal stochasticity (CV = 1 for dispersal rates of the upstream spawning migration). Data points are the probability (and the associated 95% confidence interval) that the total abundance (one-year-old juveniles [Juv 1+] and adults) will be less than a given threshold in the following 50 years (see Results | Interval Extinction Risk). These curves were completed using 100 replications of 50 years.

Discussion

The model built with RAMAS Metapop appears to be in good agreement with the dynamics of a trout population. Abundances, spatial distribution, and age structure provide a likely description of a resident trout population living in a river system in western France. We must emphasize here the improvement of the biological realism observed in this study for the values of densities compared to the previous density-dependent models we developed (Chaumot et al., 2003b). That is because in RAMAS Metapop, the density-dependent functions are defined with carrying capacities, which directly determine the order of magnitude of the simulated equilibrium densities, while in our previous Leslie approach, the density dependence is expressed more mechanistically with density-dependent survival rates.

RAMAS Metapop clearly allows the modeler to take into account spatial heterogeneity of demographic and toxicological processes. Nevertheless, one difficulty arises in our case due to the existence of two totally disconnected age-specific migratory patterns. The modeling of this situation is not easily addressed in the software because of the mismatch between the metapopulation philosophy of the software and the running of the trout population presented here. In this study, we examined a population that is spatially structured but that is not a metapopulation *sensu stricto* (Levin, 1970; Hanski and Simberloff, 1997). Here, dispersal mechanisms highly influence the dynamics of each subpopulation each year (the isolated analysis of local demographic matrices is totally inefficient to translate each local dynamics); on the other hand, in a metapopulation *sensu stricto*, dispersal slightly influences the dynamics in each living area at a short time scale. In our case study, spatial structure corresponds mainly to an age structure; dispersal is thus associated with the transitions between age classes. Dispersal is stage and direction specific, which allows the different age classes living in levels of the river network with good age-specific conditions of habitat. Because the software offers only the possibility of a common qualitative pattern of dispersal for all age classes with homothetic coefficients, from a metapopulation point of view, the demographic dynamics of local populations is equivalent in the different spatial locations. In other words, in a metapopulation *sensu stricto*, local dynamics are independent at a short time scale; this independence implies that reproduction, recruitment, and survival processes take place in each local population. In order to extend the use of RAMAS Metapop to the case of a well-connected set of subpopulations, where spatial fragmentation is mainly related to the segregation of age classes, we used artificial population management actions in the model building, which allowed us to define age-specific migratory behaviors.

When we compare the assessment of the impact of the pollution on the population features, this study confirms the trends observed with analytical deterministic Leslie models (Chaumot et al., 2003a, 2003b): decrease in densities and aging for increasing concentrations of contaminant. Thus, taking stochasticity into account in the definition of population endpoints does not qualitatively modify the expected consequences of pollution. But with a more quantitative point of view, we show here that deterministic endpoints—such as equilibrium abundances—seem to seriously underestimate the endangering effect of pollution on population levels. The analysis of stochastic features such as extinction risk appears to be of broad interest to avoid this pitfall.

In this study, we consider effects of cadmium only on fecundity. But for other concentrations, other species, or other compounds, this will not be adequate. Modeling can be applied by adding other dose–response curves to express these supplementary perturbations. Likewise, more complex exposure models may amend the simplifying ecotoxicological assumptions formulated in this case study. Modeling by simulation (as in RAMAS Metapop) would still be achievable, but this would not necessarily allow us to understand outputs that may emerge as a result of the interactions among these complex input functions. Indeed, the population level would intricately integrate levels of exposure, spatial repartition of pollutant, spatial distribution of individuals, weight of each age class in the demography, thresholds, and shapes of dose–response curves. In a previous contribution (Chaumot et al., 2003a), we showed that using analytical models (e.g., Leslie models) is a way to avoid this drawback. We were able to understand where and which vital rates or age classes are involved in explaining the effect of pollution on the population. Here, we emphasize the complementarity of the two modeling strategies (simulation/analytical models), which can give rise on one hand to more realism and on the other hand to more robustness, more understanding.

Finally, we show here that random variations in the dispersal pattern adversely alter population dynamics with the increased extinction risk related to pollution perturbations. Even if the link between such stochastic variations and potential toxicant-induced perturbations is quite vague, these results corroborate the conclusion of a previous contribution focused on the potential effect of migratory disruptions due to pollution (Chaumot et al., 2003b). Unfortunately, little is known about such contaminant effect, and very few experiments have been performed (Woodward et al., 1995; Hansen et al., 1999). Therefore, increased efforts are necessary to develop knowledge relative to toxicant-induced spatial behaviors and to integrate such effects in defining environmental quality criteria.

References

Baglinière, J.L., Maisse, G., Lebail, P.Y., and Prévost, E. 1987. Dynamique de la population de truite commune (Salmo trutta L.) d'un ruisseau breton (France). II—les géniteurs migrants. *Acta Oecologica—Oecologia Applicata* 8:201–215.

Baglinière, J.L., Maisse, G., Lebail, P.Y., and Nihouarn, A. 1989. Population dynamics of brown trout, Salmo trutta L., in a tributary in Brittany (France): spawning and juveniles. *Journal of Fish Biology* 34:97–110.

Baglinière J.L., Prévost E., and Maisse, G. 1994. Comparison of population dynamics of Atlantic salmon (*Salmo salar*) and brown trout (*Salmo trutta*) in a small tributary of the River Scorff (Brittany, France). *Ecology of Freshwater Fish* 3:25–34.

Benoit, D.A., Leonard, E.N., Christensen, G.M., and Fiandt, J.T. 1976. Toxic effects of cadmium on three generations of brook trout (*Salvenilus fontinalis*). *Transactions of the American Fisheries Society* 4:550–560.

Brown, V., Shurben, D., Miller, W., and Crane, M. 1994. Cadmium toxicity to rainbow trout *Oncorhynchus mykiss* Walbaum and brown trout *Salmo trutta* L. over extended exposure periods. *Ecotoxicology and Environmental Safety* 29:38–46.

Buhl, K.J., and Hamilton, S.J. 1991 Relative sensitivity of early life stages of Arctic grayling, coho salmon, and rainbow trout to nine inorganics. *Ecotoxicology and Environmental Safety* 22:184–197.

Caswell, H. 1996. Demography meets ecotoxicology: untangling the population level effects of toxic substances. Pages 255–292 in M.C. Newman and C.H. Jagoe (eds.), *Ecotoxicology: A Hierarchical Treatment*. Lewis Publishers, Boca Raton, Florida.

Caswell, H. 2000. *Matrix Population Models*, 2nd ed. Sinauer Associates, Sunderland, Massachusetts.

Charles, S., Bravo de La Parra, R., Mallet, J.P., Persat, H., and Auger, P. 1998. Population dynamics modelling in an hierarchical arborescent river network: an attempt with *Salmo trutta*. *Acta Biotheoretica* 46:223–234.

Chaumot, A., Charles, S., Flammarion, P., Garric, J., and Auger, P. 2002. Using aggregation methods to assess toxicant effects on population dynamics in spatial systems. *Ecological Applications* 12:1771–1784.

Chaumot, A., Charles, S., Flammarion, P., and Auger, P. 2003a. Ecotoxicology and spatial modeling in population dynamics: an illustration with brown trout. *Environmental Toxicology and Chemistry* 22:958–969.

Chaumot, A., Charles, S., Flammarion, P., and Auger, P. 2003b. Do migratory or demographic disruptions rule the population impact of pollution in spatial networks? *Theoretical Population Biology* 64:473–480.

Elliott, J.M. 1994. *Quantitative Ecology and the Brown Trout.* Oxford University Press, New York.

Gouraud, V. 1999. *Etude de la dynamique de populations de truite commune (Salmo trutta L.) à l'aide d'un modèle déterministe; application sur un bassin bas-normand et sur un bassin pyrénéen.* Ph.D. thesis, Ecole Nationale du Génie Rural des Eaux et Forêts, Paris.

Gouraud, V., Baglinière, J.L., Ombredane, D., Sabaton, C., Hingrat, Y., and Marchand, F. 1997. Caractéristiques biologiques de la population de truites (*Salmo trutta* L.) de la rivière Oir (Manche, Basse-Normandie) en 1996–1997: 1. Population en place sur le bassin. 2. Dévalaison des juvéniles au printemps et origine géographique des migrants. 3. Population de géniteurs migrants et potentiel reproducteur sur le ruisseau de la Roche. 4. Premières données sur l'exploitation par pêche à la ligne. Report INRA/EDF-DER. Laboratoire d'écologie aquatique, Institut National de la Recherche Agronomique, Rennes, France.

Gouraud, V., Baglinière, J.L., Ombredane, D., Sabaton, C., Hingrat, Y., and Marchand, F. 1998. Caractéristiques biologiques de la population de truites (*Salmo trutta* L.) de la rivière Oir (Manche, Basse-Normandie) en 1997–1998: 1. Population en place sur le bassin. 2. Caractéristiques de la population de géniteurs de truites se reproduisant sur le ruisseau de la Roche et estimation du potentiel reproducteur. 3. Quelques données sur l'exploitation par pêche à la ligne. Report INRA/EDF-DER. Laboratoire d'écologie aquatique, Institut National de la Recherche Agronomique, Rennes, France.

Hansen, D.F., Woodward, J.A., Little, E.E., Delonay, A.J., and Bergman, H.L. 1999. Behavioral avoidance: possible mechanism for explaining abundance and distribution of trout species in a metal-impacted river. *Environmental Toxicology and Chemistry* 18:313–317.

Hanski, I., and Simberloff, D. 1997. The metapopulation approach, its history, conceptual domain, and application to conservation. Pages 5–26 in I.A. Hanski and M.E. Gilpin (eds.), *Metapopulation Biology.* Academic Press, San Diego, California.

Héland, M. 1991. Organisation sociale et territorialité chez la truite commune immature au cours de l'ontogenèse. Pages 121–149 in J.L. Baglinière and G. Maisse (eds.), *La truite, biologie et écologie.* INRA Editions, Paris.

Levin, R. (1970) Extinction. Pages 77–107 in M. Gesternhaber (ed.), *Some Mathematical Problems in Biology.* American Mathematical Society, Providence Rhode Island.

Liebig, H. 1998. *Etude du recrutement de la truite commune (Salmo trutta L.) d'une rivière de moyenne montagne (Pyrénées ariègeoises, 09); effets de la gestion par éclusées d'une centrale hydroélectrique; approches in situ et expérimentales.* Ph.D. thesis, Institut National Polytechnique de Toulouse, Toulouse, France.

Maisse, G., and Baglinière, J.L. 1991. Biologie de la truite commune (*Salmo trutta* L.) dans les rivières françaises. Pages 22–45 in J.L. Baglinière and G. Maisse (eds.), *La truite, biologie et écologie.* INRA Editions, Paris.

Ovidio, M. 1999. Cycle annuel d'activité de la truite commune (*Salmo trutta* L.) adulte: étude par radio-pistage dans un cours d'eau de l'Ardenne belge. *Bulletin Français de la Pêche et de la Pisciculture* 352:1–18.

R Development Core Team. 2004. *R: A Language and Environment for Statistical Computing.* R Foundation for Statistical Computing, Vienna. Available: www.R-project.org.

Schuck, H.A. 1945. Survival, population density, growth, and movement of the wild brown trout in Crystal Creek. *Transactions of the American Fisheries Society* 73:209–230.

Sherratt, T.N., and Jepson, P.C. 1993. A metapopulation approach to modeling the long-term impact of pesticides on invertebrates. *Journal of Applied Ecology* 30:696–705.

Solomon, D.J., and Templeton, R.G. 1976. Movements of brown trout *Salmo trutta* L. in a chalk stream. *Journal of Fish Biology* 9:411–423.

Sorensen, E.M.B. 1984. *Metal Poisoning in Fish.* CRC Press, Boca Raton, Florida.

Spromberg, J.A., John, B.M., and Landis, W.G. 1998. Metapopulation dynamics: indirect effects and multiple distinct outcomes in ecological risk assessment. *Environmental Toxicology and Chemistry* 17:1640–1649.

Woodward, D.F., Hansen, J.A., Bergman, H.L., Little, E.E., and DeLonay, A.J. 1995. Brown trout avoidance of metals in water characteristic of the Clark Fork River, Montana. *Canadian Journal of Fisheries and Aquatic Sciences* 52: 2031–2037.

12

Using a Spatial Modeling Approach to Explore Ecological Factors Relevant to the Persistence of an Estuarine Fish (*Fundulus heteroclitus*) in a PCB-Contaminated Estuary

DIANE E. NACCI
STEVEN WALTERS
T. GLEASON
W. R. MUNNS, Jr.

Although population (or any ecological) models do not reflect ecological realism completely, they provide an opportunity to systematically explore biological and ecological factors that affect population vitality, condition, or persistence. Their role in ecological risk assessment is further developed in this application of RAMAS GIS where models and data from a series of research studies were used to generate spatial projections reflecting effects of short-term or chronic, multigenerational exposure to toxic pollutants on resident fish populations. Specifically, real and simulated site-specific genetic, demographic, and toxicological information was incorporated into spatial population projections to explore the potential contributions to population persistence of a) connectivity among subpopulations (e.g., source–sink dynamics), and b) adaptive demographic variation among subpopulations.

Populations of a small, nonmigratory fish species, *Fundulus heteroclitus* (mummichogs), resident to a highly contaminated estuary are being studied to improve predictions of population-level effects associated with exposures to dioxin-like compounds (DLCs), including the most toxic polychlorinated biphenyl (PCB) congeners. The study site, New Bedford Harbor (NBH), Massachusetts, USA (figure 12.1), is an urban estuary highly contaminated with PCBs (e.g., Pruell et al. 1990; Lake et al. 1995; Nelson et al. 1996), with highest concentrations in its northern reaches, which have been designated as a Superfund site. Inferences based on PCB concentrations in sediment cores suggest that since the 1950s sediments at the Superfund site have been at levels toxic to this species and other biota (Monosson 1999/2000; Nacci et al. 2002a). Although the entire harbor is contaminated, there is a steep gradient of PCB concentrations, decreasing from north to south. PCB discharge ceased in 1976

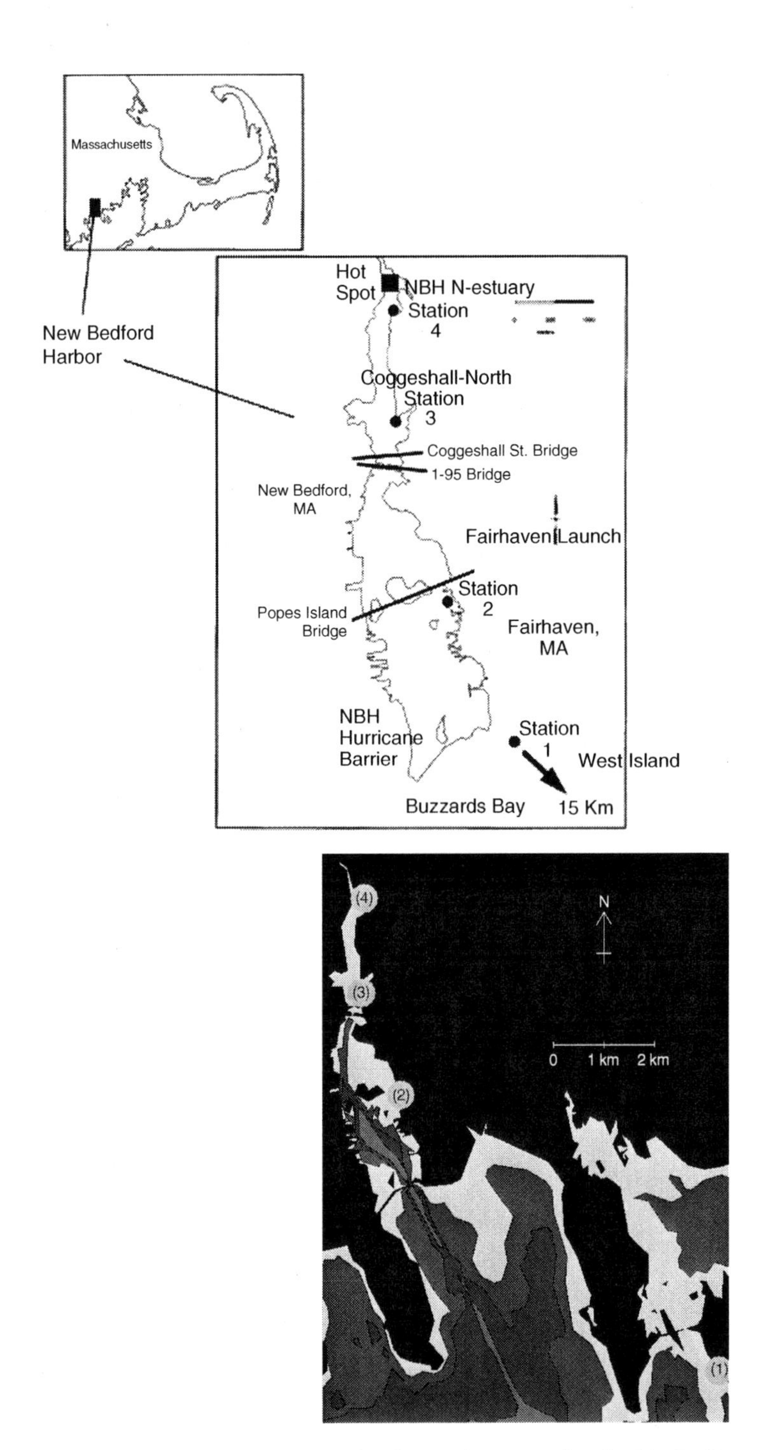

Figure 12.1. Collection stations for *Fundulus heteroclitus* associated with the New Bedford Harbor, Massachusetts, PCB-contaminated Superfund site (adapted from Munns et al. 1997), including water depth contours as represented in RAMAS GIS (bottom).

Table 12.1. PCB concentrations for sediments and toxic equivalency concentrations (TECs) for livers of *Fundulus heteroclitus* collected from four stations associated with New Bedford Harbor (MA, USA).

Source population		Sediment[a]	Fish Liver[b]
Site	Location	PCBs (μg/g dry)	TEC (pg/g dry)
41.3504N 70.4942W	West Island	165	5.72
41.637N 70.907W	Fairhaven Launch	874	132
41.39181N 70.5520W	Coggeshall Street north	3762	543
41.4017N 70.5448W	NBH northern estuary	22,666	1,560

[a] Nacci et al. (2002b).
[b] Black et al. (1998a), based on Walker and Peterson (1991).

(U.S. EPA 2007), but mummichogs from NBH are still highly contaminated with PCBs. When reported as dioxin toxic equivalency concentrations (TECs), PCB concentrations in livers of mummichogs ranged from 1,560 to 132 pg TEC/g dry weight for residents of the northernmost to southernmost NBH stations, and 6 pg/g for West Island reference station fish (Black et al. 1998a; table 12.1). To put this into perspective, the highest concentrations are over 75-fold more contaminated than the European Commission limit for DLCs in fish tissue for human consumption, about 20 pg TEC/g dry weight (Regulation EC 2375/2001, Royal Commission on Environmental Pollution 2005).

With an awareness of the NBH study area, Munns et al. (1997) developed a stage-classified matrix model for mummichogs to project population effects associated with DLC exposures. One set of models was parameterized based on laboratory-collected data (Black et al. 1998a) from mummichogs resident to NBH and nearby less-contaminated sites. These model projections suggested that exposure to PCBs at high but environmentally realistic levels for NBH could produce adverse population-level effects. However, a survey of NBH biota (Mitchell and Oviatt 1996) revealed an abundance of mummichogs of all size classes. This apparent disparity between matrix model projections and field observations suggested that simple model projections might not accurately reflect NBH mummichog population dynamics, and suggested the value of modeling approaches accounting for additional ecological complexities.

Fundulus population matrix models

The mummichog models initially used to project DLC effects (Munns et al. 1997) were constructed as simple representations of population dynamics (i.e., density independent, and without immigration or emigration) that could incorporate the toxicological responses of key developmental stages (figure 12.2). Demographic parameters were derived from the literature and from reference populations held in the laboratory, and were assumed to represent those of unstressed populations (Munns et al. 1997). Initial parameterization produced a reference model with a positive population growth rate (equal to 1.12). To produce a model against which relative toxicological effects could be evaluated, this reference model was modified (by reducing juvenile survival)

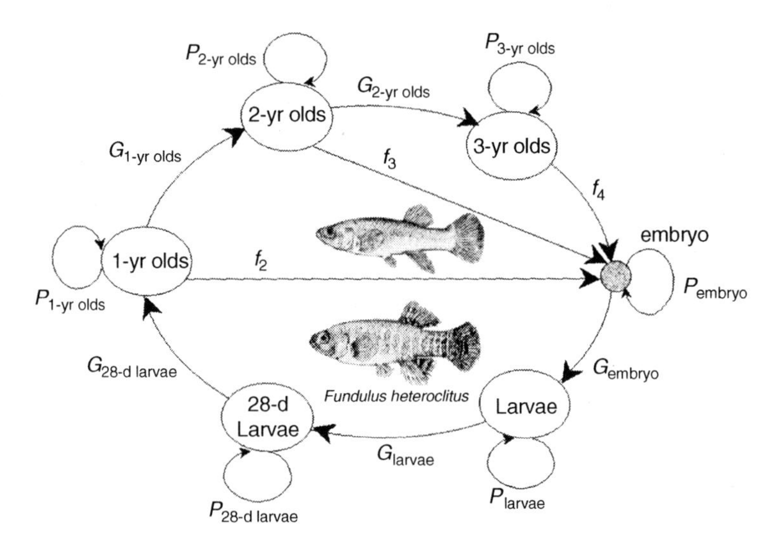

Figure 12.2. Life cycle representation of the stage-classified population model for *Fundulus heteroclitus*, where P_i is the probability of surviving and remaining in the *i*th stage, G_i is the probability of surviving and growing into stage *i*+1, and *Fi* is fertility of stage *i*.

to produce a nongrowing (stable) "base model" with population growth rate equal to 1.0 (Munns et al. 1997). To project toxicological effects, data for survivorship and reproduction were collected in five-week laboratory studies using fish sampled from NBH stations varying in PCB contamination (Black et al. 1998a). Data were used to modify transition rates in the base model and develop models representing subpopulations subject to varying, long-term exposures to DLCs. Following this same strategy, we used data from naive fish exposed in the laboratory to PCBs (Black et al. 1998b) to produce models representing populations subject to short-term exposure to DLCs of similar magnitude to the NBH sites.

Connectivity

Mummichogs are known to exhibit strong site fidelity, which has been estimated to be about 2 km of shoreline based on field observations (Lotrich 1975) and molecular genetic techniques (Brown and Chapman 1991). However, some studies suggest that the sedentary nature of this species may be overestimated (e.g., Sweeney et al. 1998; Mulvey et al. 2003). In addition, movement of mummichogs is seasonally variable (e.g., Bigelow and Schroeder 1953; Abraham 1985) and site dependent (e.g., as reviewed in Sweeney et al. 1998). Of particular concern in this study area, rocky shorelines or deep fast-moving currents (e.g., those associated with the NBH hurricane barrier) may act as hard barriers to dispersal (Lotrich 1975; Abraham 1985).

This information about movement patterns suggests that mummichogs resident to the NBH upper harbor might be isolated from less contaminated regions, including the lower and outer harbor (Buzzards Bay). Furthermore, the extreme toxicity of the upper harbor might limit effective migration (e.g., Wirgin and Waldman 1998;

Bellafiore and Anderson 2001; Mulvey et al. 2003). The potential for site fidelity and/or the presence of partial barriers to dispersal suggests that metapopulation theory would be a useful framework for investigating the effects of connectivity on local and regional population persistence (e.g., Pulliam 1988; Hanski 1998). Specifically, this framework provides a mechanism to evaluate how connectivity among local growing, stable, and declining subpopulations contributes to the maintenance of an NBH mummichog metapopulation and supports the persistence of subpopulations in the most contaminated locations.

The extent to which mummichogs have migrated historically throughout NBH can be estimated using genetic methods (e.g., Nei 1978). Specifically, McMillan et al. (2006) used amplified fragment length polymorphism (AFLP) analysis and Roark et al. (2004) used allozymes to assess population genetics of mummichogs associated with NBH. On a regional scale (>100 km), allele frequencies reflected a pattern of isolation by distance (Roark et al. 2004). However, AFLP markers distinguished subpopulations at both moderate and smaller scales, suggesting potential limitations to gene flow at distances of 2 km or less. At the harbor scale, subpopulation structure was unrelated to distance or to current PCB contamination. While many assumptions underlying the use of population genetic statistics to estimate connectivity among subpopulations may not be met in the case of NBH mummichogs (Whitlock and McCauley 1999), we used this example to demonstrate the value of genetics as a basis for dispersal estimates among sites, and to explore the potential for less contaminated sites to support subpopulation persistence at more contaminated sites.

Adaptive demography

Typically, effects of toxic contaminants have been assessed in laboratory studies using animals from naive (previously unexposed) populations. However, short-term responses of naive individuals may not reflect population outcomes associated with long-term exposures. In fact, recent research has shown that NBH mummichogs, subject to multigenerational PCB exposures, are dramatically insensitive to some (or all) of the toxic effects of DLCs (Nacci et al. 1999; Bello et al. 2000; Van Veld and Nacci in press). Results of laboratory exposures showed that concentrations of DLCs similar to those measured in NBH mummichog eggs are lethal to reference embryos and larvae (Nacci et al. 1999). These data suggest that few mummichogs from historically uncontaminated populations could survive and reproduce in NBH. Furthermore, similar comparisons between the sensitivities of laboratory-raised progeny of mummichogs collected at different sites demonstrated that DLC tolerance in NBH fish is inherited (Hahn 1998; Nacci et al. 1999, 2002b). These findings are consistent with the conclusion that multigenerational exposure to DLCs in NBH has contributed to the selection of DLC-adapted fish at NBH stations (Nacci et al. 2002b).

We used laboratory-based estimates of demographic rates to explore the realized and potential benefits of genetic adaptation to mitigate the population-level effects of intense PCB contamination on NBH mummichogs. Data measuring fecundity and survival during early (Nacci et al. 1999, 2002b) and adult (Black et al. 1998b) life stages of laboratory-exposed reference and field-contaminated fish were used to estimate demographic rates reflecting effects of PCB exposures on naive and adapted fish, respectively. Simulations were also conducted to assess the impact of hypothesized changes in demographic rates on metapopulation persistence and occupancy at the most contaminated stations.

A spatial modeling approach

RAMAS GIS was used here to generate alternative models that incorporate spatially explicit information to explore ecological mechanisms by which resident fish populations persist in chemically polluted habitats. Specifically, we evaluated the relative importance of population connectivity and adaptive demography to metapopulation persistence and site occupancy among NBH mummichog subpopulations. First, we characterized the study site spatially as mummichog habitat, defined by shoreline water depth, and described subpopulations using published demographic matrix models (Munns et al. 1997). We estimated subpopulation migration rates based on genetic relatedness (or inversely related to shoreline distance) and evaluated the relationship between the direction and strength of connectivity, metapopulation persistence, and site occupancy. The benefits to metapopulation persistence of additional hypothesized nearby stable and growing (uncontaminated) subpopulations were also evaluated. Finally, RAMAS GIS was used to explore how PCB effects could be mitigated by compensatory demography, incorporated into models as increased fecundity or survival derived from laboratory or simulated data.

Methods

Spatial data

A cell length (dimension of cells, in km) of 0.005 was used in calculating patch coordinates. However, shoreline (nonlinear) distances between populations were calculated outside of RAMAS GIS, as described below. Input maps including water depth were created from bathymetric points obtained from the National Ocean Service. These data are historic and were published in years ranging from 1935 to 1976. Coastline coverage was also constructed using a 1:25,000 scale datalayer extracted from the MassGIS hydrography data sets (Office of Geographic and Environmental Information (MassGIS), 2005), where both data sets are in UTM zone 18, NAD83 projection. Two other maps were created to show sediment PCBs values in 1993 and 1995, that is, before and after sediment dredging was used as the remedy for site cleanup. The data used to create these maps were derived from Nelson et al. (1996).

Habitat relationships were defined solely by water depth. Specifically, habitat suitability (HS) was defined as a binary function such that HS = 1.0 for depths from 0.1 to 3.0 m, and zero otherwise. Regions where HS = 0 are defined in RAMAS GIS as space that can be used for dispersal between subpopulations but not for reproduction. To exclude land as dispersal space, functional (vs. Euclidean) distances between aquatic subpopulations were calculated through water of specified depth using ArcInfo (ESRI 2004). A functional distance matrix was then used as the foundation for distance-based dispersal estimates in RAMAS GIS.

Stage structure

Female-only models for this species were based on Munns et al. (1997). Stage-specific parameters were calculated for six stages (figure 12.2): embryo (E), larvae (L), >28-day larvae (J), one-year adults (A1), two-year adults (A2), and three-year adults (A3). Time steps were set to 14 days (as in Munns et al. 1997), and simulations were run for 500 time steps (about 20 years, or about 20 generations) with 500 replications.

Subpopulation stage matrices

A reference model reflecting a growing population (population growth rate = 1.12) and a base model constrained to population growth rate = 1 (Munns et al. 1997) were used to represent hypothesized, local uncontaminated populations. Site-specific models representing geographic subpopulations were parameterized as described in Munns et al. (1997), using data acquired during laboratory studies (Black et al. 1998a). Site-specific subpopulations are identified by residence stations (from north to south; figure 12.1): NBH northern estuary and Coggeshall Street–north (both in the NBH Superfund area), Fairhaven boat launch site (in NBH, but south of the Superfund area), and West Island in the adjacent Buzzards Bay, bounded from NBH by a hurricane barrier. Using similar processes as those used to develop the site-specific models, models were also developed incorporating effects of laboratory exposures to PCBs for naive fish (Black et al. 1998b).

Information on variation in vital rates among field and laboratory contemporary replicates was available or could be calculated from published studies (Munns et al. 1997; Black et al. 1998b; Nacci et al. 1999, 2002a). Therefore, demographic stochasticity was invoked during RAMAS simulations. However, no data were available to estimate environmental stochasticity (i.e., time series data) for these subpopulations and treatments. Absent these data, variations in matrix parameters were included based on average values of specific matrices and coefficients of variation of 0.1 for survival and 0.2 for fecundity.

Density dependence was population specific. Density-independent models were used with most field- and laboratory-exposed subpopulations (population growth rate <1), except that scramble-type density dependence was used for the base and reference models (population growth rates ≥1). To explore demographic modifications, fecundity, survival, and carrying capacity (for density-dependent subpopulations) were treated as uncorrelated.

Abundances

Most simulations were initiated with stage abundances reflecting stable age structures and total abundances equal to 100,000. Out of this total abundance, about 500 were adult stages (i.e., sum A1, A2, A3). For simulations to assess outcomes associated with the addition of hypothetical uncontaminated subpopulations, a geographic "station 0" subpopulation was defined using the base or growth reference matrix model to represent the subpopulations. For these simulations, the initial total abundance of the hypothesized subpopulation was set to 0.01, 0.1, or 1 times the initial abundance for site-specific subpopulations. Final metapopulation abundance was calculated using adult stages only and included field- or laboratory-exposed subpopulations but not additional hypothesized base or reference populations.

Dispersal

For some simulations, no dispersal was permitted; in others, all migration/dispersal rates were symmetrical (i.e., the same magnitude in both directions), while in others migration was constrained to be unidirectional. This last constraint was set specifically to represent the limitation to effective migration of toxic conditions in the harbor. For these simulations, migration was permitted only from more to less contaminated stations, for example, from the subpopulation at station 4 (NBH northern estuary) to station 1 (West Island), but not vice versa. This was modeled by setting the lower diagonal values for the dispersal matrix to 0. Dispersal between subpopulations was first estimated as the inverse of average shoreline distances between stations (see "Spatial data" above). These distance-based dispersal rates ranged between 0.0001 and 0.001 (data not shown).

Magnitudes of estimated dispersal rates were similar whether calculated by distance-based or genetically based methods. Therefore, simulations reported here used dispersal estimates based on genetic relatedness to remain consistent with their use describing connectedness among subpopulations for this species (i.e., Mulvey et al. 2003). McMillan et al. (2006) analyzed AFLP data to describe genetic subpopulation structure for NBH mummichogs and report inbreeding statistics, F_{ST}. Here (as in Mulvey et al. 2003), F_{ST} is used to estimate subpopulation migration rates:

$$F_{ST} = 1/(1 + 4 \times N_e \times m),$$

where N_e is the effective population size and $N_e \times m$ is the effective migration rate (i.e., Nei 1978). Migration was calculated based on $N_e = 0.1 \times$ census population abundance. Dispersal per time step was calculated by dividing m by the number of time steps per generation (26). Genetically based dispersal rates ranged from 0.00017 to 0.0008 per time step. Dispersal between the base or reference subpopulation and field stations was set lower than the smallest calculated dispersal rate (0.0001). To examine the effect of connectivity among populations on meta- and subpopulation persistence, we also varied the dispersal rate by multiplying the per-time-step dispersal estimates by a factor ranging from 0.01 to 1,000.

Results

Subpopulations

Population matrices were used to obtain population growth rates for independent, unconnected populations subject to chronic or short-term contamination by PCBs, or uncontaminated conditions (figure 12.3). Matrices from Munns et al. (1997) represent hypothetical stable (base) or growing (reference) subpopulations and four field subpopulations associated with PCB-contaminated NBH. As described in Munns et al. (1997), subpopulation annual growth rates for the field stations were projected to range from 0.92 to 0.98. Annual growth rate projections from subpopulation matrices

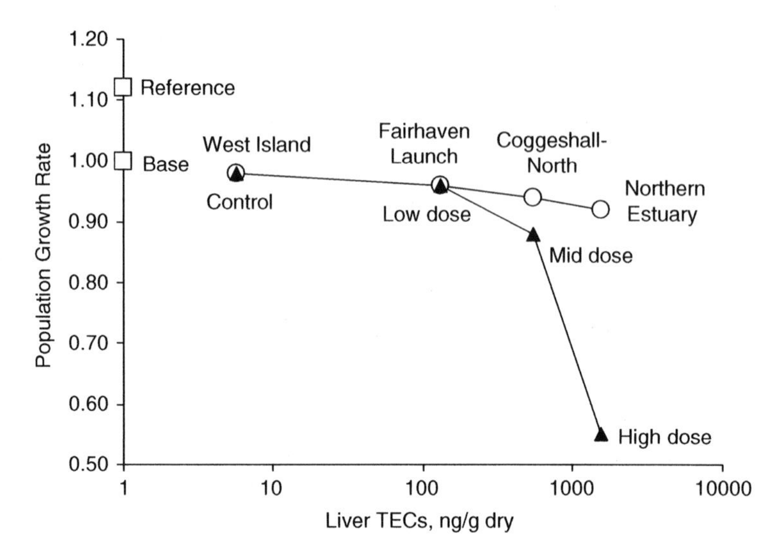

Figure 12.3. Population growth rates projected from population matrices representing unstressed *Fundulus heteroclitus* (squares) under base (constrained to population growth rate = 1) or reference (unconstrained population growth rate) conditions, or incorporating responses from PCB exposures (circles and triangles). Effects of PCB contamination were based on data collected during laboratory studies of fish contaminated with PCBs from their residence sites (Black et al. 1998a; circles) or from laboratory dosing studies (Black et al. 1998b; triangles). PCB contamination is expressed as liver dioxin TEC units or "dioxin equivalents," as described in Black et al. (1998a, 1998b). Population growth rates were calculated from those data as described in Munns et al. (1997).

representing short-term effects of PCB contamination (constructed for this study from results of laboratory dosing of naive fish, Black et al. 1998b) ranged from 0.55 to 0.98. Population growth rates were inversely correlated with concentrations of DLCs in livers of exposed fish. However, much lower population growth rates were projected for similar exposures to high levels of PCBs in naive versus field-adapted fish populations (figure 12.3).

Connecting subpopulations

Matrices from four NBH field subpopulations were spatially associated within RAMAS GIS, and the outcomes were assessed as viability of the metapopulation (reported as probability of persistence) and the most contaminated subpopulation (reported as probability of site occupancy for the NBH northern estuary). In initial simulations, subpopulations were set to initial abundances, and no dispersal was permitted. Under these conditions, there was a modest likelihood of metapopulation persistence during the simulation period of 20 years (39%), and occupancy of the NBH northern estuary was relatively unlikely (30%; figure 12.4).

The inclusion of genetically based estimates of one- or two-way dispersal among field subpopulations had a relatively small effect on the likelihood of metapopulation persistence and site occupancy for the NBH northern estuary (figure 12.4). High levels

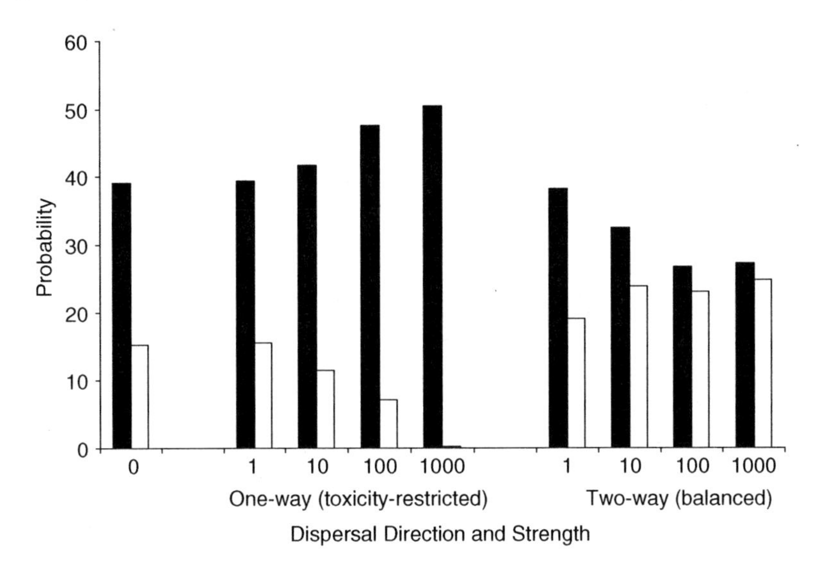

Figure 12.4. Results of simulations using population matrices representing *Fundulus heteroclitus* for four field sites associated with New Bedford Harbor unconnected (dispersal factor = 0) or connected by dispersal of varying strength and direction. Probabilities of metapopulation persistence (solid bars) and site occupancy for the most contaminated site (NBH northern estuary; open bars) are shown are shown for one-way (from high to low contamination, only, or toxicity restricted) and two-way (balanced) dispersal at 1 to 1,000 times dispersal rates estimated as described in the text.

of dispersal among field subpopulations (10–1,000 times estimated rates) were necessary to affect the likelihood of metapopulation persistence and site occupancy for the NBH northern estuary. Connecting subpopulations using increasing levels of two-way dispersal caused small incremental decreases in the probability of metapopulation persistence and small incremental increases in the probability of occupancy at the NBH northern estuary. Increasing the level of one-way migration (from more to less toxic stations, "toxicity-restricted" migration, only) reduced the probability of occupancy at NBH northern estuary but increased the probability of metapopulation persistence (slightly). However, even by including an optimal effect for connectivity on the NBH northern estuary subpopulation (two-way migration at 1,000-fold estimated rates), persistence of the metapopulation was improbable within a 20-year time period, and the probability of occupancy for the NBH northern estuary was low.

The addition of a hypothetical, nearby uncontaminated subpopulation increased the probability of metapopulation persistence, but effects on the occupancy of the most contaminated site varied dependent on the subpopulation growth rate and on the magnitude and direction of dispersal among subpopulations. When migration toward more contaminated subpopulations was completely prohibited (one-way migration), the probabilities of NBH metapopulation persistence and NBH northern estuary site occupancy were unaffected by the addition of a stable (base population growth rate = 1) or growing (reference population growth rate = 1.12) subpopulation. However, two-way dispersal that included either a base or reference subpopulation increased the probability of metapopulation persistence and the probability of site occupancy for the NBH northern estuary (figure 12.5). Moreover, an additional growing subpopulation connected by two-way dispersal had a strong positive effect on site occupancy: A magnitude of dispersal even 10 times lower than that estimated for field populations was sufficient to project an increased probability of occupancy at the NBH northern estuary. These results suggest that even if toxicity or physical barriers limit effective migration, dispersal into the most contaminated sites has a beneficial effect on site and metapopulation persistence.

Projections of unconnected populations using data from laboratory exposure studies indicated a modest likelihood of metapopulation persistence over the 20-year simulation (47%) but a very low probability of occupancy at the NBH northern estuary (2%). As with field subpopulations, probabilities of persistence were low for highly exposed subpopulations, even when connected by two-way dispersal from less-exposed populations. Consistent with simulations representing field subpopulations, migration from a growing source improved viability. However, even high levels of two-way dispersal from a nearby, growing population (which were necessary to produce a large increase in the probability of metapopulation persistence) produced only a small increase in the occupancy of a subpopulation exposed in the laboratory to PCB concentrations as high as those measured in fish from the NBH northern estuary (figure 12.5).

Modifying demographic rates

The probability of persistence could be increased without adding hypothesized subpopulations to the system by modifying subpopulation-specific demographic

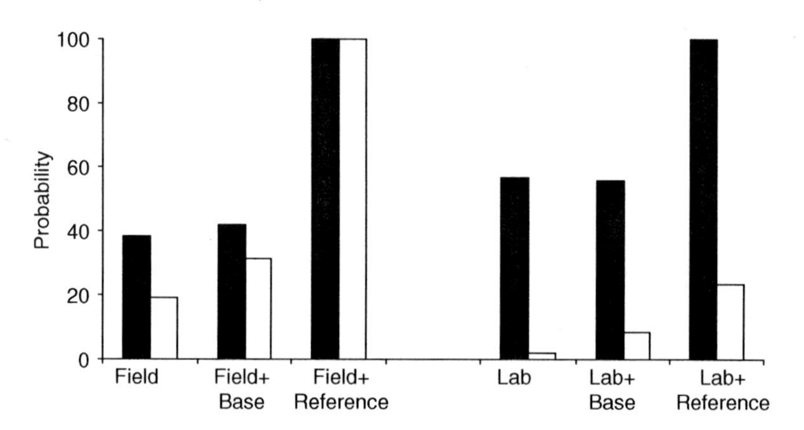

Figure 12.5. Results of simulations using population matrices representing *Fundulus heteroclitus* subpopulations for four field sites (Field) or subpopulations exposed in the laboratory to PCB concentrations similar to New Bedford Harbor sites (Lab) and an additional hypothesized subpopulation at base (population growth rate = 1) or reference (population growth rate = 1.12) conditions. Probability of metapopulation persistence (solid bars) and probability of site occupancy for the most contaminated site (NBH northern estuary, open bars) are shown for balanced two-way rates, estimated as described in the text.

rates for the most contaminated subpopulations (i.e., those resident to the Superfund stations: NBH northern estuary and Coggeshall Street–north). For unconnected field subpopulations, metapopulation persistence and high site occupancy could be achieved by large increases in demographic rates for NBH northern estuary and Coggeshall–north subpopulation, that is, increasing survival rates by >10% or fecundity rates by 20-fold (data not shown). Similarly, metapopulations composed of field subpopulations connected by two-way dispersal were persistent throughout the simulation period and had a high probability of occupancy at all sites when Superfund subpopulation survivals were increased by about >5% or fecundity was increased by about 10-fold (figure 12.6).

Additional simulations suggested that a metapopulation composed only of subpopulations resembling those in laboratory exposure experiments (i.e., growth rates for highly exposed subpopulations <<1), whether connected or unconnected, had a relatively low likelihood of persistence. Persistence could be affected by connecting subpopulations and modifying demographic rates for the most highly exposed populations, that is, those exposed at levels similar to the Superfund NBH northern estuary and Coggeshall–north sites (figure 12.6). For example, the likelihood of subpopulation persistence was increased when laboratory-exposed subpopulations were connected by two-way dispersal and survival was increased by 8% for the Superfund site subpopulations. However, even these relatively large changes in demographic rates had a relatively modest effect on increasing the probability for persistence for the metapopulation.

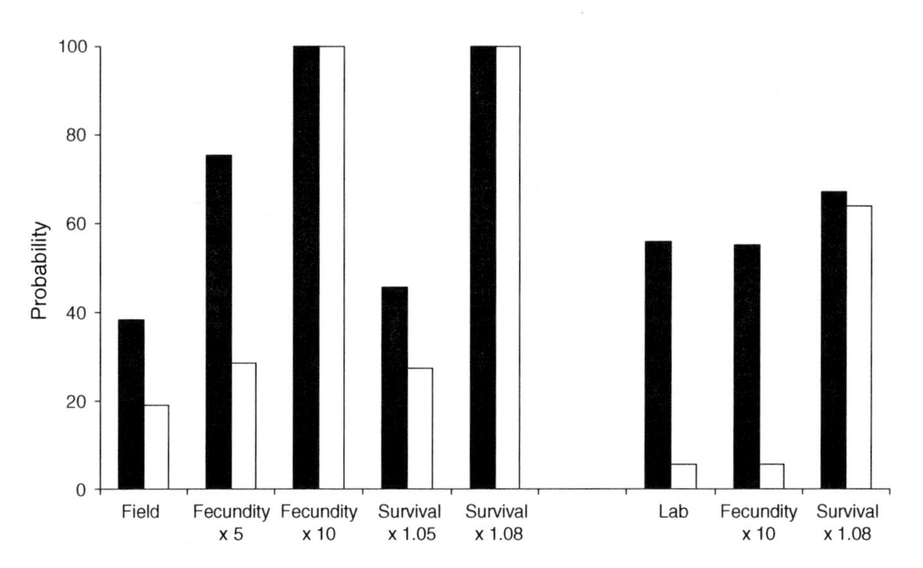

Figure 12.6. Results of simulations using population matrices representing *Fundulus heteroclitus* for four field sites associated with New Bedford Harbor (Field) or laboratory exposures to similar PCB levels (Lab) connected by balanced two-way dispersal. Probability of metapopulation persistence (solid bars) and probability of site occupancy for the most contaminated site (NBH northern estuary; open bars) are shown for changes in fecundity and survival.

Discussion

RAMAS GIS provided a mechanism to explore ecological factors by which *Fundulus heteroclitus* persists in a PCB-contaminated estuary. The New Bedford Harbor includes a Superfund site where exposures are highly toxic to nonresidents of this fish species (Black et al. 1998a, 1998b; Nacci et al. 1999, 2002a). Using a nonspatial population modeling approach and laboratory-collected data, Munns et al. (1997) suggested that independent (closed) subpopulations at the most contaminated sites were unlikely to persist. However, persistence could be influenced by unaccounted for ecological factors such as immigration to declining subpopulations and/or by demographic compensation. By incorporating real and simulated data into a spatial population modeling context, we were able to evaluate the impact on persistence of connectivity and variation in demographic rates among fish subpopulations exposed to a gradient of PCB contamination.

Consistent with theoretical expectations, simulations demonstrated that metapopulation persistence was more probable with higher rates of migration from faster to slower growing subpopulations. Metapopulation persistence depends on a balance between rates of localized extinction within subpopulations and colonization from occupied patches (Hanski and Gilpin 1997; Hanski 1998). As demonstrated in these simulations, dispersers from source populations can counteract declines in nearby patches, thereby "rescuing" more sink-like populations from extinction (Pulliam 1988). Our results suggest that migrants from nearby stable or growing populations could provide some support to subpopulations in the highly contaminated NBH sites. The results also indicate, however, that only higher rates of dispersal (and/or high initial population sizes in source subpopulations) could successfully mitigate declines in contaminated subpopulations: Lower rates were not sufficient to offset the declining growth rates characteristic of subpopulations at the contaminated sites. Given the relatively high degree of historical connectivity suggested by genetic data, high abundance, and proximity of subpopulations, dispersal may have been important for population establishment and stability, even for this "most sedentary of species" (Abraham 1985). However under current NBH conditions, one-way "toxicity restricted" migration probably contributes little to the support of the mummichog populations residing at the most contaminated sites.

Subpopulation-specific demographic rates were also varied to explore their effects on metapopulation persistence and site occupancy. Consistent with theoretical expectations, even large increases in fecundity had small effects on outcomes. Also consistent with expectations for a species with this life-history strategy (e.g., Stearns 1976), relatively small increases in survival (here, affected across all life stages) had profound effects on persistence. Survival was higher in adult (Black et al. 1998a, 1998b) and early life stages (Nacci et al. 1999, 2002a) of mummichogs from the NBH northern estuary versus those exposed in the laboratory to similarly high concentrations of PCBs. This increase contributed to higher growth rates for field-exposed (annual population growth rate = 0.92) versus lab-exposed subpopulations (annual population growth rate = 0.55), and higher probabilities of site occupancy and persistence for the field- than lab-exposed metapopulations. Thus, the increased survival (but not altered reproduction, Nacci et al. 2002a) observed in chemically-tolerant NBH mummichogs functions as effective compensatory demography and probably contributes substantially to population persistence at the most highly contaminated sites.

However, the importance of small increases in survival also suggests that survival estimates based on short-term laboratory studies should be used with caution. These "snapshots" may not capture important differences in lifelong survival, such as those related to overwintering and disease vulnerability. Furthermore, even for relatively well-studied aquatic species such as *Fundulus heteroclitus*, few data are available to develop good estimates of rates and variations in demography, especially those related to (environmental) stochasticity, which are critical for understanding and predicting population dynamics.

Conclusions

Models and data derived from a series of research studies were used to represent effects of short- and long-term exposures to toxic pollutants on fish populations. RAMAS GIS provided a mechanism to incorporate demographic and toxicological data into a spatially explicit framework and to explore ecological mechanisms by which wildlife populations persist in chemically polluted habitats.

We used this mechanism to evaluate the relative importance of population connectivity of varying strengths, sources, and directions to metapopulation persistence and site occupancy. We also explored the extent to which toxicity could be mitigated by compensatory demography, incorporated into models as increased fecundity or survival derived from laboratory or simulated data. Our results highlight the value of field observations and laboratory studies to inform and test population models intended to project stressor effects. This application demonstrates a useful role for spatially explicit modeling in the assessment of ecological risks and the development of management strategies to minimize effects of heterogeneously distributed stressors and wildlife populations.

Acknowledgments We acknowledge the valued assistance of Amy McMillan (Buffalo State University) and Jane Copeland (CSC), and the helpful advice and reviews provided by Jason Grear, Anne Kuhn, and Sandy Raimondo of the U.S. Environmental Protection Agency, and our editor, Reşit Akçakaya. This is contribution AED-06-044 of the U.S. EPA Office of Research and Development, National Health and Environmental Effects Research Laboratory, Atlantic Ecology Division. Although the research described in this contribution has been funded partially by the U.S. EPA, it has not been subjected to agency-level review. Therefore, it does not necessarily reflect the views of the agency. Mention of trade names, products, or services does not constitute endorsement or recommendation for use.

References

Abraham, B.J. Species profiles: Life histories and environmental requirements of coastal fishes and invertebrates (mid-Atlantic)—mummichog and striped killifish. U.S. Fish and Wildlife Service Biological Report 82 (11.40). U.S. Army Corps of Engineers, TR EL-82-4, 1985.

Bellafiore, N., and Anderson, S. Effects of contaminants on genetic patterns of aquatic organisms. Mutat. Res., 489: 97–122, 2001.

Bello, S.B., et al. Acquired resistance to aryl hydrocarbon receptor agonists in a population of *Fundulus heteroclitus* from a marine superfund site: In vivo and in vitro studies on the induction of xenobiotic metabolizing enzymes. Toxicol. Sci., 60: 77–91, 2000.

Bigelow, H.B., and Schroeder, W.G. Fishes of the Gulf of Maine. Fishery Bulletin of the Fish and Wildlife Service, Vol. 53. U.S. Government Printing Office, Washington, D.C., 1953.

Black, D.E., et al. Reproduction and polychlorinated biphenyls in Fundulus heteroclitus (Linnaeus) from New Bedford Harbor, Massachussetts, USA. Environ. Toxicol. Chem., 17: 1405, 1998a.

Black, D.E., et al. Effects of a mixture of non-ortho and mono-ortho-polychlorinated biphenyls on reproduction in Fundulus heteroclitus (Linnaeus). Environ. Toxicol. Chem., 17: 1396, 1998b.

Brown, B.L., and Chapman, R.W. Gene flow and mitochondrial DNA variation in the killifish, Fundulus heteroclitus. Evolution, 45: 1147, 1991.

ESRI. ArcInfo, version 9.0. Environmental Systems Research Institute, Inc., Redlands, California, 2004.

Hahn, M.E. Mechanisms of innate and acquired resistance to dioxin-like compounds, Rev. Toxicol. Ser. B, Environ. Toxicol., 2: 395, 1998.

Hanski, I.A. Metapopulation dynamics. Nature 396: 41–49, 1998.

Hanski, I.A., and Gilpin, M.E. (editors). Metapopulation Biology: Ecology, Genetics, and Evolution. Academic Press, London, 1997.

Lake, J.L., et al. Comparisons of patterns of polychlorinated biphenyl congeners in water, sediment, and indigenous organisms from New Bedford Harbor, Massachusetts. Arch. Environ. Contam. Toxicol., 29: 207, 1995.

Lotrich, V.A. Summer home range and movements of *Fundulus heteroclitus* (Pisces: Cyprinodontidae) in a tidal creek. Ecology, 56: 191, 1975.

McMillan, A., Bagley, M., Jackson, S., and Nacci, D. Genetic diversity and structure of an estuarine fish (*Fundulus heteroclitus*) indigenous to sites associated with a highly contaminated urban harbor. Ecotoxicology, 15: 539–548, 2006.

Mitchell, E.L., and Oviatt, C.A. Comparison of fish biomass, abundance and community structure between a stressed estuary, New Bedford Harbor, MA, and a less-stressed estuary, the Slocums River, MA. Graduate School of Oceanography Report, University of Rhode Island, Kingston, RI, 1996.

Monosson, E. Reproductive and developmental effects of PCBs in fish: A synthesis of laboratory and field studies. Rev. Toxicol., 3: 25, 1999/2000.

Mulvey, M., Newman, M.C., Vogelbein, W.K., Ynger, M.A., and Ownby, D.R. Genetic structure and mtDNA diversity of Fundulus heteroclitus populations from polycyclic aromatic hydrocarbon-contaminated sites. Environ. Toxicol. Chem., 22(3): 671–677, 2003.

Munns, W.R., Jr., et al. Evaluation of the effects of dioxin and PCBs on Fundulus heteroclitus populations using a modeling approach. Environ. Toxicol. Chem., 16: 1074, 1997.

Nacci, D., et al. Adaptation of wild populations of the estuarine fish *Fundulus heteroclitus* to persistent environmental contaminants. Mar. Biol., 134: 9, 1999.

Nacci, D., Gleason, T., Gutjahr-Gobell, R., Huber, M., and Munns, W.R., Jr. Effects of environmental stressors on wildlife populations. In: Coastal and Estuarine Risk Assessment: Risk on the Edge, Newman, M.C. (editor). CRC Press/Lewis Publishers, Washington, D.C., 2002a.

Nacci, D., Coiro, L., Champlin, D., Jayaraman, S., and McKinney, R. Predicting the occurrence of genetic adaptation to dioxinlike compounds in populations of the estuarine fish *Fundulus heteroclitus*. Environ. Toxicol. Chem., 21(7): 1525–1532, 2002b.

Nei, M. Estimation of average heterozygosity and genetic distance from a small number of individuals. Genetics 89: 583–590, 1978.

Nelson, W.G., et al. New Bedford Harbor long-term monitoring assessment report: Baseline sampling. EPA/600/R-96/097. U.S. Environmental Protection Agency, National Health and Environmental Effects Research Laboratory, Atlantic Ecology Division, Narragansett, Rhode Island, 1996.

Office of Geographic and Environmental Information (MassGIS), Commonwealth of Massachusetts Executive Office of Energy and Environmental Affairs. MassGIS.

Datalayers/GIS database Hydrography (1:25,000) Datalayer. http://www.mass.gov/mgis/hd.htm, 2005.

Pruell, R.J., et al. Geochemical study of sediment contamination in New Bedford Harbor, Massachusetts. Mar. Environ. Res., 29: 77, 1990.

Pulliam, H.R. Sources, sinks and population regulation. Am. Nat. 132(5): 652–661, 1988.

Roark, S.A., Nacci, D., Coiro, L., Champlin, D., and Guttman, S.I. Population genetic structure of a non-migratory marine fish Fundulus heteroclitus across a strong gradient of PCB contamination. Environ. Toxicol. Chem., 24(3): 717–725, 2004.

Royal Commission on Environmental Pollution. 25th Report. Turning the tide—addressing the impact of fisheries on the marine environment. Health issues linked to eating fish, Appendix G, pp 445. http://www.rcep.org.uk/fishreport.htm, 2005.

Stearns, S.C. Life-history tactics: A review of ideas. Q. Rev. Biol., 51: 3, 1976.

Sweeney, J., Deegan, L., and Garritt, R. Population size and site fidelity of Fundulus heteroclitus in a macrotidal saltmarsh creek. Biol. Bull., 195: 238–239, 1998.

U.S. Environmental Protection Agency (US EPA). Imprint of the past: the ecological history of New Bedford Harbor. http://www.epa.gov/nbh, 2007.

Van Veld, P.A., and Nacci, D. Chemical tolerance: Acclimation and adaptions to chemical stress. In: The Toxicology of Fishes, Di Giulio, R.T., and Hinton, D.E. (editors). Taylor and Francis, Washington, D.C., in press.

Walker, M.K., and Peterson, R.E. Potencies of polychlorinated dibenzo-p-dioxin, dibenzofuran, and biphenyl congeners, relative to 2,3,7,8-tetrachlorodibenzo-p-dioxin, for producing early life stage mortality in rainbow trout (Oncorhynchus mykiss). Aquat. Toxicol., 21: 219, 1991.

Whitlock, M.C., and McCauley, D.E. Indirect measures of gene flow and migration: Fst < > 1/(2Nm + 1). Heredity, 82: 117–125, 1999.

Wirgin, I., and Waldman, J.R. Altered gene expression and genetic damage in North American fish populations. Mutat. Res., 399: 193, 1998.

13

Demographic Effects of the Polycyclic Aromatic Hydrocarbon Fluoranthene on Two Sibling Species of the Polychaete *Capitella capitata*

ANNEMETTE PALMQVIST
VALERY E. FORBES

Assessing the ecological risks of toxic chemicals is generally based on individual-level responses such as survival, reproduction, or growth, even though the aim of ecological risk assessment is most often to prevent adverse effects of chemicals at population, community, and ecosystem levels. In this context, life table response experiments can provide a powerful tool for analyzing the responses of life-history variables to pollutants and to explore demographic consequences of sublethal toxic effects on populations (Levin et al. 1987; Caswell 1989, 1996; Hansen et al. 1999). Specifically, impacts on population dynamics can be estimated by integrating the life-history variables (or vital rates) via an appropriate demographic model to calculate, for example, population growth rate, λ. Changes in λ in response to changes in individual life-history traits depend on the life-history type of the study organism, on the severity of the toxicant effect on each life-history trait, and on the sensitivity of λ to changes in the life-history traits contributing to it (e.g., Levin et al. 1987, 1996; Caswell 1989; Kammenga et al. 1996; Hansen et al. 1999).

Species belonging to the *Capitella capitata* species complex provide an ideal model system for exploring linkages between individual-level and population-level effects of toxic chemicals. The species are small and have relatively short life spans (i.e., several months), and at least some of the species can be bred for many generations in the laboratory. *Capitella capitata* represents a cryptic species complex, in which adult morphology is so similar among species that no effort has been made to separate this complex with proper taxonomical species descriptions. Nevertheless, the so-called *Capitella* sibling species are reproductively isolated and differ substantially in karyotypes and enzyme patterns (Grassle and Grassle 1976; Grassle et al. 1987; Wu et al. 1991); in ultrastructure of eggs and ovarian follicle cells and in larval, genital spine, and sperm morphology (Eckelbarger and Grassle 1983, 1987); in ecophysiological characters (Gamenick et al. 1998); in brood and egg size, larval dispersal mode, and the occurrence of hermaphrodites (Grassle and Grassle 1976; Grassle et al. 1987); and in tolerance to various types of environmental stressors (Gamenick and Giere 1994; Linke-Gamenick et al. 2000a, 2000b).

We based the simulations presented in this chapter on previously published data comparing effects of the polycyclic aromatic hydrocarbon fluoranthene (Flu), a widespread contaminant of marine sediments, on two *Capitella* sibling species. The purpose of the present analysis was to explore the potential effects of patchiness in sediment contamination on the population dynamics of these species, paying particular attention to interactions between population density and sediment patchiness and the influence of life history. We used the simulations to generate hypotheses that we will employ in the design of future experimental studies addressing these issues.

Methods

Study species and area

Two previously identified, non-interbreeding sibling species from different geographical regions and habitat types were used in this study: *Capitella* sp. I (Grassle and Grassle 1976) originally obtained from Setauket Harbor, New York, USA, and *Capitella* sp. S (Gamenick and Giere 1994) from intertidal sediments of the Island of Sylt, North Sea, Germany. These species reproduce readily in culture and have been reared under identical conditions in our laboratory for many generations.

Capitella sp. I has been described as highly opportunistic and is found at high densities in organically contaminated sediments (Grassle and Grassle 1974, 1976). This species exhibits rapid population explosions following oil spills, heavy loads of organic inputs, and anoxic events (Bridges et al. 1994). In contrast, *Capitella* sp. S inhabits oxygen-rich low-sulfide habitats and is physiologically intolerant of low oxygen, sulfide, and Flu (Gamenick and Giere 1994; Linke-Gamenick et al. 2000a).

Morphologically, the two species are indistinguishable, but they differ markedly in reproductive and larval modes. *Capitella* sp. S matures into females and males and hatches benthic juveniles that emerge from the parental brood tube after metamorphosis (= direct development) in low numbers of 16–50 (Gamenick and Giere 1994; Méndez et al. 2000). *Capitella* sp. I develops into females, males, and hermaphrodites and reproduces via free-swimming metatrochophore larvae (= lecithotrophic development), with between about 200 and 500 larvae per brood (Grassle and Grassle 1976; Hansen et al. 1999; Méndez et al. 2000).

Experimental methods

Details of the experimental methods are described in Linke-Gamenick et al. (2000b). Briefly, experiments were initiated with 3- to 6-day-old juveniles. A total of 180 *Capitella* sp. S and 144 *Capitella* sp. I juveniles were randomly assigned to six Flu treatment groups (30 *Capitella* sp. S per group and 24 *Capitella* sp. I per group). Each group was transferred into a petri dish containing experimental sediment (with nominal concentrations of 0, 5, 10, 20, 40, or 80 μg/g Flu) and seawater (30‰ Salinity). The 18 dishes were placed in a moisture chamber to reduce evaporation and kept in the dark at 18°C. Sediments were spiked with Flu as described in Linke-Gamenick et al. (2000b), and the nominal concentrations were checked by high-performance liquid chromatography using methods described in Linke-Gamenick et al. (1999). Results are presented in terms of measured sediment Flu concentrations (0, 5, 10, 20, 50, and 95 μg/g Flu; for mean ± standard deviations, see Linke-Gamenick et al. 2000b).

When juveniles became sexually mature (female, appearance of ovaries; male, appearance of genital spines), a female and a male were paired and transferred to new separate experimental dishes (up to nine replicate pairs of worms per Flu concentration). All dishes were monitored once a week for a duration of 175 days (*Capitella* sp. S) or 161 days (*Capitella* sp. I). At each census day, worms were removed from the sediment, and data on worm survivorship, body size, wet weight and reproduction were recorded. Worms were returned to their corresponding treatments with new experimental sediment and seawater. The following life table parameters were measured: age-specific survival, body volume over time until sexual maturity, juvenile specific growth rate, age at maturity, age at first reproduction (appearance of brood tubes), percentage of reproducing females and hermaphrodites, numbers of broods per female or hermaphrodite, age-specific fecundity, numbers of offspring per brood, time between broods, and size-specific sediment processing rate (feeding rate).

Population model and simulations

We used the laboratory data to generate age-classified matrices for each species and exposure treatment. For the present analysis, we focused particularly on the 0, 50, and 95 μg Flu/g treatments. The time steps in the model are defined in terms of weeks.

We assumed Scramble competition for most simulations with the exception of the *Capitella* sp. S 95 μg/g Flu matrix, since the starting λ value was less than 1 (see RAMAS manual: 74, Akçakaya 2002). For this simulation, we assumed ceiling-type density dependence. Scrample-type density dependence is defined by R_{max} and K; ceiling-type density dependence, only by K. For both *Capitella* sp. I and sp. S, maximal theoretical K was set at 200,000 individuals for the whole metapopulation and was divided equally among all patches (subpopulations) within the metapopulation. We assumed that density dependence affects all age classes. To model dispersal for the two species, considering that *Capitella* sp. I has a swimming larval stage (high dispersal) whereas *Capitella* sp. S has direct development (low dispersal), in the Stage dialogue we set the percentage dispersing as 1.0 for *Capitella* sp. I age class 0, and 0.1 for all other age classes. For comparison, we set the percentage dispersing for all age classes of *Capitella* sp. S as 0.1. Dispersal is defined to be a function of distance between patches according to

$$\begin{aligned} &\text{dispersal rate} = a^*\exp(-\text{distance}^c/b), && \text{if distance} \le D_{max} \\ &\text{dispersal rate} = 0, && \text{if distance} > D_{max} \end{aligned} \qquad (13.1)$$

For all experiments, D_{max}, the maximal distance an individual can migrate, was set to 2, and the parameters *a* and *c* were set to 1, giving a dispersal rate that depends only on the parameter *b* and the distance between patches [dispersal rate = exp(–distance/*b*)].

We defined three types of patches to simulate sediment containing 0, 50, or 95 μg Flu/g sediment. We arranged the landscape to consist of either a mixture of clean (i.e., 0) and highly contaminated (i.e., 95) patches or a homogeneous sediment of average contaminant level (i.e., 50).

Three sets of simulations were performed. For all simulations, the individuals for the metapopulation always originated from the patch in the upper left corner, that is, modeling a "recolonization" situation, and all simulations were deterministic. The metapopulation dynamics were described by the realized carrying capacity, *K*, and the time it took to reach 95% of *K*. The realized K was always less than or equal to the maximum theoretical K of 200.000 individuals. Both values were calculated by assuming logistic population growth and using the equation

$$N_t = (N_0 \times K) \times [N_0 + (K - N_0) \times \exp(-r_0 \times t)]^{-1},$$

where N_t is the abundance at time *t*, N_0 is the initial abundance, *K* is the carrying capacity, r_0 is the initial growth rate, and *t* is time. *K* and r_0 were estimated by fitting the equation to the data from the trajectory summary using the NONLIN module of SYSTAT (version 10.0; SYSTAT Software, Inc., Richmond, CA, USA) using a least squares loss function and Gauss-Newton estimation method. Time to reach 95% of *K* was calculated from the equation by setting $N_t = 0.95 \times K$.

Simulation 1: Hotspot effect

In this simulation we explored how a single contaminated patch ("hotspot") in an otherwise clean habitat influenced population dynamics compared to a habitat that was moderately and homogeneously contaminated. We initially compared a scenario with two patches such that the average Flu concentration in the hotspot habitat was equal to the average Flu concentration in the homogeneously contaminated habitat. By increasing the total number of patches while maintaining only one contaminated patch in the hotspot habitat, we explored whether and at what degree of habitat patchiness the population in the hotspot habitat reached the same carrying capacity, *K*, and time to reach 95% of *K* as the population in the homogeneous habitat. Figure 13.1 shows how patchiness (number of patches) was defined. We performed this scenario using both *Capitella* sp. I and *Capitella* sp. S.

Simulation 2: Dispersal effect

In this simulation, we explored how varying the dispersal rate of *Capitella* sp. S (the low dispersal species) influenced *K* and time to reach 95% of *K*. Here we defined a habitat consisting of six patches so that the overall average Flu concentration in the hotspot habitat was only one-third of the Flu concentration in the homogeneous habitat (see figure 13.1). We varied the dispersal rate by varying *b* in the dispersal–distance function. As described above, dispersal rate is a function of distance between patches, and increasing the value of *b* will increase dispersal/migration between

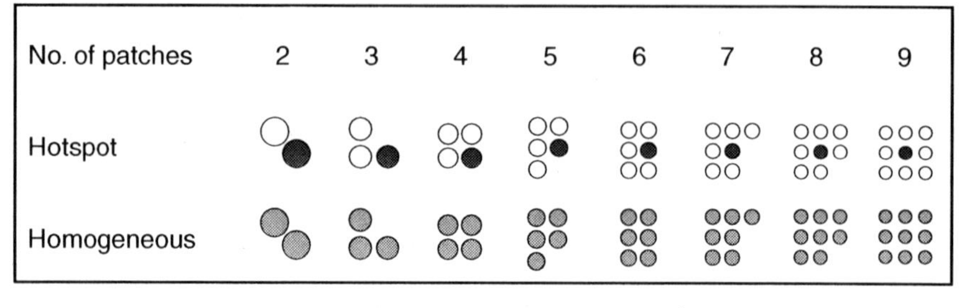

Figure 13.1. Definitions of the heterogeneous (hotspot) and homogeneous habitats used in the simulations. The hotspot habitat was simulated as one highly contaminated patch (100% = 95 μg Flu/g dry weight sediment) associated with one to eight clean patches, whereas the homogeneous habitat was simulated by two to nine moderately contaminated patches (50% = 50 μg Flu/g dry weight sediment). The total habitat area was defined to be the same regardless of patch number (i.e., increasing patch number means decreasing patch size). Average Flu concentration in the habitat decreased with increasing patch number in the hotspot habitats, whereas average Flu concentration was constant with patch number in the homogeneous habitats. Average Flu concentration in the hotspot habitat was approximately equal to the average Flu concentration in the homogeneous habitat in simulations with two patches.

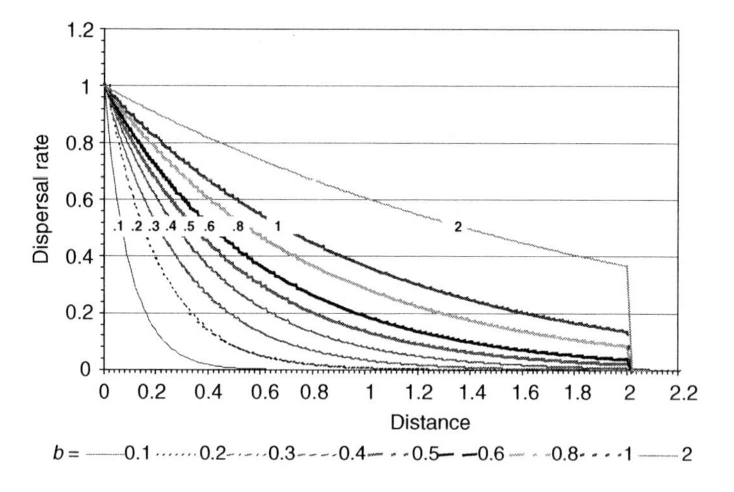

Figure 13.2. Dispersal–distance function. Dispersal rate was calculated as a function of distance according to equation 13.1. By adjusting the parameter b, the dispersal–distance curve can be shifted toward a higher or lower dispersal rate.

patches (figure 13.2). We performed a number of simulations to explore how varying the degree of dispersal would influence the patchiness effect (compared to the results for *Capitella* sp. S from simulation 1). In addition, a series of simulations was run for all of the patch combinations shown in figure 13.1 at two different dispersal rates, that is, high dispersal ($b = 1$) and medium dispersal ($b = 0.5$).

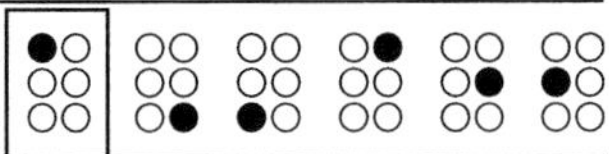

	●○ ○○ ○○	○○ ○○ ○●	○○ ○○ ●○	○● ○○ ○○	○○ ○● ○○	○○ ●○ ○○
Position of contaminated patch	0	1	2	3	4	5
Number of "donor" patches	4	4	4	4	5	5
Distance from source patch	0	2.2	2	1	1.4	1

Figure 13.3. Contaminated patch position. The position of the contaminated patch in relation to the source patch (upper left corner) was defined by a combination of number of donor patches from which the contaminated patch could receive individuals and the distance from the source patch to the contaminated patch. See text for details.

Simulation 3: Patch distribution effect

In this simulation, we explored the importance of position/location of the contaminated hotspot in relation to the source patch (i.e., the patch from which all individuals in the metapopulation originated). For this simulation, we used *Capitella* sp. S in a hotspot scenario with six patches and defined the contaminated patch position as follows:

1. The source patch in all scenarios is placed in the upper left corner (see figure 13.3).
2. The position of the contaminated patch = 0 when the contaminated patch is the same as the source patch.
3. We defined as "donor" patches all those patches from which a contaminated patch could receive individuals, and we defined "position" on the basis of the number of donor patches from which a contaminated patch could receive individuals (given $D_{max} = 2$) and the distance between the source patch and the contaminated patch.
4. When there were equal numbers of donor patches, the distance between the source patch and the contaminated patch was defined such that longer distance = lower position number.

The patch position scenarios in figure 13.3 were simulated for high, medium, and low dispersal ($b = 1$, $b = 0.5$, and $b = 0.3$, respectively).

Results

Simulation 1: Hotspot effect

Figure 13.4 shows trajectory summaries of the simulated metapopulations for both *Capitella* sp. S and *Capitella* sp. I living either in a clean environment containing one hotspot or in a moderately and homogeneously contaminated environment. For *Capitella* sp. S, the metapopulation showed rhythmic oscillations around K (indicated where the curves become horizontal), whereas for *Capitella* sp. I the metapopulation showed damped oscillations. For *Capitella* sp. S, the oscillations were larger for worms living in the hotspot habitat compared to worms living in the homogeneous habitat, whereas the opposite was the case for *Capitella* sp. I. For *Capitella* sp. S, a hotspot in an otherwise clean environment had a pronounced effect on the metapopulation's K (figures 13.4A and 13.5A), in that the resulting K

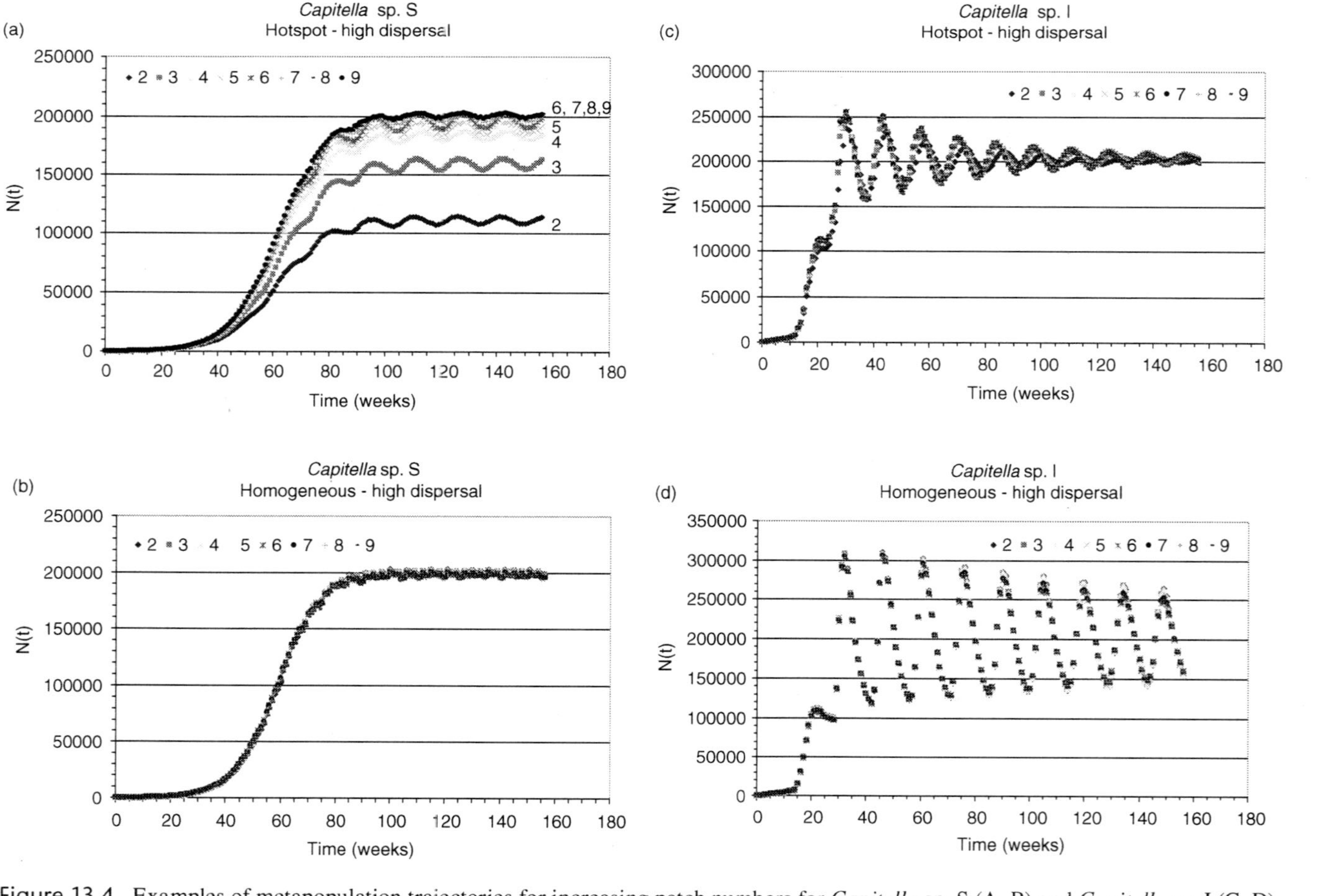

Figure 13.4. Examples of metapopulation trajectories for increasing patch numbers for *Capitella* sp. S (A, B) and *Capitella* sp. I (C, D). Numbers (2–9) relate to the total number of patches in either the heterogeneous (A, C) or the homogeneous (B, D) habitats. All simulations were made at high dispersal (*b* = 1; see text for further explanation).

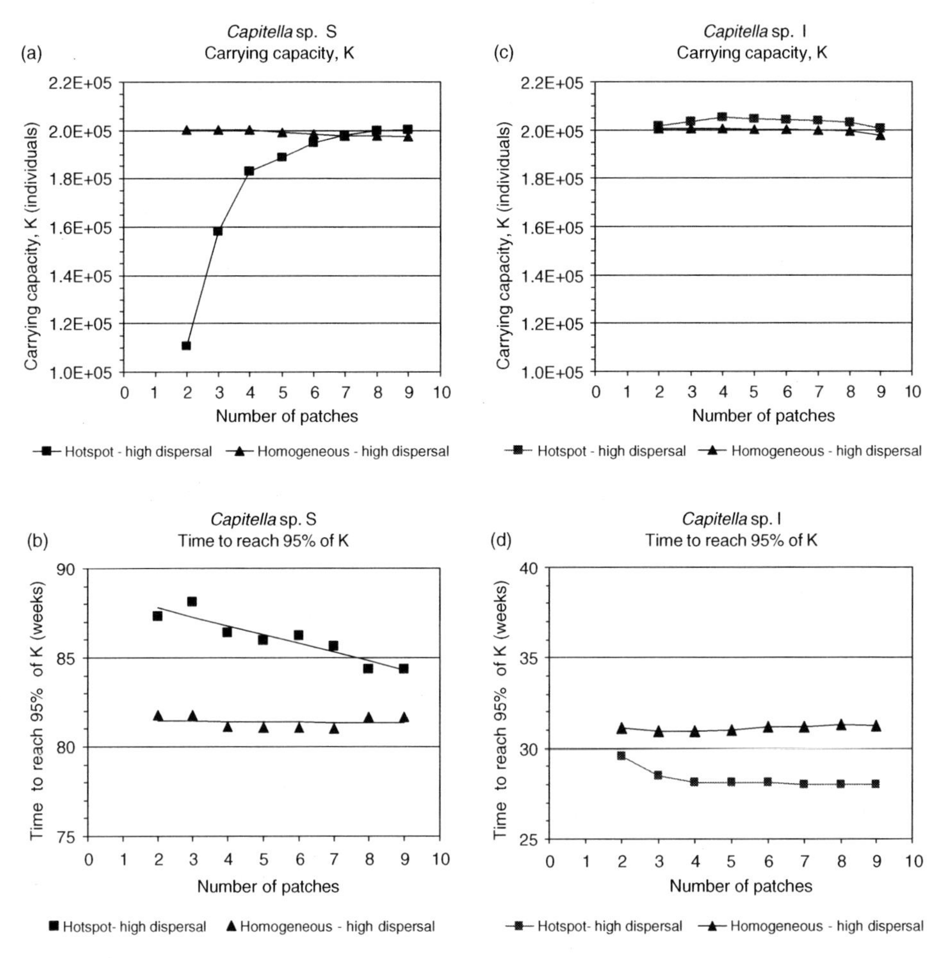

Figure 13.5. *Capitella* sp. S (A, B) and *Capitella* sp. I (C, D) carrying capacity, *K* (A, C), and time to reach 95% of *K* (B, D) as a function of patch number. Carrying capacity and time to reach 95% of *K* were calculated from the metapopulation trajectory, by fitting a logistic growth curve to the data. Simulations were made for worms exposed to either the hotspot habitat or the homogeneous habitat, and both endpoints (*K* and time to reach 95% of *K*) were related to the number of patches.

was lower for the metapopulation in the hotspot habitat than in the homogeneous habitat up to a patch number of seven (note that the average Flu concentrations were equal in the hotspot and the homogeneous habitat at a patch number of 2 and decreased in the hotspot habitat with increasing number of patches). In addition to reducing *K* (figure 13.5B), longer time to reach 95% of *K* was needed for metapopulations of *Capitella* sp. S in the hotspot habitats compared to metapopulations in the homogeneous habitats. The difference in time to reach 95% of *K* between populations in the hotspot habitat and the homogeneous habitat decreased with increasing number of patches (i.e., increasing difference in average Flu concentration), from approximately six weeks (at equal average Flu concentrations) to two

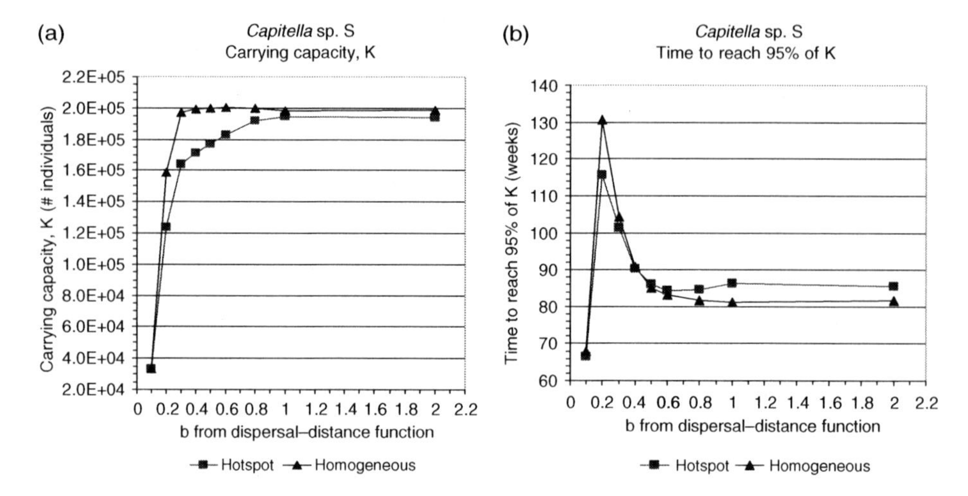

Figure 13.6. *Capitella* sp. S carrying capacity, *K* (A), and time to reach 95% of *K* (B) as a function of dispersal ability. Dispersal ability was varied by varying the parameter *b* (see text and figure 13.2 for further explanation). Simulations were made for worms exposed to either the hotspot habitat or the homogeneous habitat, and *K* and time to reach 95% of *K* were calculated by fitting a logistic growth curve to the metapopulation trajectories.

weeks in habitats with nine patches. In contrast, for *Capitella* sp. I there were only minor differences, if any, in the metapopulation *K* between worms living in the hotspot habitat and worms living in the homogeneous habitat (figure 13.5C), though time to reach 95% of *K* was generally a few weeks shorter for metapopulations in the hotspot habitat compared to metapopulations in the homogeneous habitat. Generally, time to reach 95% of *K* was considerably longer for *Capitella* sp. S than for *Capitella* sp. I (figure 13.5B,D).

Simulation 2: Dispersal effect

Figure 13.6 shows how increasing dispersal rate (increasing *b* results in increasing dispersal rate; see figure 13.2) affects *K* and time to reach 95% of *K* in metapopulations of *Capitella* sp. S living in either a hotspot habitat or a homogeneous habitat. At the lowest *b*(0.1), the source subpopulation is not able to spread to other subpopulations, so the metapopulation does not reach a *K* higher than *K* for the source subpopulation in either the hotspot habitat or the homogeneous habitat. However, the metapopulations reach *K* relatively quickly (~67 weeks) for both habitat types (since no individuals are removed from the source population due to migration). The metapopulation in the homogeneous habitat reaches a *K* of 200,000 individuals (equal to the maximum *K*) at a lower dispersal rate than does the metapopulation in the hotspot habitat. This indicates that the hotspot has a larger negative impact on the metapopulation as dispersal ability decreases (figure 13.6A). Except for the situation in which *b* is so low that there is no dispersal between patches, time to reach 95% of *K* decreases with increasing dispersal for worms living in both the

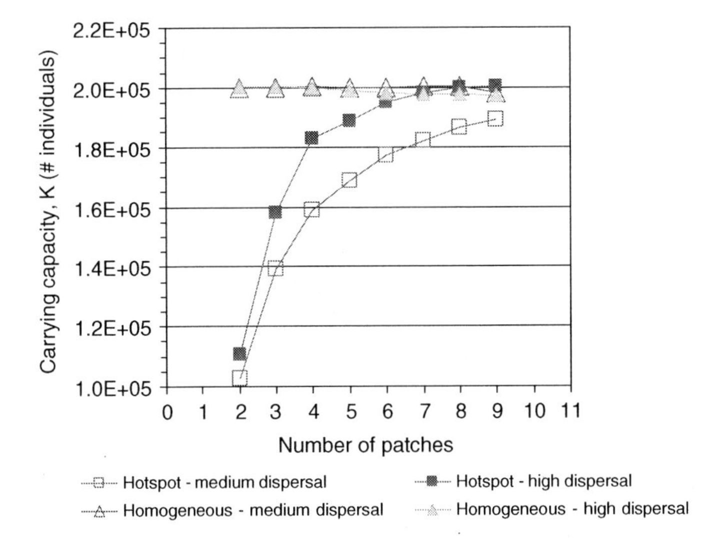

Figure 13.7. *Capitella* sp. S: Influence of dispersal on the patchiness effect. Carrying capacity, K, is shown as a function of number of patches in the habitat. Simulations were made with either high (b = 1) or medium (b = 0.5) dispersal for worms living in either the hotspot habitat or the homogeneous habitat. Notice that the homogeneous-medium dispersal curve is almost completely covered by the homogeneous-high dispersal curve. K was calculated by fitting a logistic growth curve to the metapopulation trajectories.

hotspot habitat and the homogeneous habitat until a certain dispersal rate, above which the time to reach 95% of K is constant (figure 13.6B). The dispersal rate above which both K and time to reach K are constant may be considered to be full dispersal between patches. According to this definition, we observed full dispersal in simulation 1, since worms from both hotspot and homogeneous habitats reached this dispersal rate at b = 1. The effect of dispersal on the patchiness effect is shown in figure 13.7: The metapopulation K, for all patch numbers, is decreased with lower dispersal ability in the hotspot habitat, whereas it is not affected in the metapopulations living in homogeneous habitats. The hotspot metapopulation K did not reach the same value as the homogeneous metapopulation K for up to nine patches simulated in the present study.

Simulation 3: Patch distribution effect

Figure 13.8 shows that the position/location of the contaminated patch in relation to the source patch influences both K and time to reach 95% of K, but that dispersal ability also influences the effect of contaminated patch position on the two endpoints. Generally, lowering dispersal ability decreases K and increases time to reach 95% of K. Figure 13.8 shows that as position of the contaminated patch increases (= more donor patches and shorter distances to source patch), the effect on carrying capacity is opposite for medium and high dispersal compared to low dispersal.

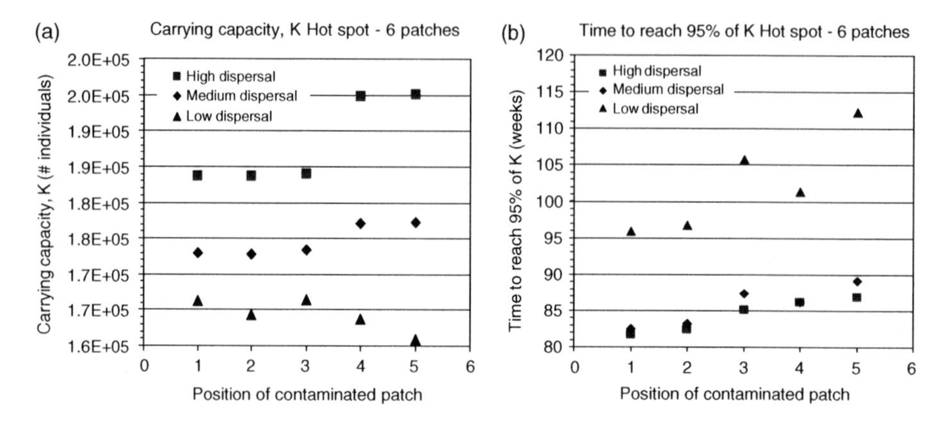

Figure 13.8. Importance of contaminated patch location for *Capitella* sp. S: Carrying capacity, K (A), and time to reach 95% of K (B) as a function of the position of the contaminated patch in relation to the source patch (see figure 13.3 and text for definitions of patch positions). Simulations were made with high (b = 1), medium (b = 0.5), or low (b = 0.3) dispersal for worms living in the hotspot habitat.

Discussion

The results of the simulations performed here have implications for extrapolating effects of toxic chemicals, measured in the laboratory, to field conditions, for understanding recolonization of contaminated habitats and the recovery potential of affected populations.

Typically, sediment (and other) toxicity tests are performed under homogeneous exposure conditions (with a single exposure concentration per treatment group), and rarely, if ever, are test organisms intentionally exposed to spatially patchy contamination. Yet sediment contaminant concentrations in the field often vary over a spatial scale such that members of resident invertebrate populations could experience very different exposures. It is therefore relevant to ask whether averaging exposure for the population as a whole may tend to over- or underestimate risk to the exposed population. Our simulations demonstrate that for *Capitella* sp. S, heterogeneous exposure (i.e., a hotspot) had a "sink effect" on the metapopulation, because λ for the hotspot subpopulation was lower than 1. To remove the sink effect, the hotspot had to be diluted by adding more clean patches until the overall average concentration in the hotspot habitat was considerably lower than the average concentration in the homogeneously contaminated habitat. This implies that extrapolation of responses from homogeneous laboratory exposures to spatially heterogeneous field conditions could underestimate the population risk, particularly for species with limited dispersal. With respect to recolonization, both *Capitella* sp. S and *Capitella* sp. I showed a delay (2–6 weeks and ~3 weeks, respectively) in recolonization (time to reach 95% of K) in the hotspot habitat compared to the homogeneous habitat. Thus, patchiness may reduce the rate at which recolonization of contaminated sediments occurs, and the degree to which patchiness slows down recolonization will depend on, among other things, the dispersal abilities of resident species.

Previous studies have emphasized the importance of considering life-history differences when extrapolating toxicant (or other stressor) effects to the population level from measures of individual performance (Calow et al. 1997; Forbes and Calow 1999, 2002; Forbes et al. 2001), although they have not considered differences in dispersal ability. Liess and Von der Ohe (2005) explicitly include dispersal or migration potential as an important variable characterizing "species at risk" from, for example, pesticide exposure. Our simulations confirm that dispersal ability is a very important parameter to consider when estimating the effect of contaminant patchiness on the dynamics of exposed populations. We observed that a decrease in dispersal rate seriously increased the degree of dilution needed to remove the sink effect of a contaminant hotspot in an otherwise clean sediment (figure 13.7). In addition, we observed that the extent to which the spatial location of the contaminated patch relative to the source patch influenced the effect on the metapopulation (i.e., K and time to reach K) was crucially dependent on the dispersal between patches (figure 13.8). Furthermore, we found that not only the degree of patchiness but also the spatial distribution of patches with respect to each other (i.e., relative position of donor vs. contaminated patches) influenced the dynamics of the metapopulation.

Extrapolating the effects of chemicals measured on organisms under controlled laboratory conditions to the field is extremely challenging. Current approaches ignore many of the complexities present in natural ecological systems and therefore add an unknown degree of uncertainty to ecological risk assessments. Population models provide powerful and cost-effective tools for exploring the relationships between organism- and population-level responses to chemicals. They can be used to generate testable hypotheses about the roles of biological (e.g., life history, dispersal ability) and environmental (e.g., spatial and temporal variation) factors in influencing the risks of chemicals to exposed populations and in focusing experimental studies and monitoring strategies.

Acknowledgments We thank Richard Sibly for valuable feedback on the analyses. This work was funded by a Centre of Excellence Grant from the Danish Natural Sciences Research Council to Valery Forbes.

References

Akçakaya HR. 2002. *RAMAS GIS: Linking Spatial Data with Population Viability Analysis (version 4.0).* Applied Biomathematics, Setauket, New York, USA.

Bridges TS, Levin LA, Cabrera D, Plaia G. 1994. Effects of sediment amended with sewage, algae, or hydrocarbons on growth and reproduction in two opportunistic polychaetes. *Journal of Experimental Marine Biology and Ecology* 177: 99–119.

Calow P, Sibly R, Forbes VE. 1997. Risk assessment on the basis of simplified population dynamics' scenarios. *Environmental Toxicology and Chemistry* 16: 1983–1989.

Caswell H. 1989. Analysis of life table response experiments I. Decomposition of effects on population growth rate. *Ecological Modelling* 46: 221–237.

Caswell H. 1996. Analysis of life table response experiments II. Alternative parameterizations for size- and age-structured models. *Ecological Modelling* 88: 73–82.

Eckelbarger KJ, Grassle JP. 1983. Ultrastructural differences in the eggs and ovarian follicle cells of *Capitella* (Polychaeta) sibling species. *Biological Bulletin* 165: 379–393.

Eckelbarger KJ, Grassle JP. 1987. Interspecific variation in genital spine, sperm, and larval morphology in six sibling species of *Capitella. Bulletin of the Biological Society of Washington* 7: 62–76.

Forbes VE, Calow P. 1999. Is the per capita rate of increase a good measure of population-level effects in ecotoxicology? *Environmental Toxicology and Chemistry* 18(7): 1544–1556.

Forbes VE, Calow P, Sibly RM. 2001. Are current species extrapolation models a good basis for ecological risk assessment? *Environmental Toxicology and Chemistry* 20(2): 442–447.

Forbes VE, Calow P. 2002. Population growth rate as a basis for ecological risk assessment of toxic chemicals. *Philosophical Transactions of the Royal Society Series B* 357: 1299–1306.

Gamenick I, Giere O. 1994. Population dynamics and ecophysiology of *Capitella capitata* from North Sea intertidal flats. Evidence for two sibling species. *Polychaete Research* 16: 44–47.

Gamenick I, Vismann B, Grieshaber MK, Giere O. 1998. Ecophysiological differentiation of *Capitella capitata* (Polychaeta). Sibling species from different sulfidic habitats. *Marine Ecology Progress Series* 175: 155–166.

Grassle JF, Grassle JP. 1974. Opportunistic life histories and genetic systems in marine benthic polychaetes. *Journal of Marine Research* 32: 253–284.

Grassle JP, Grassle JF. 1976. Sibling species in the marine pollution indicator *Capitella* (Polychaeta). *Science* 192: 567–569.

Grassle JP, Gelfman CE, Mills SW. 1987. Karyotypes of *Capitella* sibling species, and of several species in the related genera *Capitellides* and *Capitomastus* (Polychaeta). *Bulletin of the Biological Society of Washington* 7: 77–88.

Hansen FT, Forbes VE, Forbes TL. 1999. The effects of chronic exposure to 4-n-nonylphenol on life-history traits and population dynamics of the polychaete *Capitella* sp. I. *Ecological Applications* 9: 482–495.

Kammenga JE, Busschers M, Van Straalen NM, Jepson PC, Bakker J. 1996. Stress induced fitness reduction is not determined by the most sensitive life-cycle trait. *Functional Ecology* 10: 106–111.

Levin LA, Caswell H, DePatra KD, Creed EL. 1987 Demographic consequences of larval development mode: planktotrophy vs. lecitotrophy in Streblospio benedicti. *Ecology* 68: 1877–1886.

Levin LA, Caswell H, Bridges T, DiBacco C, Cabrera D, Plaia G. 1996. Demographic responses of estuarine polychaetes to pollutants: life table response experiments. *Ecological Applications* 6: 1295–1313.

Liess M, Von der Ohe P. 2005. Analyzing effects of pesticides on invertebrate communities in streams. *Environmental Toxicology and Chemistry* 24: 954–965.

Linke-Gamenick I, Forbes VE, Sibly RM. 1999. Density-dependent effects of a toxicant on life-history traits and population dynamics of a capitellid polychaete. *Marine Ecology Progress Series* 184: 139–148.

Linke-Gamenick I, Vismann B, Forbes VE. 2000a. Effects of fluoranthene and ambient oxygen levels on survival and metabolism in three sibling species of *Capitella* (Polychaeta). *Marine Ecology Progress Series* 194: 169–177.

Linke-Gamenick I, Forbes VE, Méndez N. 2000b. Effects of chronic fluoranthene exposure on sibling species of *Capitella* with different development modes. *Marine Ecology Progress Series* 203: 191–203.

Méndez N, Linke-Gamenick I, Forbes VE. 2000. Variability in reproductive mode and larval development within the *Capitella capitata* species complex. *Invertebrate Reproduction and Development* 38: 131–142.

Wu BL, Qian PY, Zhang SL. 1991. Morphology, reproduction, ecology and allozyme electrophoresis of three *Capitella* sibling species in Qingdao (Polychaeta: Capitellidae). *Ophelia* 5 (Suppl): 391–400.

14

Application of Population Modeling to a Causal Analysis of the Decline in the Cherry Point Pacific Herring (*Clupea pallasi*) Stock

WAYNE G. LANDIS

A recent Pellston workshop sponsored by the Society of Environmental Toxicology and Chemistry (SETAC) focused on the issue of risk assessments at the population scale (Barnthouse, Muns and Sorenson, 2007) and discussed the issues and problems of performing risk assessments upon valued populations. Populations provide important ecological services and are often the focus of watershed and regional-scale ecological risk assessments (Landis 2006). Recent discussions have underlined the importance of populations in risk assessment and have attempted to define a population for this application (Landis 2002; Munns et al. 2002). Populations can also be regarded as the stressor in a risk assessment, especially if an invasive species or disease organism is under consideration (Deines et al. 2005).

A fundamental question is how to evaluate risks upon populations. Extrapolations using models and concentration responses from laboratory toxicity tests have been conducted (Forbes et al. 2001; Calow and Forbes 2003; Lin et al. 2005; Stark and Banks 2003). Stark et al. (2004) used extrapolations from laboratory chronic toxicity tests to evaluate the importance of life-history characteristics at the population scale.

In an elegant and detailed paper, Spromberg and Meador (2005) looked specifically at the effects of chronic toxicity upon the characteristics of salmon populations. A critical finding is the effect of contaminants upon the age structure of chinook populations, and this pattern as a source of signal about cause and effect.

Collier (2003) and Adams (2003) have described the application of a weight of evidence (WoE) approach to the establishment of causality in ecological systems. Seven criteria are important to meet in establishing causality: (1) strength of association, (2) consistency of association, (3) specificity of association, (4) time order or temporality, (5) biological gradient, (6) experimental evidence, and (7) biological plausibility. These are described in more detail in table 14.1.

Table 14.1. Criteria for causality and the application of population modeling: Application of modeling to the analysis of the CPPHS (based on Adams 2003; Collier 2003).

Causal criteria	Description	Application of population modeling	Application to CPPHS analysis
1. Strength of association	Cause and effect coincide.	Investigation of the lag time between onset of the cause and the observability of the effect at a population scale.	The hypothesized cause, changing of the ocean climate as part of the PDO, does precede the decline of the rates of increase and the alteration of age structure that preceded the detectable decline of the 1980s.
2. Consistency of association	The association between a particular stressor or stressors and an effect has been observed by other investigators in similar studies and at other times and places.	Use of models from prior studies to estimate the magnitude and type of effect from the suspected causes.	The population model did not specifically address this issue, but the switching of fisheries because of changes in ocean conditions is well documented (see text).
3. Specificity of association	The effect is diagnostic of exposure.	Calculation of the range of effects that can be observed in a population from different types of stressors (e.g., endocrine disruption, general mortality, reduction in fecundity).	The collapse of the age structure that ultimately was observed in the CPPHS was consistently predicted in the normalized equilibrium age structures calculated from surveys in the late 1970s and early 1980s. The consistency of pattern indicates that the same type of stressor was in place during the entire period. The effects seen in the field or in the model are not consistent with chemical stressors that affect reproduction or development that would cause a decline of the younger age classes.
4. Time order or temporality	The cause precedes the effect in time, and also the effect decreases when the cause is decreased or removed.	Modeling can be useful in setting the likely onset of measurable effects resulting from of a stressor agent. Application of models incorporating stochastic components can aid in setting a range of expected variability so that fluctuations of populations are not misinterpreted.	The compression of the normalized equilibrium age classes observed in the field surveys points to a causative agent that reduces the survivorship of older age classes being present since the 1970s. Although the population did demonstrate increases during the mid-1990s, the calculated intrinsic rates of growth and the normalized equilibrium age structure indicated that the stressor was still in place.

5. Biological gradient	There is a dose–response relationship either spatially or temporally within the system. The risk of an effect is a function of magnitude of exposure.	Population models can provide an indication of the shape of the dose–response relationship (linear, sigmoid). Spatially explicit models can also provide a prediction of the spatial distribution of the effect, which may appear distant from the initial cause (Spromberg et al. 1998).	
6. Experimental evidence	Valid experimental studies support the proposed cause–effect relationship.	Modeling can assist in the design of experiments, predict outcomes, and provide a tool for extrapolation.	Because of the length of the survey program, data generated from one period can be used to examine its predictive capability. For example, the decline of the CPPHS from 1994 to 2004 is very similar to that of the model output based on survivorship and growth data from 1990 to 1998. Deviations from model predictions at this point would lead to suggestions that other stressors are now coming into play or that ocean conditions have changed.
7. Biological plausibility	There is a credible or reasonable biological and/or toxicological basis for the proposed mechanism linking the proposed cause and effect.	A population model can provide a framework for connecting disparate lines of evidence into a causal pathway for examining the plausibility of a cause and effect.	The modeling supports the conclusions reached from other studies that the change in ocean conditions has depressed the CPPHS fishery.

In this chapter I apply population modeling within a causality framework to examine causes of the decline of the Cherry Point Pacific herring (*Clupea pallasi*) stock (CPPHS) at Cherry Point, Washington, USA (Landis et al. 2004). The age structure, intrinsic rate of increase, and variability in these features are used to examine potential causes of the decline and the timing of the impact. I estimate future dynamics of the CPPHS and compare this to data collected since the origination of the original data set. This extends the approach of Landis et al. (2004, 2005) by using the RAMAS computational tools to examine the features of this age-structured population in detail.

The next sections present the application and results of this analysis. First, the characteristics of the CPPHS are reviewed and updated to the spring of 2004. Next, the process of applying the tools in RAMAS to the causal analysis is described. The results of the modeling are presented and applied to the causal framework. The conclusion is then presented that the causal factor was present by at least the late 1970s to early 1980s and is present to this day. As previously suggested by Landis et al. (2004) and Hershberger et al. (2005), there are indications that the causal agent is not related to Cherry Point but operates on a larger scale. Finally, the application of population modeling to a WoE is discussed, specifically within the context of meeting the criteria for causality.

Background on the Cherry Point Pacific herring stock

The characteristics of the CPPHS and the Cherry Point region have been presented previously (EVS Environmental Consultants 1999; Landis et al. 2004, 2005; Hart Hayes and Landis 2005; Markiewicz 2005) and are summarized below.

The Pacific herring fishery in Puget Sound first began in the early 1890s. The bulk of the herring catch was taken during spawning when the herring moved into the nearshore areas. The historical catch records indicate that the population dynamics and harvest of Pacific herring have been highly variable. In 1957, the General Purpose Fishery was opened, and annual landings of herring in Washington State waters increased from an average of 305 short tons (240,157 kg) per year to 3,259 short tons (2.567×10^6 kg) per year up through 1976. In 1977, a general decline began in most of the herring stocks in Washington waters. Subsequent annual landings decreased to an average of 1,278 short tons (1.006×10^6 kg) in 1996.

Two clear trends appear from plotting the numbers of adult fish of the CPPHS since the early 1970s. The first is a general decline in the population, although the total number of fish fluctuates widely (figure 14.1). Simultaneously, there has been a change of the age structure of the population (figure 14.2). The population of the 1970s consisted of older age class fish, up to ages 8 and 9. In the late 1990s, the population consisted of fish younger than age 4. Fishing was halted in the mid-1990s.

Although reproduction rates declined, no decrease in survivorship from egg to age 2 was observed in the data. This is based upon a number of assumptions about egg production of different age classes and the catchability of age 2 fish. The only exception was an increase in survivability during 1991 and 1992.

During the early 1970s, there was a surplus of age 3 fish compared to age 2 fish counted the previous year. This could have been due to a delay in age 2 fish coming to shore to spawn or migration from another population. Recent genetic analysis by Small et al. (2005) makes the immigration hypothesis unlikely.

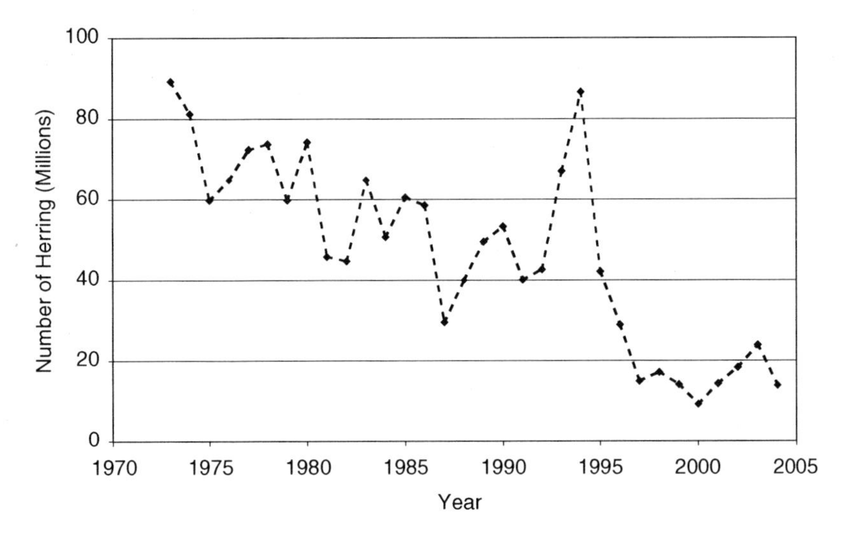

Figure 14.1. Decline of the Cherry Point Pacific herring stock (CPPHS) at Cherry Point, 1973–2004. Data from EVS Environmental Consultants (1999) and Stick (2005).

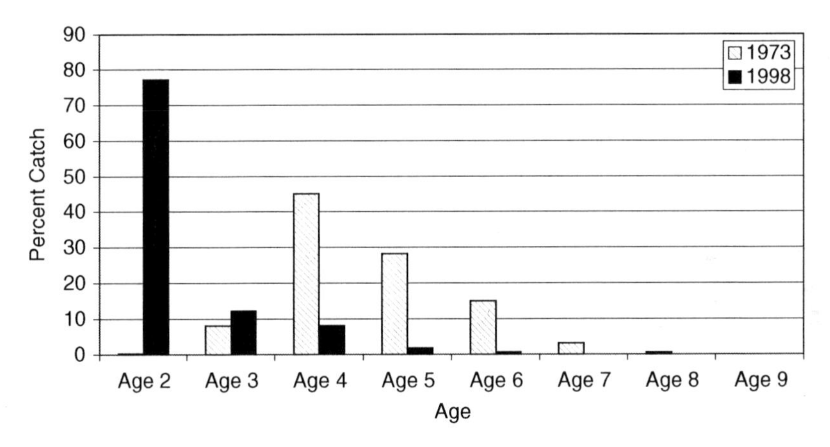

Figure 14.2. Comparison of age structures of the 1973 CPPHS compared to 1998. In 1998, no age class 7–9 fish were identified in the survey samples.

Hershberger et al. (2005) have demonstrated that CPPHS has a high rate of larval abnormalities compared to Puget Sound stocks. Eggs incubated in water from other sites exhibited similar rates of abnormalities. Eggs from other stocks incubated in water from Cherry Point do not exhibit a corresponding increase in abnormalities.

Stick (2005) has reviewed the status of Pacific herring in Washington State, including the CPPHS. This report includes the 2004 spawning data and age structure that are compared to the model output in the analysis presented here. CPPHS is still at low numbers; however, the age structure may be showing signs of aging in the last several spawning seasons.

One of the important features of the Cherry Point region is the Pacific Decadal Oscillation (PDO). This is an important driver of the marine community and therefore of the fisheries in the region.

The Pacific Decadal Oscillation

Mantua and Hare (2002) have reviewed the presence and effects of the PDO. It is a decades-long, 20- to 30-year shift in ocean temperature regime with periods of cooler ocean water followed by a warming trend that has been documented since 1900. Paleoclimatic reconstructions have detected the pattern of the PDO back into the seventeenth century.

A shift in temperature occurred in 1977 from a colder to a warmer ocean regime (Hare and Mantua 2000). A subsequent shift occurred in 1989 that was more subtle and was signaled by shifts in the biological community. This shift was not a return to pre-1977 conditions and was not widespread (Hare and Mantua 2002). The current warm period has lasted at least until 1998 (Mantua and Hare 2002). The current records compiled by the Joint Institute for the Study of the Atmosphere and Ocean (http://jisao.washington.edu/pdo/) do not show a clear shift by 2006 to a cold PDO.

It is now apparent that regime shifts in climate alter fisheries. Prior to the recognition of the PDO, several researchers discovered links between climate and fisheries. Mysak (1986) documented the relationships between El Niño events and the interannual variability of climate upon the fisheries in the Pacific waters of the North American West Coast. He predicted broad changes in community structure and fisheries such as Pacific herring, bluefin tuna (*Thunnus thynnus*), and Fraser River sockeye salmon (*Oncorhynchus nerka*). Beamish and Bouillon (1993) document the importance of climate on the production of Pacific salmon. Mantua et al. (1997) demonstrated the correspondence to the reversals of the PDO in 1925, 1947, and then 1976/1977 with shifts in salmon production in the Pacific. Mantua and Hare (2002) have compiled a wide variety changes in marine and terrestrial systems due to the PDO regime changes.

Landis et al. (2004) suggested that a large-scale feature was responsible for the change in the productivity of the CPPHS. The PDO has the features as described in the retrospective analysis.

Cherry Point Pacific herring retrospective assessment

Landis et al. (2004, 2005) described the use of WoE and path analysis to evaluate the decline of the Cherry Point Pacific herring. The WoE approach was based upon a risk assessment type of conceptual model in order to link the paths of potential sources of stressors to the effects seen in the population. Ranking criteria and regressions were used to assign weights to the potential sources and stressors. A Monte Carlo analysis was applied to represent the uncertainty in each of the ranks, correlations, and filters and to estimate the uncertainty of the analysis. This technique resulted in a series of multinomial distributions representing the likelihood of a stressor causing an impact. The conclusions of the Landis et al. (2004, 2005) analysis concerning the CPPHS were as follows:

> The population declines observed at Cherry Point and the compressed age structure in common to Puget Sound stocks are due to large-scale events, such as habitat loss,

the Pacific Decadal Oscillation and the resultant changes in the community structure and dynamics of the Georgia Basin. Using alternative and independent analyses, Hershberger et al. (2005) and Stick (2005) have reached similar conclusions.

Contaminants are possibly an important stressor, but there is considerable uncertainty in the linkage of toxicity to changes in age structure and population decline at large scales.

One of the sources of uncertainty in the Landis et al. (2004, 2005) analysis was the difficulty linking the impacts of contaminants to a specific pattern of changes in population dynamics and structure observed in the CPPHS. The lack of predictive capability also made it difficult to differentiate the patterns caused by potential contaminants from those of disease or other factors.

The timing of the 1976–1977 shift of the PDO to a warm regime is clearly suggestive of a link between regime change and the decline of the CPPHS. In spite of the warm PDO, the CPPHS has exhibited a number of fluctuations, with some of the population numbers of the early 1990s rivaling that of the 1970s. Modeling of the characteristics of the CPPHS and combining that with a causal analysis or a WoE approach should be informative. The following section presents the sources of data and the application of RAMAS to the analysis of the CPPHS.

Population-Level Analysis of the CPPHS

Model construction

Data on the sampling of the Pacific herring were obtained from EVS Environmental Consultants (1999) and updated by Stick (2005). These data include estimates of biomass for each age class, and these are converted to estimated numbers. Survivorship tables from 1974 to 2004 are generated from these estimates (table 14.2), as well as age structure. Mean survivorships and the standard deviations are calculated from these transformed values. Fertility data are from Chapman et al. (1941). These data sets are used to generate survivorship from egg to age 2, the stage matrix, and initial population sizes used in the modeling exercises.

A model was constructed in RAMAS GIS, based upon the CPPHS counts from 1990 and a transition matrix generated from the averages and standard deviations from 1982–1998. The model was run for 50 years with 30 replications. Constraints were in effect, and density dependence was set at the exponential, meaning that no density dependence was assumed for the vital rates. Demographic stochasticity was used. Equilibrium age structure, intrinsic rate of increase of the population (λ), and sensitivity matrices were calculated for the CPPHS for 1974–2004.

Sources of uncertainty

The data for the CPPHS were collected over a 30-year period, with slightly different methods and personnel during this time span. However, there were significant efforts to ensure the reliability of the data set over this time period (Stick 2005). Estimates are made of biomass and size, and those numbers are converted to counts of individuals of a particular age class.

Table 14.2. Estimated survivorship matrix for the CPPHS from 1974 to 1998 from Landis et al. (2004) and from 1999 to 2004 derived from data in Stick (2005) (boxes: survivorships > 1.0).

Year	Age 2–3	Age 3–4	Age 4–5	Age 5–6	Age 6–7	Age 7–8	Age 8–9
1974	79	2.32	0.64	0.6	0.46	0.37	0.25
1975	30.5	0.44	0.62	0.54	0.5	0.36	0.2
1976	44	0.95	1.14	0.86	0.76	0.57	0.55
1977	3.96	0.69	0.44	0.63	0.76	0.6	0.93
1978	2.89	0.58	0.8	0.61	0.52	0.46	0.64
1979	7.36	0.65	0.79	0.69	0.96	0.65	0.81
1980	1.85	0.63	0.38	0.71	0.62	0.3	0.65
1981	0.55	0.84	0.69	0.57	0.21	0.19	0.38
1982	1.5	0.51	0.64	0.4	0.18	0.2	0
1983	0.76	0.96	1	0.82	0.5	0.9	0.67
1984	0.26	0.46	0.64	0.4	0.36	0.43	0.33
1985	0.91	1.07	0.25	0.17	0.13	0.2	0.33
1986	0.6	0.24	0.31	0.87	0.3	0.33	0
1987	0.33	0.27	0.21	0.17	0.15	0.33	0.5
1988	1.3	0.6	0.48	0.42	0.25	0	0
1989	0.51	0.31	0.29	0.05	0	0	
1990	0.52	0.4	0.65	0.68	1		
1991	0.61	0.43	0.9	0.56	0.15	1	0
1992	0.78	0.47	0.24	0.15	0.11	0	0
1993	0.28	0.27	0.18	0.47	0.75	0	
1994	0.17	0.32	0.32	0.36	0.11	0	
1995	0.25	0.25	0.23	1	0.2	0	
1996	0.53	0.43	0.48	0.4	0.43	0	
1997	0.7	0.41	0.05	0.09	0	0	
1998	0.54	0.23	0.07	0.25	0		
1999	0.85	0.38	0.01	0.06	0.00		
2000	0.92	0.24	0.04	0.00			
2001	0.79	0.49	0.15	0.00			
2002	0.72	0.18	0.09	0.52			
2003	0.54	0.18	0.16	0.24	0.00		
2004	0.51	1.27	3.74	3.30	0.65	0.00	

Unlike many population models, the analysis of the CPPHS is on those fish that arrive to spawn at the Cherry Point site. Age 1 animals are not counted since the fish do not spawn in the region until at least age 2. Rates of survivorship from egg are therefore based on egg to age 2. In order to accommodate the RAMAS environment, the calculated survivorship was input as if it were to age 1, and the transition value for age 1–2 fish was set at 1. No reproductive rate was included for age 1 fish.

Since only the spawning population is counted, fish not identified in the surveys may be part of the population. This may account for the survivorship being greater than 1 for several of the populations observed in the 1970s and the 2004 collection. Most often, this occurred in the transition from age 2 to age 3 fish, but also occasionally occurred in the transition from age 3 to 4 and age 4 to 5. This finding suggests that there is a cohort that skips spawning seasons. The genetics data of Small et al. (2005) indicate that the occasional surplus in an age class is not due to immigration from other populations.

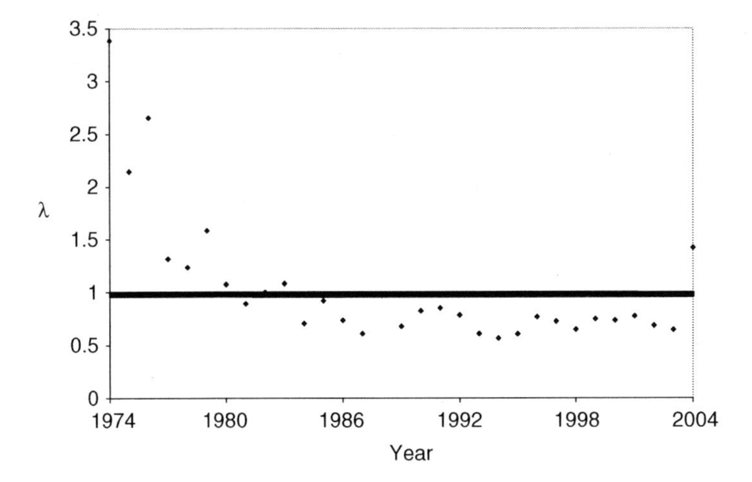

Figure 14.3. Plot of λ as calculated by RAMAS for the CPPHS. Since the early 1980s, the value of λ has been generally less than 1.0. The switch to the warm PDO occurred in 1977. A switch to a cooler period may have begun in 1998, but it is not clear that this date is correct.

A critical component is an estimate of fecundity within the population, and this was estimated as in Landis et al. (2004). Age-specific fecundity calculations were based on Chapman et al. (1941), which counted eggs for age 2–8 Pacific herring from Puget Sound. A regression was used to fit a line to describe the relationship between age and fecundity. The regression was significant ($p < 0.05$), and the residuals displayed no apparent bias. Since the CPPHS is a genetically unique stock, there is uncertainty in these estimates of fecundity. However, the similarity of the patterns seen in the model output and the dynamics of the CPPHS (see below) demonstrates that these estimates reflect the characteristics of this population.

Results

There are two segments to the remainder of the present analysis. The first segment is an analysis of the CPPHS from 1973 until 1998, the period of analysis covered by Landis et al. (2004, 2005). The second segment is the period from 1999 until 2004, as stock appears to be changing its age structure and the PDO enters a colder phase.

The calculated intrinsic rate of increase (λ) is plotted during the study period in figure 14.3. During the 1970s, λ was at a very high value with a decline throughout the decade. After 1980, λ is generally below 1.0, indicating that the population growth rate is negative during this period. The value ranges from approximately 1.08 (1980, 1983) to a low of 0.57 for 1994.

Also calculated as part of the RAMAS process is the normalized equilibrium age structure. Examples of the normalized age structure for 1978, 1988, and 1998 are presented in figure 14.4. The normalized equilibrium age structure for 1978 is broad and includes age 9 fish. This pattern mimics the pattern observed in the surveys of CPPHS

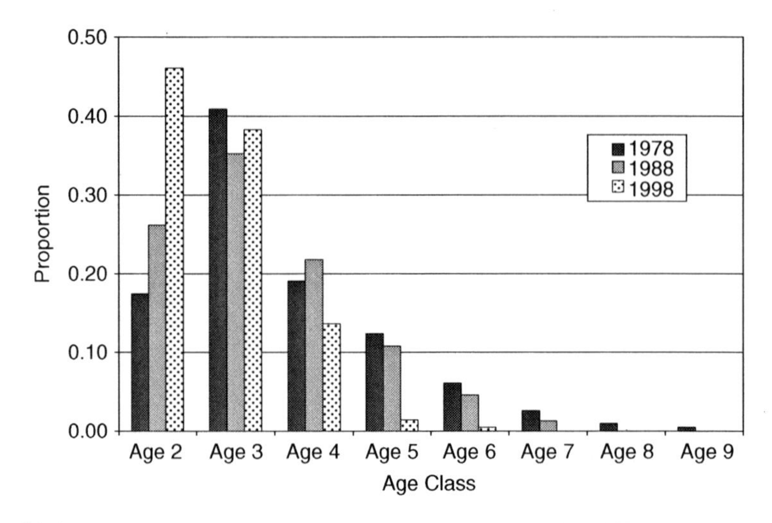

Figure 14.4. Normalized equilibrium age structure for 1978, 1988, and 1998 CPPHS populations.

of that period. In 1988, the pattern shows a lack of age 8 and 9 fish, but the age 3 class is still largest. By 1998, the calculated normalized equilibrium is even more compressed, with age 2 fish the largest component and no age 7 fish. This calculation is very similar to the 1998 survey (compare figures 14.2 and 14.4).

The means and standard deviations of survivorship were used to estimate the population dynamics of the CPPHS from two points. The first time period is the maximum that occurred in 1990, using the population numbers for each age class and the total. The second time period is the maximum in 1994 that rivaled the populations recorded for the early 1970s but with a very different age structure. The results are presented in figure 14.5.

Using 1990 as a starting point, the model underestimated the number of fish, especially from 1994 onward. The number of fish observed at Cherry Point was larger than any of the simulations, and this increase in population size resulted in the model underestimating population size (figure 14.5A). When 1994 was used as the starting population, the model did a better job of predicting the population dynamics of the CPPHS. The observed population was within the maximums and minimums observed in the simulations and was often within the range of the standard deviations (figure 14.5B).

Discussion

Criteria for causality

At this point, it is most useful to examine the criteria for causality and discuss what characterizes the stressor agent as demonstrated by the field surveys and the modeling analysis (table 14.1). Each of the seven criteria is examined below regarding the PDO shift and the application of the results of the RAMAS modeling effort.

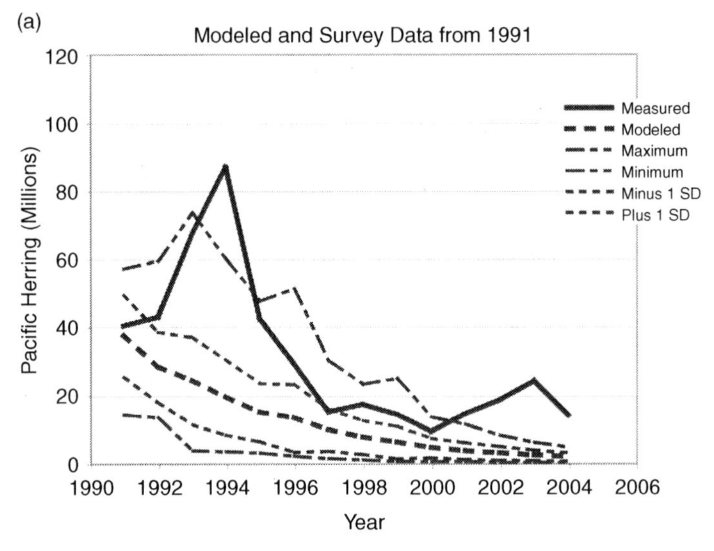

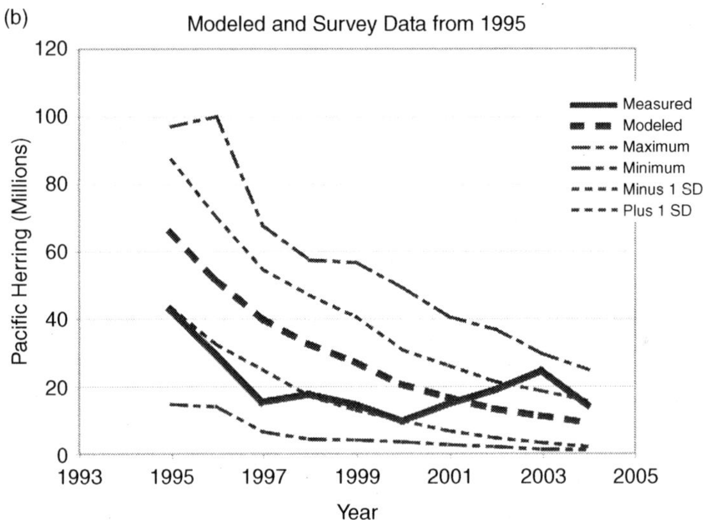

Figure 14.5. Model results and survey data compared. (A) Summary of 30 simulations of the fate of the CPPHS using the transition matrix and standard deviations, starting with the 1990 population counts. (B) Summary of the simulation from 1994 to 2004 compared to the survey data. Minus 1 SD, the mean less than one standard deviation; Plus 1 SD, the mean plus one standard deviation. The maximum and minimum are those calculated for the 30 simulations performed for each modeling series.

1. Strength of association

The hypothesized cause, changing of the ocean climate as part of the PDO, does precede the decline of the rates of increase and the alteration of age structure that preceded the detectable decline of the 1980s. The λ values started to decrease in the late 1970s compared to the highs of the early 1970s. By 1980, the value of λ was near 1 or below for the remainder of the data set, except for 2004.

2. Consistency of association

The population model did not specifically address this issue, but the switching of fisheries because of changes in ocean conditions is well documented (see Mantua and Hare 2002).

3. Specificity of association

The collapse of the age structure that ultimately was observed in the CPPHS was consistently predicted in the normalized equilibrium age structures calculated from surveys in the late 1970s and early 1980s. The consistency of pattern indicates that the same type of stressor was in place during the entire period of the decline of the CPPHS.

The effects seen in the field or in the model are not consistent with chemical stressors that affect reproduction or development that would cause a decline of the younger age classes (Spromberg and Meador 2005). A persistent bioaccumulative pollutant may affect the survivorship of the older age classes, but reproductive effects are known for PCBs and dioxins, two of the expected classes of agents. However, Landis et al. (2004) did not document a decrease in survivorship from egg to age 2 in the CPPHS.

4. Time order or temporality

A change in λ and the compression of the normalized equilibrium age classes that began in the late 1970s and eventually stabilized in the 1980s correspond to the regime shift to a warm PDO. The compression of the age structure was then seen in the field surveys from the 1980s and especially in the late 1990s to early 2000s.

Although the population did demonstrate increases during the early to mid-1990s, the calculated intrinsic rates of growth and the normalized equilibrium age structure indicated that the stressor was still in place. Again, this corresponds to the documented period of the warm regime of the PDO.

5. Biological gradient

Because the scale of the PDO is larger than the range of the CPPHS, a clear biological gradient will be difficult to document. Analysis of other Pacific herring stocks in Puget Sound and Georgia Straits may provide a north–south gradient along which the effects of the PDO could be documented.

6. Experimental evidence

Direct experimental manipulation at the scale of this region is not possible. However, the length of the survey program allows data generated from one period to be used to examine the predictive capability of the modeling effort. For example, the decline of the CPPHS from 1994 to 2004 is very similar to that of the model output based on survivorship and growth data from 1990 to 1998. As the natural experiment of the

PDO shift to a colder regime occurs, the modeling should be able to detect changes in equilibrium age structure and λ that would predict an increase in the CPPHS.

7. Biological plausibility

The modeling supports the conclusions reached from other studies that the change in ocean conditions has depressed the CPPHS fishery. As discussed above, changes in regime shift affect a wide number of fisheries and the coastal ecological structure.

The PDO as the causal factor

The preceding analysis is consistent with the change of ocean regime being the cause of the decline of the CPPHS since the late 1970s. The intrinsic rate of increase and the change in the equilibrium age structure of the CPPHS revealed the signal before it was clear that a decline was occurring. Even during the fluctuations of the early 1990s, the features of the population derived from the field data indicated that the impact was still present.

The greater unknown is the specific pathway that creates the decline in the CPPHS. Oceanographic conditions, including temperature, influence mass of the adults and consequently the fertility of the females. Changes in prey density due to change in community structure and temperature may cause an overall decrease in size and fecundity of adult fish.

Disease may also be a causal mechanism related to the PDO and its effects. Hershberger et al. (2002) has found that the incidence of the parasite *Ichthyophonus hoferi* increases with the age of Pacific herring. A connection between disease prevalence and changes in ocean conditions for the Georgia Strait–Puget Sound region has not been made, but the pathway should be investigated.

What to measure: λ, normalized equilibrium age structure?

One of the questions in population-scale ecological risk assessment is the appropriate endpoint. The intrinsic rate of increase or λ has been proposed by Lin et al. (2005) and did indicate that the CPPHS was in decline starting in the late 1970s to early 1980s. Small changes in λ can result in large differences in the dynamics of populations (Stark et al. 2004; Spromberg and Meador 2005). Estimated λ values from the survey data at Cherry Point varied widely even after the onset of the decline. The variability in these estimates is likely due to the challenge of measuring the numbers of the CPPHS as well as real changes in the value. It is useful to know if the population is increasing or decreasing; however, this measure does not provide diagnostic information regarding the cause of the change in rate of increase.

The more informative attribute of the CPPHS was the compression of the age structure to the younger age classes. This feature was apparent both in the survey data and in the calculated equilibrium age structure. Both indicated that the causative agent reduced the survivorship of the older age classes. Because it may take several years for the survey data to demonstrate a change in age structure, the calculated equilibrium can provide an early indication of the direction. Spromberg and Meador (2005) have also modeled the impacts of different types of toxicants upon age structure and found that this is an important diagnostic variable at the population scale.

Application of population models to risk assessment

Can population models be applied to risk assessment? In the case of causality for the decline of the CPPHS, the answer is yes. Population modeling could be used to examine the causality criteria as presented in table 14.2 to examine likely scenarios. In this instance, the scale and time period of the PDO has led this feature of the eastern Pacific to be implicated even if a complete causal pathway is not clear.

It is also apparent that there are limitations to what is possible to measure. A value such as λ is going to have significant variation compared to the magnitude of difference that is known to change population dynamics. Knowing λ with confidence to the third or forth decimal place is an unlikely goal with populations such as the Pacific herring.

Survivorship also has an associated variability (table 14.2). In several instances, the standard deviation is as large as the mean. So if λ and the survivorship values have such high uncertainty, what is an appropriate measure?

The most robust attribute of the CPPHS appears to be the pattern found in the age structure. The compression of the age structure was an early symptom and is persistent during a 20-year period. The specific values may change, but the loss of older age classes and the dominance of age 2 is a clear signal that an impact that persistently alters the population size has occurred.

References

Adams, S.M. 2003. Establishing causality between environmental stressors and effects on aquatic ecosystems. *Human and Ecological Risk Assessment* 9:17–35.

Barnthouse, L., Munns W.R. and Sorenson, M. 2007. *Population Level Ecological Risk Assessment.* SETAC Press, Pensacola, Florida.

Beamish, R.J., and Bouillon, D.R. 1993. Pacific salmon production trends in relation to climate. *Canadian Journal of Fisheries and Aquatic Sciences* 50:1002–1016.

Calow, P., and Forbes, V.E. 2003. Does ecotoxicology inform ecological risk assessment? *Environmental Science and Technology* 37:146A–151A.

Chapman, W.M., Katz, M., and Erickson, D.W. 1941. The Races of Herring in the State of Washington. Biological Report No. 38A. State of Washington Department of Fisheries, Olympia, Washington.

Collier, T.K. 2003. Forensic ecotoxicology: Establishing causality between contaminants and biological effects in field studies. *Human Ecological Risk Assessment* 9:259–266.

Deines, A.M., Chen, V., and Landis, W.G. 2005. Modeling the risks of non-indigenous species introductions using a patch-dynamics approach incorporating contaminant effects as a disturbance. *Risk Analysis* 6:1637–1651.

EVS Environmental Consultants. 1999. Cherry Point Screening Level Ecological Risk Assessment. EVS Project No. 2/868-01.1, prepared for the Washington Department of Natural Resources. EVS Environmental Consultants, Seattle, Washington.

Forbes, V.E., Calow, P., and Sibly, R.M. 2001. Are current species extrapolation models a good basis for ecological risk assessment? *Environmental Toxicology and Chemistry* 20:442–447.

Hare, S.R., and Mantua, N.J. 2000. Empirical evidence for North Pacific regime shifts in 1977 and 1989. *Progress in Oceanography* 47(2–4):103–146.

Hart Hayes, E., and Landis, W.G. 2005. The ecological risk assessment using the relative risk model and incorporating a Monte Carlo uncertainty analysis. Pages 257–290 in W.G. Landis (ed.), *Regional Scale Ecological Risk Assessment Using the Relative Risk Model.* CRC Press, Boca Raton, Florida.

Hershberger, P.K., Stick, K., Bui, B., et al. 2002. Incidence of *Ichthyophonus hoferi* in Puget Sound fishes and its increase with age of adult Pacific herring. *Journal of Aquatic Animal Health* 14:50–56.

Hershberger, P.K., Elder, N.E., Wittouck, J., Stick, K., and Kocan, R.M. 2005. Abnormalities in larvae from the once-largest Pacific herring population in Washington State result primarily from factors independent of spawning location. *Transactions of the American Fisheries Society* 142:326–337.

Landis, W.G. 2002. Population is the appropriate unit of interest for a species-specific risk assessment. *SETAC Globe Newsletter* 3:31–32.

Landis, W.G. 2006. Population scale assessment endpoints in ecological risk assessment part 1: Reflections of stakeholders' values. *Integrated Environmental Assessment and Management* 2:86–91.

Landis, W.G., and Wiegers, J.A. 1997. Design considerations and a suggested approach for regional and comparative ecological risk assessment. *Human and Ecological Risk Assessment* 3:287–297.

Landis, W.G., Duncan, P.B., Hart Hayes, E., Markiewicz, A.J., and Thomas, J.F. 2004. A regional assessment of the potential stressors causing the decline of the Cherry Point Pacific herring run and alternative management endpoints for the Cherry Point Reserve (Washington, USA). *Human and Ecological Risk Assessment* 10:271–297.

Landis, W.G., Hart Hayes, E., and Markiewicz, A.M. 2005. Retrospective regional risk assessment predictions and the application of a Monte Carlo analysis for the decline of the Cherry Point herring stock. Pages 245–256 in W.G. Landis (ed.), *Regional Scale Ecological Risk Assessment Using the Relative Risk Model.* CRC Press, Boca Raton, Florida.

Lin, B., Tokai, A., and Nakanishi, J. 2005. Approaches for establishing predicted-no-effect concentrations for population-level ecological risk assessment in the context of chemical substances management. *Environmental Science and Technology* 39:4833–4840.

Mantua, N.J., Hare, S.R., Zhang, Y., Wallace, J.M., and Francis, R.C. 1997. A Pacific interdecadal climate oscillation with impacts on salmon production. *Bulletin of the American Meteorological Society* 78:1069–1079.

Mantua, N.J., and Hare, S.R. 2002. The pacific decadal oscillation. *Journal of Oceanography* 58:35–44.

Markiewicz, A.J. 2005. The use of regional risk assessment in the management of the marine resources. Pages 227–239 in W.G. Landis (ed.), *Regional Scale Ecological Risk Assessment Using the Relative Risk Model.* CRC Press, Boca Raton, Florida.

Munns, W.R., Jr., Nelson Beyer, W., Landis, W.G., and Menzie, C. 2002. What is a population? *SETAC Globe Newsletter* 3:29–31.

Mysak, L.A. 1986. El Niño, interannual variability and fisheries in the northeast Pacific Ocean. *Canadian Journal of Fisheries and Aquatic Sciences* 43:464–497.

O'Neill, S.M., and West, J.E. 2001. Exposure of Pacific herring (*Clupea pallasi*) to persistent organic pollutants in Puget Sound and the Georgia Basin. In T. Droscher (ed.), *Proceedings of the 2001 Puget Sound Research Conference.* Puget Sound Water Quality Action Team, Olympia, Washington. Available: www.psat.wa.gov/Publications/01_proceedings/sessions/sess_2b.htm.

Small, M.P., Loxterman, J.L., Frye, A.E., Von Bargen, J.F., Bowman, C., and Young, S.F. 2005. Temporal and spatial genetic structure among some Pacific herring populations in Puget Sound and the southern Strait of Georgia. *Transactions of the American Fisheries Society* 134:1329–1341.

Spromberg, J.A., and Meador, J.P. 2005. Relating results of chronic toxicity responses to population-level effects: Modeling effects on wild chinook salmon populations. *Integrated Environmental Assessment and Management* 1:9–21.

Stark, J.D., and Banks, J.E. 2003. Population-level effects of pesticides and other toxicants on arthropods. *Annual Review of Entomology* 48:505–519.

Stark, J.D., Banks, J.E., and Vargas, R. 2004. How risky is risk assessment: The role that life strategies play in susceptibility of species to stress. *Proceedings of the National Academy of Sciences of the USA* 101:732–736.

Stick, K.C. 2005. 2004 Washington State Herring Stock Status Report. Washington Department of Fish and Wildlife SS05-01. Available: wdfw.wa.gov/fish/papers/herring_status_report/index.htm

15

Endocrine Disruption in Eelpout (*Zoarces viviparus*) on the Swedish Baltic Coast

Population-Level Effects of Male-Biased Broods

NIKLAS HANSON

Humans are, and have always been, dependent on the earth's ecosystems. As the human population grows, the needs for ecosystems services (e.g., food production) increase. However, the growth of the human population in combination with increased per capita consumption has also led to large-scale degradation of ecosystems (Vitousek et al. 1997; Hughes and Tanner 2000; Tilman and Lehman 2001; Myers and Worm 2003; Penn 2003). The release of environmental toxicants is closely linked to the consumption of resources and has historically had severe effects on several ecosystem services, for example, restrictions on the consumption of fish from polluted waters. Besides health effects on humans, there is also the risk of a decline, or even extinction, of important species. Ever since the publication of Rachel Carson's *Silent Spring* in 1962, environmental toxicants have been on the political agenda. With increasing knowledge of environmental effects, restrictions and bans have been applied to some of the most examined toxicants, for example, DDT (dichlorodiphenyltrichloroethane) and PCBs (polychlorinated biphenyls).

When a human-made chemical affects the endocrine system of a species, the term "endocrine disruption" is used (European Commission 1996). Many endocrine-disrupting chemicals (endocrine disruptors) can interact with hormone receptors directly and thereby mimic or antagonize endogenous hormones (Witorsch 2002). Environmental toxicants that have indirect effects on the hormone system, for example, by affecting synthesis, storage, release, transport, metabolism and elimination of natural hormones, also fall under the definition of endocrine disruptors (Kavlock et al. 1996). The complexity of the hormone system and the multiple routes of action for endocrine disruption often make it difficult to link an observed effect to a specific route of action or to a specific compound.

The presence of endocrine-disrupting chemicals in the aquatic environment has been shown in several studies. Jobling et al. (1998) demonstrated a high incidence of intersexuality in wild populations of riverine fish in the United Kingdom, something that could be associated with the discharges of hormonally active substances from sewage treatment works. In Sweden, findings of endocrine disruption downstream from a sewage treatment plant has been associated with the synthetic hormone ethinyl estradiol, which is used in contraceptives (Larsson et al. 1999). Some of the classical environmental toxicants (e.g., DDT and PCB) have also been shown to have endocrine-disrupting properties. One example of endocrine disruption coupled to DDT is the reproductive abnormalities found in reptiles in Lake Apopka (FL, USA) after a DDT spill in 1980 (Guillette et al. 1995). There are also examples of suspected endocrine disruption where the causes are not known, for example, the findings of an extremely high proportion of immature female perch (*Perca fluviatilis*) close to a public refuse dump in Sweden (Noaksson et al. 2001, 2003).

One of the first known examples of masculinization in the aquatic environment was the development of male characters in female dog whelk (*Nucella lapillus*), caused by tributyltin used in antifouling marine paints (Gibbs and Bryan 1986). Development of male secondary sex characters in the presence of pulp mill effluents is another example where females have developed malelike phenotypes. This has been shown in a number of fish species, for example, fathead minnow (*Pimephales promelas*), mosquitofish (*Gambusia affinis*), and guppies (*Poecilia reticulata*) (Kovacs et al. 1995; Cody and Bortone 1997; Larsson et al. 2002). The effects on secondary sex characters can be serious, because the reproductive capacity of the population may be affected. Even more alarming, however, is that hormonal treatments during embryonic development can affect sexual differentiation, hence the functional sex, of many fish species (Hunter et al. 1983; Piferrer et al. 1994; Piferrer 2001). Therefore, there is a risk that endocrine-disrupting substances may affect the sex structure of wild fish populations.

Sex reversal at embryonic stages will cause a bias in the sex ratio of recruits (primary sex ratio). A decrease in the proportion females has been recorded in life cycle tests on fathead minnows exposed to different concentrations of pulp mill effluents (Kovacs et al. 1995). In wild populations, however, this effect may be difficult to disentangle from biased sex ratios due to other causes, for example, sex-dependent mortality or a sex bias of catches at sampling. The best way, although difficult, to determine the primary sex ratio is therefore to examine the recruits. The viviparous (i.e., gives birth to live young) eelpout (*Zoarces viviparous*) offers a unique opportunity to examine the sex ratios of whole broods because the embryos undergo sexual differentiation before birth. Studies of the eelpout have shown significantly smaller proportions of female embryos close to a pulp mill on the Swedish Baltic coast during several years of sampling (Larsson et al. 2000; Larsson and Förlin 2002). In reference areas, primary sex ratios were close to 50%. Furthermore, in 1999, the mill had a planned shutdown period just before and during the expected onset of sexual differentiation, and no effect was seen that year (Larsson and Förlin 2002). Altogether, the results of these studies give strong support to the hypothesis that pulp mill effluents can affect the primary sex ratios of nearby eelpout populations.

In this chapter, the population-level response to male-biased sex ratios is examined using population modeling in RAMAS GIS. A reduction in the proportion of females clearly represents a reduction in the fertility of the population, something that may lead to a decrease in population growth and population size. Furthermore, due to stochastic variations in the environment, there is always a risk of local extinction or decline of a species. In this chapter, the effects on predicted population size as well as on the risk of falling below a threshold level are analyzed for different primary sex ratios.

Methods

The population model presented here is partly based on a single-sex model developed and described previously (Hanson et al. 2005). Major changes are the introduction of density dependence and the inclusion of males in the model. Furthermore, minor changes have been made to fit the RAMAS GIS software.

Study species and area

The eelpout is a common fish species in coastal areas of the North Atlantic as well as the Baltic Sea. The eelpout is suitable for environmental monitoring because it has stationary behavior throughout its entire life cycle and is viviparous. Embryological development is two to three weeks long, and the larvae stay in the ovary for three to four months before birth (Vetemaa 1999).

The fecundities and survival rates in this study were parameterized from a data set of catches from Kvädöfjärden Bay, on the Baltic coast of Sweden. The area has been used in Swedish environmental monitoring programs as a reference area because it is located far from known point sources of pollution (Balk et al. 1996; Sandström et al. 2005). The data were provided by the National Board of Fisheries Institute of Coastal Research, Öregrund, Sweden and the sampling has been financed by the Swedish Environmental Protection Agency. Although sex ratios have not been sampled in Kvädöfjärden Bay, the absence of point sources suggests that the population is undisturbed and that sex ratios are close to 50%. Biased sex ratios, as found by Larsson and co-workers (Larsson et al. 2000; Larsson and Förlin 2002), were then applied to the model of an undisturbed population.

Experimental methods

The data set was based on annual catches of female eelpout for the years 1994–2001. The sampling took place in early November each year, which is during the pregnancy period of the females. Among the parameters available in the data set, catch per unit effort (CPUE), age, and the number of fry per female were used for model development. The structure of the data set, and the parameters available, led to the development of an age-structured model with time steps of one year. Due to the type of data available for parameterization, population size in this chapter is expressed in terms of CPUE (i.e., a relative rather than absolute number). The eight years of catch data available enabled estimations of the variability among years.

Estimation of survival

The survival rates could be estimated from data of CPUE and age of the females. Survival (S) of age class x at time t can be calculated from the population size (N) of age classes x and $x + 1$ at times t and $t + 1$:

$$S_x^t = \frac{N_{x+1}^{t+1}}{N_x^t} \tag{15.1}$$

From the data set, it could be seen that age classes 0, 1, and 2 were underrepresented in the catches, something that would lead to an overestimation of the corresponding survival rates when using equation 15.1. Missing information on the number of individuals in age class 0 also made it impossible to calculate the survival of recruits; that is, the probability for a fry to reach age class 0. To estimate the survival of these groups, survival rates were assumed to increase linearly with age up to age class 2. This is described in more detail in Hanson et al. (2005).

Estimation of fecundity

In this model, maternity is the number of living fry (male and female) in the ovary of a female at sampling in November, and fecundity is the number of surviving class 0 daughters, from one specific female, one year later. Fecundity thus requires knowledge of the age-dependent maternity (m_x), proportion of breeding females of the age class (P_x), primary sex ratio (α), and survival rate of recruits (S_{rec}). The fecundity of age class x can then be calculated:

$$F_x = m_x \cdot P_x \cdot \alpha \cdot S_{rec} \tag{15.2}$$

Density dependence

For the age classes with most data available (2 and 3), significant ($p_2 = 0.030$, $p_3 = 0.035$) density dependence was found on survival rates. For the rest of the age classes, trends were similar but not significant ($0.073 < p < 0.288$). For maternity values, no density dependence was found ($0.109 < p < 0.85$). Weighting of the age classes based on size or fecundity values did not improve the correlation values, nor did exclusion of one or several age classes.

The sex ratios found by Larsson and co-workers were recorded during the period from late October to early November of the years 1997–2000. At a site 1.7 km north of the discharge, 45.5%, 42.2%, and 38.7% female embryos were recorded for the years 1997, 1998, and 2000, respectively (Larsson et al. 2000; Larsson and Förlin 2002). Based on these findings, primary sex ratios in the range of 30–50% were used in the population model.

Estimation of vital rates for males

When the primary sex ratio is male biased, the proportion of males in the whole population will increase. It can therefore be assumed that males are not limiting in a scenario with decreased primary sex ratio, something that could justify the exclusion of males from the model. However, the extra proportion of males will affect the population density. Because of the density dependence, an exclusion of males would lead to a conservative model. This is true because the population size for male-biased populations would be underestimated and hence the population growth rate would be overestimated. Because the data set used to parameterize the model is based on females, a few assumptions have to be made. In this model, males are assumed to have the same stages and survival rates as females. To calculate the number of surviving class 0 sons, per average female of age x, the sex ratio in equation 15.2 has to be adjusted:

$$F_x = m_x \cdot P_x \cdot (1 - \alpha) \cdot S_{\text{rec}} \tag{15.3}$$

Population model

All population data discussed below can be found in the file Eelpout.MP on the CD-ROM accompanying this book.

Stage matrix

The eelpout population was divided into nine age classes for each sex, numbered 0 through 8. Age class 0 is assumed to be immature (Vetemaa 1999). Figure 15.1 shows the life cycle graph for females. In the model, five stage matrices each represents one primary sex ratio (50%, 45%, 40%, 35%, and 30%). Comparing the stage matrices shows that the survival rates remain constant in the different scenarios while the fertility values change (in accordance with equations 15.2 and 15.3). The different stage matrices are set up as five populations at five different locations (Model → Populations), something that allows for easy comparison of the results. The five populations can be considered to be located along an effluent gradient, with population 5 (primary sex ratio, 30%) closest to the pulp mill. However, the populations are idealized in the sense

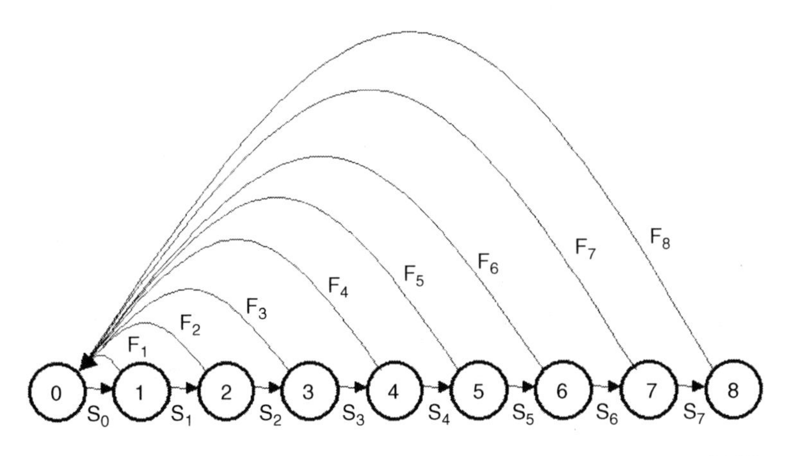

Figure 15.1. Life cycle graph for female eelpout. Modified from Hanson et al. (2005).

that all populations have the same initial abundance, carrying capacity, and so forth. Therefore, the populations can just as well be considered treatments to analyze the factor primary sex ratio at five levels. Exchange of individuals between the populations is set to zero (Model → Dispersal).

Environmental variability

To simulate the effects of natural fluctuations in vital rates, a stochastic model is needed. Standard deviations were calculated from yearly means of survival rates and maternity values. This estimation of variability from catch data also included demographic stochasticity, something that was adjusted for using a procedure suggested in Akçakaya (2002). Based on earlier work (Hanson et al. 2005), lognormal distribution for environmental stochasticity and no correlation between survival and fecundity values were chosen (Model → Stochasticity).

Density dependence

The finding of density dependence on survival rate, but not on fecundity values, was incorporated in the model (Model → Density Dependence). Furthermore, because weighting or exclusion of age classes did not improve the correlation value in the density dependence, all stages were assumed to affect density dependence. The RAMAS GIS software allows for several types of density dependence that are based on carrying capacity (K) and maximum population growth (R_{max}). Carrying capacity was calculated from catch data of females as the population size where the population growth rate (N_{t+1}/N_t) equals 1 (figure 15.2). To account for the male part of the population, however, this value had to be doubled. This is consistent with the assumption that males have the same number of age classes and survival rates as females, and that the primary sex ratio is 50% in an undisturbed area. The maximum annual population growth rate (R_{max}) for the undisturbed population was set at the y-intercept of the linear trend for density dependence (2.0). Because the stage matrix for the undisturbed population is based on mean survival rates and fecundity values of the population during 1994–2001, the eigenvalue (λ) of the matrix and the mean population size from the catch data should fit the density dependence curve resulting from the R_{max} and K values. This was best achieved using the Contest-type density dependence (Model → Density Dependence).

When the sex ratio changes and fecundity values are reduced, the R_{max} value will be reduced, as well. The new R_{max} values for the male-biased populations were set so that the eigenvalues of the stage matrices (with reduced fecundity) fitted to the density dependence curve. Carrying capacity (K) was assumed to be unaffected by the treatment (sex ratio). The K and R_{max} values can be seen in the Populations window (Model → Populations) under the "Density dep" tab. Initial population size was set as the mean of catch data, once again doubled to account for males. The density dependence in relation to the initial population size and eigenvalues of the four stage matrices can be studied under Model → Populations → Display → Density Dependence in R. The eigenvalues of the different stage matrices can also be found under Model → Populations → Display → Finite Rate of Increase (λ). The initial population size was set to 308 individuals (per unit effort) for all treatments (sex ratios).

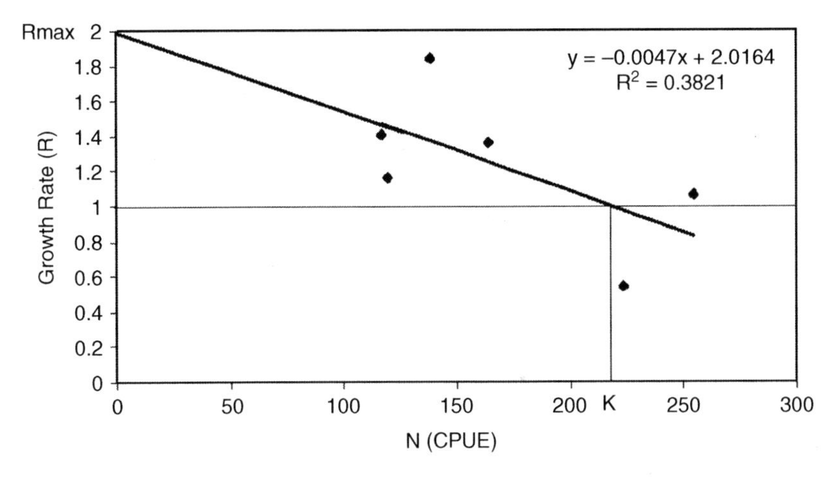

Figure 15.2. Growth rate as a function of population size in Kvädöfjärden Bay: Data from 1994 to 2000. Based on data from the Swedish Environmental Protection Agency.

Results

After running the population model (Simulation → Run), the results can be examined for the different options of the Results menu. In this chapter, results from predicted population size and risk of quasi extinction are presented. Because of the stochasticity in the model, results will vary some between simulation runs. However, by using the maximum amount of replications (10,000), these variations can be kept at a minimum. Here, a warning is appropriate. Running the model with five populations and 10,000 replications will take some time. Reducing the number of populations (Model → Populations) or the number of replications (Model → General Information) will reduce the time needed to complete the simulation.

Population size

To analyze the predicted population size for the different primary sex ratios, simulation runs with 100 time steps (= 100 years) and 10,000 replications were performed. The predicted population size over time can be seen in the Trajectory Summary window (Results → Trajectory Summary). Here, the population number can be changed to analyze the different primary sex ratios. A comparison in predicted population size between an undisturbed population (50%) and a male-biased population (30%) is shown in figure 15.3: The predicted population size of the undisturbed population stabilizes after a few years, while the male-biased population keeps falling for the whole simulation period. By analyzing the other proportions of female fry (change the population number in the Trajectory Summary window), it is obvious that this trend is less pronounced (or nonexistent) when the primary sex ratio is less male biased. (Note that the graph represents an average of 10,000 simulation runs, hence the relatively straight line after adjusting for the initial gap to carrying capacity.) Obviously, each individual

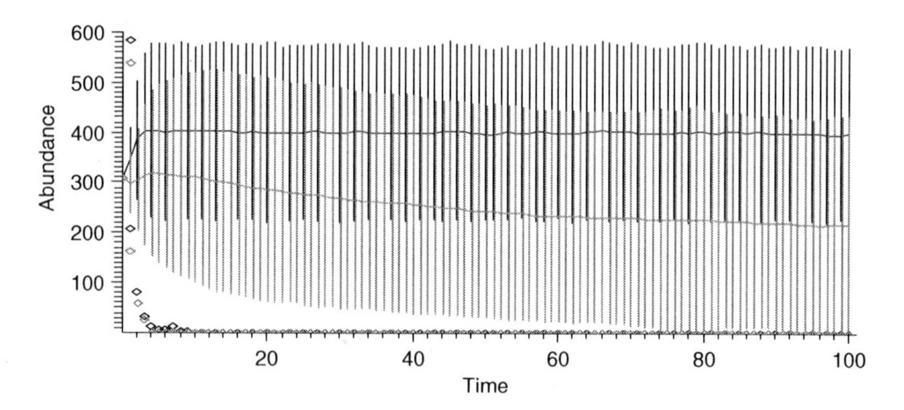

Figure 15.3. Predicted population size when the primary sex ratio is 50% (solid line) or 30% (dashed line) females. Error bars are standard deviation and the diamonds show the minimum and maximum abundances.

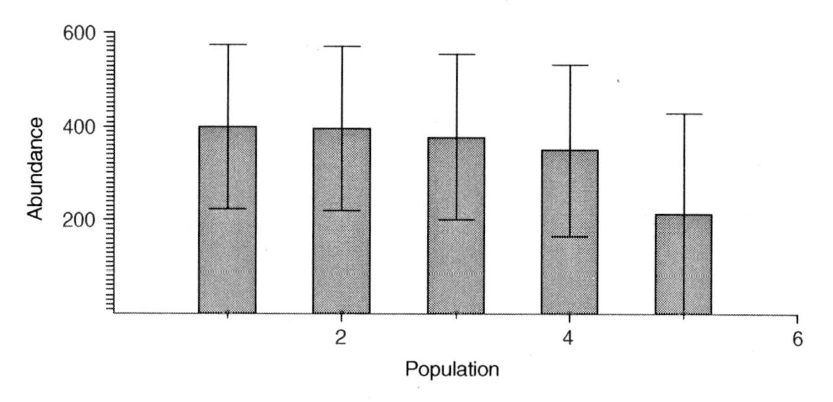

Figure 15.4. Average population size after 100 years. Primary sex ratios are, from left to right, 50%, 45%, 40%, 35%, and 30% females. Error bars are standard deviation.

trajectory will be a lot more variable over time, something that can be seen during the simulation run (Simulation → Run → Trajectories).

To compare the predicted population size of the five treatments, population structure can be used (Results → Population Structure). Figure 15.4 shows the average predicted population size after 100 years. The population size at other times can be examined by changing the time step in the Population Structure window.

Table 15.1 shows the predicted population size as a proportion of the expected size of an undisturbed population at the same time step. The decrease over time during the first 100 years is, once again, most apparent for the population with a primary sex ratio of 30%. For a longer time horizon (500 years), however, this effect can also be seen for a primary sex ratio of 35%.

Table 15.1. Predicted population size as proportions of an undisturbed population.

Time	Primary sex ratio			
	45%	40%	35%	30%
0	1	1	1	1
10	0.98	0.94	0.90	0.77
20	0.99	0.95	0.88	0.71
30	0.99	0.96	0.89	0.67
50	0.98	0.94	0.87	0.60
100	0.99	0.95	0.88	0.53
500	0.99	0.95	0.85	0.21

Quasi extinction

The variability in population growth means that there always is a certain risk that the population size will fall below a certain lower limit. Several options are available under the Result menu to analyze this risk. Time to quasi extinction is the time it takes the population size to fall below the extinction threshold, specified in the Stochasticity dialog box (Model → Stochasticity). For the results presented here, the threshold was set to 43 individuals, which is 10% of the carrying capacity and about 11% of the predicted size of an undisturbed population (~400; figure 15.3). To analyze the time to quasi extinction for one of the five treatments, only the corresponding population should have the "Include in summation" box checked at the start of the population run (Model → Populations → General). The results presented here are thus based on five individual simulation runs, all with 100 time steps and 10,000 replications. Figure 15.5 shows the accumulated probability of falling below 43 individuals as a function of time.

The mean time for the population to fall below 43 individuals was 81.5 and 358 years for the populations with a primary sex ratio of 30% and 35%, respectively. In the scenarios with less male-biased primary sex ratios, the mean time for quasi extinction was greater than 500 years and could thus not be calculated. To compare treatments, it is therefore better to look at the accumulated probability of quasi extinction within a certain time horizon. The accumulated probability to fall below 43 individuals within 10, 100, or 500 years is shown in figure 15.6.

Discussion

The results from the population model show that the male-biased broods found by Larsson and co-workers could have a significant impact on the eelpout population. For primary sex ratios of 40% females or more, the density dependence partly compensates for the male-biased broods by increasing the survival rates when the population size falls. Therefore, the major effect in the 45% or 40% scenarios is a reduction in predicted population size. This can be seen clearly in table 15.1, where the predicted population size is almost stable over time. For populations with a primary sex ratio

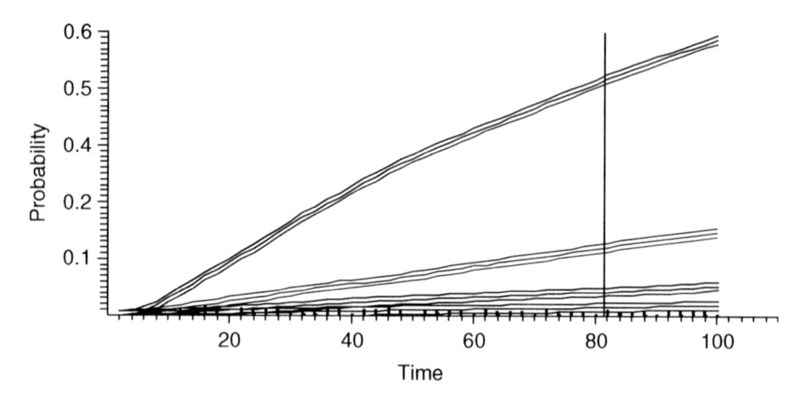

Figure 15.5. Probability of quasi extinction as a function of time for primary sex ratios of (from top to bottom) 30%, 35%, 40%, and 50% females. Each curve shows the probability that the population will decline to 10% of its carrying capacity at, or before, the time step given in the x-axis. The vertical line marks the median time of quasi extinction in a scenario with 30% female fry (81.5 years).

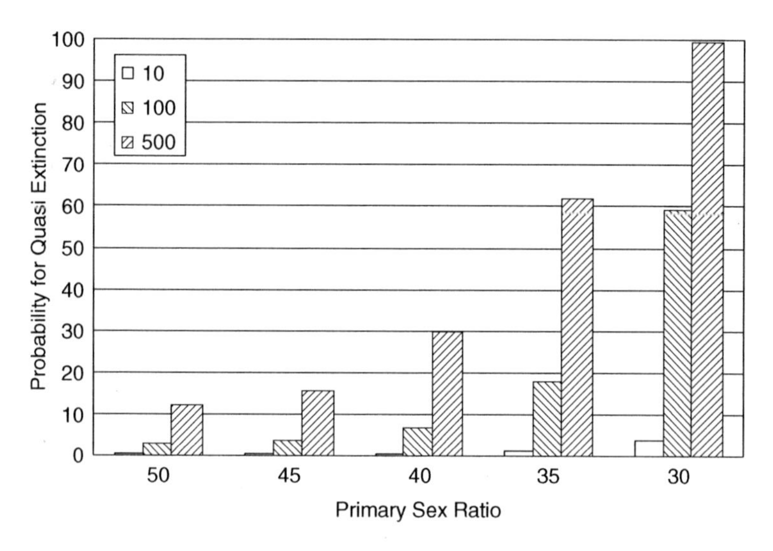

Figure 15.6. Probability of quasi extinction (risk of falling to or below CPUE = 43, which is 10% of the carrying capacity) for three different time horizons: 10, 100, and 500 years.

of 30% or 35% females, however, the results are somewhat different. Figure 15.3 and table 15.1 show that the predicted population size does not stabilize as it does for primary sex ratios of 40% females or more. Instead, the predicted population size keeps falling, perhaps because the density dependence is weaker when the R_{max} value is smaller; hence, the reduction in fecundity is not fully compensated by increased survival rates. This effect can easily be understood by imagining a density dependence with R_{max} = 1, that is, no density dependence (Model → Populations → Display → Density Dependence in *R*). The lower predicted population size for more male-biased populations is also reflected in the probability for quasi extinction. The risk of the population falling below 10% of the carrying capacity within 100 years is about 23% for a primary sex ratio of 35% females, which is not far from the 38.7% found by Larsson and Förlin in 2000 (Larsson and Förlin 2002).

All simulation runs have an initial population size that is lower than the carrying capacity; thus, the average population size increases over the first few years. The fact that the population size is fixed at the start of the simulation run also explains the lower risk of quasi extinction during the first years (Results → Time to Quasi extinction). Because the average population size is lower at the start, one could believe that the risk of quasi extinction would be higher. However, the average population size has little to do with the probability for quasi extinction at any given time step. Instead, it is the variability of the population size that is important. Because the population size at the start of each trajectory is fixed, there is no variation at year zero. This can be seen clearly from the error bars in figure 15.3 (Results → Trajectory Summary).

The average population size from 10,000 simulation runs was higher for populations with higher primary sex ratio. For the case with 50% female fry, the average population size after 100 years was 400 CPUE, that is, somewhat under the carrying capacity of 430. In the most extreme case (30% female fry), the average population size after 100 years was only 211. The larger gap to carrying capacity for the male-biased populations can be explained by the lower R_{max} values. This means that the capacity to recover after one or more unfavorable years is reduced, resulting in a population size under the carrying capacity. This effect of reduction in population growth rate is expected after the introduction of stochastic variability, something that is compensated by the density dependence at a certain population size. It is important to remember that the average population size is calculated from many (in this case 10,000) replications that all will differ from the predicted population size at any given time step. Replications that differ upward will increase the mean population size; those who differ downward will decrease it. Because of the density dependence, there is a limit to how much a replication is likely to differ in any direction, and at a certain level it will start to approach the equilibrium population size. However, because there is a substantial risk of extinction, some replications will not be able to approach the equilibrium again. This means that the average abundance is declining toward zero, which will be reached when the last replication is extinct. Although this does not happen (within 500 years) for any of the scenarios described here, the trend is obvious for all sex ratios (Results → Trajectory Summary → Text). In a deterministic (no stochasticity) model, the population would stabilize at 430 CPUE, because this is where the growth rate equals 1. Hence, the stochastic growth rate of the undisturbed population could be said to be equal to one at a population size of about 400 CPUE. From this, the undisturbed population could be described as varying around an equilibrium population size and that at any time a sequence of unfavorable years could threaten the population. When the primary sex ratio is male biased, the ability to recover is lower, hence the increased risk of quasi extinction. The reduction in predicted population size for the male-biased populations further increases the probability of quasi extinction as the "safety margin" is reduced.

In this model, the extra proportion of males (the sex-reversed females) is assumed to become either normal or nonreproductive. However, if the sex determination in eelpout is based on male heterogameity (XY), genetic females (XX) that have been sex-reversed due to endocrine disruption could give broods consisting of 100% females. If this is the case, the masculinization at the individual level could render feminization on the population level. This is something that has been applied in aquaculture to produce all female stocks (Piferrer 2001). It is, however, unclear if this would happen in a wild population where factors such as behavior and reproductive competition between males could be important.

References

Akçakaya, H. R. 2002. Estimating the variance of survival rates and fecundities. *Animal Conservation* 5: 333–336.

Balk, L., Larsson, Å., and Förlin, L. 1996. Baseline studies of biomarkers in the feral female perch (*Perca fluviatilis*) as tools in biological monitoring of anthropogenic substances. *Marine Environmental Research* 42: 203–208.

Cody, R. P., and Bortone, S. A. 1997. Masculinization of mosquitofish as an indicator of exposure to kraft mill effluent. *Bulletin of Environmental Contamination and Toxicology* 58: 429–436.

European Commission. 1996. European Workshop on the Impact of Endocrine Disrupters on Human Health and Wildlife, 2–4 December, Weybridge, U.K. Report Reference 17549. Brussels: European Commission.

Gibbs, P. E., and Bryan, G. W. 1986. Reproductive failure in populations of the dog-whelk, *Nucella lapillus*, caused by imposex induced by tributyltin from antifouling paints. *Journal of the Marine Biological Association of the United Kingdom* 66: 767–777.

Guillette, L. J., Crain, D. A., Rooney, A. A., and Pickford, D. B. 1995. Organization versus activation—the role of endocrine-disrupting contaminants (EDCs) during embryonic-development in wildlife. *Environmental Health Perspectives* 103: 157–164.

Hanson, N., Åberg, P., and Sundelöf, A. 2005. Population-level effects of the male-biased broods in eelpout (*Zoarces viviparus*). *Environmental Toxicology and Chemistry* 24: 1235–1241.

Hughes, T. P., and Tanner, J. E. 2000. Recruitment failure, life histories, and long-term decline of Caribbean corals. *Ecology* 81: 2250–2263.

Hunter, G., Donaldson, E., Stoss, J., and Baker, I. 1983. Production of monosex female groups of chinook salmon (*Onchorhynchus tsawytscha*) by the fertilisation of normal ova with sperm from sex-reversed females. *Aquaculture* 33: 355–364.

Jobling, S., Nolan, M., Tyler, C. R., Brighty, G., and Sumpter, J. P. 1998. Widespread sexual disruption in wild fish. *Environmental Science and Technology* 32: 2498–2506.

Kavlock, R. J., Daston, G. P., DeRosa, C., et al. 1996. Research needs for the risk assessment of health and environmental effects of endocrine disruptors: A report of the US EPA-sponsored workshop. *Environmental Health Perspectives* 104: 715–740.

Kovacs T. G., Gibbons J. S., Tremblay L. A., O'Connor B. I., Martel P. H., and Voss R. H. 1995. The effects of a secondary-treated bleached kraft mill effluent on aquatic organisms as assessed by short-term and long-term laboratory tests. *Ecotoxicology and Environmental Safety* 31: 7–22.

Larsson, D. G. J., and Förlin, L. 2002. Male-biased sex ratios of fish embryos near a pulp mill: Temporary recovery after a short-term shutdown. *Environmental Health Perspectives* 110: 739–742.

Larsson, D. G. J., Adolfsson-Erici, M., Parkkonen, J., Pettersson, M., Berg, A. H., Olsson, P.-E., and Förlin, L. 1999. Ethinyloestradiol—an undesired fish contraceptive? *Aquatic Toxicology* 45: 91–97.

Larsson, D. G. J., Hallman, H., and Förlin, L. 2000. More male fish embryos near a pulp mill. *Environmental Toxicology and Chemistry* 19: 2911–2917.

Larsson, D. G. J., Kinnberg, K., Sturve, J., Stephensen, E., Skön, M., and Förlin, L. 2002. Studies of masculinization, detoxification, and oxidative stress responses in guppies (*Poecilia reticulata*) exposed to effluent from a pulp mill. *Ecotoxicology and Environmental Safety* 52: 13–20.

Myers, R., and Worm, B. 2003. Rapid worldwide depletion of predatory fish communities. *Nature* 423: 280–283.

Noaksson, E., Tjärnlund, U., Bosveld, A. T. C., and Balk, L. 2001. Evidence for endocrine disruption in perch (*Perca fluviatilis*) and roach (*Rutilus rutilus*) in a remote Swedish lake in the vicinity of a public refuse dump. *Toxicology and Applied Pharmacology* 174: 160–176.

Noaksson, E., Linderoth, M., Bosveld, A. T. C., Norrgren, L., Zebuhr, Y., and Balk, L. 2003. Endocrine disruption in brook trout (*Salvelinus fontinalis*) exposed to leachate from a public refuse dump. *Science of the Total Environment* 305: 87–103.

Penn, D. J. 2003. The evolutionary roots of our environmental problems: Toward a Darwinian ecology. *Quarterly Review of Biology* 78: 275–301.

Piferrer, F. 2001. Endocrine sex control strategies for the feminization of teleost fish. *Aquaculture* 197: 229–281.

Piferrer, F., Zanuy, S., Carrillo, M., Solar, I. I., Devlin, R. H., and Donaldson, E. M. 1994. Brief treatment with an aromatase inhibitor during sex-differentiation causes chromosomally female salmon to develop as normal, functional males. *Journal of Experimental Zoology* 270: 255–262.

Sandström, O., Larsson, Å., Andersson, J., Appelberg, M., Bignert, A., Ek, H., Förlin, L., and Olsson, M. 2005. Three decades Swedish experience stresses the need of integrated long-term monitoring in marine coastal areas. *Water Quality Research Journal of Canada* 40:233–250.

Tilman, D., and Lehman, C. 2001. Human-caused environmental change: Impacts on plant diversity and evolution. *Proceedings of the National Academy of Sciences of the USA* 98: 5433–5440.

Vetemaa, M. 1999. Reproduction biology of the viviparous blenny (*Zoarces viviparus* L.). *Fiskeriverket Rapport* 2: 81–96.

Vitousek, P. M., Mooney, H. A., Lubchenco, J., and Melillo, J. M. 1997. Human domination of Earth's ecosystems. *Science* 277: 494–499.

Witorsch, R. J. 2002. Endocrine disruptors: Can biological effects and environmental risks be predicted? *Regulatory Toxicology and Pharmacology* 36: 118–130.

16

Leptocheirus plumulosus in the Upper Chesapeake Bay

Sediment Toxicity Effects at the Metapopulation Level

TODD S. BRIDGES
H. REŞIT AKÇAKAYA
BARRY BUNCH

Although it has been known for some time that spatial heterogeneity of the environment (including the spatial distribution of toxicity) may have important effects on the quantification of ecotoxicological impacts, practical methods for incorporating spatial structure into ecotoxicological assessment has only recently been developed (Hallam and Lika 1997; Ares 2003). Among the spatially explicit methods of population-level ecotoxicological assessment are metapopulation models of target or sensitive species (Spromberg et al. 1998; Johnson 2002; Landis 2002; Chaumot et al. 2003). The metapopulation concept (and spatial structure in general) is important because species that exist in a metapopulation face particular issues related to environmental impacts and have management and mitigation options that can be evaluated more completely, or only, in a metapopulation context (Akçakaya et al. 2007).

The main goal of our analysis was to assess the sediment quality and analyze the potential sediment toxicity on the study species, *Leptocheirus plumulosus*, which is considered an indicator species. We also aimed to analyze the effects of spatial structure (especially source–sink dynamics between the main river and the tributaries) and to demonstrate the feasibility of using realistic, spatially structured models in ecotoxicological risk analysis at the population level. We demonstrate this approach with a stochastic metapopulation model for *L. plumulosus*, based on the distribution and life-history traits of the species in the Gunpowder River in the upper Chesapeake Bay. A unique feature of this model is the dispersal rates, which were based on a hydrological model of the Gunpowder River. This is the first model in which a stochastic metapopulation model is linked to a hydrological model for ecological risk analysis.

Methods

Study species and area

Leptocheirus plumulosus (Shoemaker) is an infaunal gammaridean amphipod found in estuaries along the East Coast of the United States, ranging from Cape Cod, Massachusetts, to northern Florida. *Leptocheirus plumulosus* is generally accepted as an appropriate indicator species for sediment toxicity tests (U.S. Environmental Protection Agency 2001). Several field studies have shown a correlation between amphipod distribution and sediment contamination, and laboratory experiments have shown that amphipods are more sensitive to sediment toxicants than are many other infaunal groups (Lotufo et al. 2001a, 2001b; Ferraro and Cole 2002).

The study area was in the Gunpowder River and its tributaries, in the vicinity of the Aberdeen Proving Ground in Maryland, where the Gunpowder River joins the Chesapeake Bay estuary.

Experimental methods

This analysis is based on data from three studies: a 30-week population study (data used to estimate the age-specific survival and fecundity values), a long-term field study (data used to correct the parameters for field conditions and to estimate seasonal and stochastic variability), and a series of 28-day bioassays (data used to add spatial variability by adjusting these parameters to local conditions).

In the population study, experiments on the organisms were carried out under six different treatments, each defined by a sediment concentration (0%, 3%, or 6%) and a food ration (normal or double): 0% sediment and normal food ration, 0% sediment and double food ration, 3% sediment and normal food ration, 3% sediment and double food ration, 6% sediment and normal food ration, and 6% sediment and double food ration. For each treatment, six replicates were executed, each with 20 animals at the initiation of the experiment. Each replicate was executed in a 1-L beaker containing the medium and the amphipods. In all cases, the animals were estimated to be between one and two weeks old at the start of the experiment. An additional experiment was conducted to estimate survival in the first week under the first treatment (0% sediment and normal food ration). This first-week survival estimate was then used for all treatments. Weekly data were collected on each replicate of each treatment, including the number of the original individuals still alive and the number of offspring alive at the end of each week. The offspring were removed from each beaker at the time of census. The experiment continued for 30 weeks or until all the animals died, whichever came first. These data were used to calculate age-specific survival rates and fecundities by pooling the data over the six replicates of each treatment.

The field study consisted of counts of *L. plumulosus* taken in an embayment of the Magothy River, a western shore subestuary of Chesapeake Bay considered to be little affected by toxic chemicals (McGee and Spencer 1997; see also Spencer and McGee 2001; McGee and Spencer 2001). The data were in the form of censuses taken at irregular intervals during the years 1995 and 1996, and the censuses were broken down into four life stages: eggs, recruits, juveniles, and adults. The vital rates were then calculated by several different methods, one of which was linear regression.

In the 28-day bioassays, sediment from several locations in the Gunpowder River were used (Neubauer et al. 1997, appendix G). Sediment was taken from each site, and *L. plumulosus* were reared in each type of sediment. The total number of individuals that survived until the end of the experiment and the total number of offspring were recorded for each sediment sample.

Population model

We used RAMAS Metapop (Akçakaya 2005) to develop an age-structured model with one-week age classes, and incorporated both demographic and environmental stochasticity. Environmental stochasticity was modeled with a set of standard deviations, each corresponding to an element (survival or fecundity) of the Leslie matrix. Each vital rate (survival or fecundity) was then selected at each time step from a random distribution with its mean (as specified in the transition matrix) and its standard deviation. Given the sampled survival rate and fecundity for a time step, the number of survivors were sampled from a binomial distribution, and the number of offspring were sampled from a Poisson distribution, to model demographic stochasticity (Akçakaya 1991). The model included 12 subpopulations, each with its own set of vital rates, based on field data. The dispersal among these subpopulations was based on a hydrological model, which is a "particle tracer" used for predicting the movement of discrete items in the water. We developed two models: a "control" model, based on laboratory data with 0% sediment, and an "impact" model, based on laboratory data with 3% sediment. In both models, the Leslie matrix based on the laboratory data was modified according to field data from Magothy River (a western shore subestuary of Chesapeake Bay), and according to bioassay data from the Gunpowder River to reflect the local conditions. The details of the model and the estimation of its parameters are described in the following sections.

The Leslie matrix

The age-specific survival rates and fecundities of the model were estimated from various sources, including laboratory and field studies. The Leslie matrix was based on the matrix calculated under the laboratory conditions, and modified as described in this section. We used data from the normal food ration treatment and created two Leslie matrices for "control" model (based on 0% sediment treatment) and for "impact" model (3% sediment).

In defining the Leslie matrix, we used one-week age classes and thus a one-week time step. We pooled individuals 19 or more weeks of age, for a total of 20 age classes. The survival rate of age class x individuals is defined as the probability of an individual in age class x reaching age class $x + 1$ after one time unit has passed. Age class x is defined as individuals between x and $x + 1$ weeks old. The survival rate of the composite age class was calculated as the weighted average of the week-specific survival rates of the age classes composing the composite class, weighted by the number of individuals in each age class. The fecundity of age class x individuals, F_x, is defined as the number of age class 0 individuals alive at time $t + 1$ per age class x individual at time t. The fecundity of the composite class was calculated as the weighted average of the week-specific fecundities of the classes composing the composite class.

Table 16.1. Age-specific survival rates and fecundities based on laboratory data.

Age (weeks)	Survival rate	Fecundity
0–1	0.970	0.00
1–2	0.950	0.00
2–3	0.956	0.00
3–4	0.991	0.01
4–5	0.972	1.92
5–6	0.981	3.66
6–7	0.990	4.50
7–8	1.000	5.12
8–9	0.971	4.70
9–10	0.929	6.45
10–11	0.957	7.28
11–12	0.943	6.93
12–13	0.904	9.02
13–14	0.920	10.2
14–15	0.913	6.07
15–16	0.841	6.63
16–17	0.962	5.98
17–18	0.902	3.98
18–19	0.957	4.24
19^{+}	0.854	0.07

We corrected these matrices for field conditions, based on field counts of *L. plumulosus* by McGee and Spencer (1997). These researchers calculated seven transition matrices, each corresponding to a different time of year. Thus, application of all seven matrices in sequence represents one full year of population fluctuation of *L. plumulosus*. Because we are modeling the population dynamics of the *L. plumulosus* population under natural conditions, we modified our original Leslie matrix such that the eigenvalue (λ) of the modified Leslie matrix was equal to λ of the stage-based transition matrices of McGee and Spencer. They calculated seven transition matrices, so seven corrections were needed. This was done by proportionally decreasing all elements of our Leslie matrix such that the λ values would coincide. We had to decrease the vital rates in all cases because the laboratory-based λ was higher than all field-based λ values. We ran the model for 52 time steps, where each time step was one week, correcting the Leslie matrix at each time step in order to remain consistent with the field data. Tables 16.1 and 16.2 show survival rates and fecundities of the Leslie matrix we calculated from laboratory data and the multipliers we used to proportionally decrease the survival rates and fecundities.

Environmental stochasticity

The standard deviations of the survival rates and fecundities were estimated by matching the variation in the simulated population abundance to the variation in field counts of *L. plumulosus* in the Magothy River (McGee and Spencer 1997). The coefficient of variation of the field data was about 0.8. We increased the variation in survival rates

Table 16.2. Multipliers used to correct for field conditions.

Multiplier	Number of weeks applied
0.639	5
0.736	8
0.587	13
0.785	8
0.637	6
0.492	6
0.811	6

and fecundities until the model output, with the same initial number of individuals as in the field data, gave a similar amount of variability in abundance. A coefficient of variation of 0.22 (for all vital rates) resulted in variation in abundance that was similar to the variation observed in the field. We ran 1,000 replicates of each model.

Metapopulation structure

We created a metapopulation model for *Leptocheirus plumulosus*, to simulate its metapopulation in the Gunpowder River. Subpopulations were defined based on spatial proximity and homogeneity of survival rates and fecundities. Using these criteria, we defined 12 distinct subpopulations in the Gunpowder area (see figure 16.1). Each subpopulation was modeled as a structured population with age-specific survival rates and fecundities.

The basis of the metapopulation model is the age-structured model described above, in which seven different Leslie matrices are used to model the population dynamics of *Leptocheirus plumulosus* over a one-year period. Each matrix was then modified for each subpopulation in order to correct for differential survival rates and fecundities in different populations, as described below.

Population-specific survival rates and fecundities

We modified the time series of weekly population matrices (see above) for each subpopulation, to account for spatial differences in population dynamics. For this modification, we first calculated survival rates and fecundities for each subpopulation, using data from the 28-day sediment bioassays (Neubauer et al. 1997). Most of the subpopulations we defined in our model covered several of the sites from which the sediments were taken and tested. We pooled the data over all sites comprising a subpopulation, taking the total number of individuals that survived and dividing by the total number of individuals at the initiation of the experiment. Each bioassay experiment was conducted for four weeks; thus, we took the fourth root of the four-week survival rate to convert to a one-week survival rate. The total number of offspring was also recorded for each replicate (i.e., for each site). We also pooled replicates within each subpopulation in order to calculate fecundity. Weekly fecundity (F) was calculated for each subpopulation based on

$$N \cdot F + (N \cdot S)F + (N \cdot S^2)F + (N \cdot S^3)F = \text{total number of offspring},$$

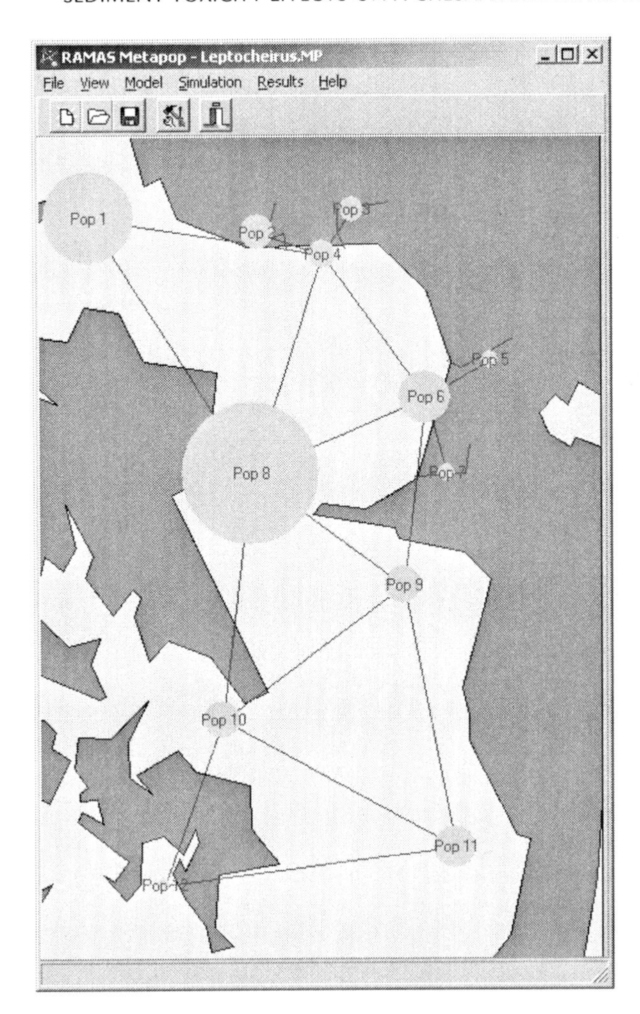

Figure 16.1. The spatial structure of the metapopulation model for *Leptocheirus plumulosus* in the Gunpowder River. Twelve distinct subpopulations were defined based on spatial proximity and homogeneity of survival rates and fecundities.

where N is the total number of animals at the beginning of the experiment over all sites included in the subpopulation. We then divided the weekly survival rate and fecundity of each subpopulation with the overall, weighted average survival and fecundity, respectively, of all the subpopulations, obtaining a "relative survival" and a "relative fecundity" value for each subpopulation (see table 16.3). We proportionally changed each survival rate and each fecundity for each population by multiplying by the relative survival rate or fecundity. This resulted in a population-specific time series of vital rates. Thus, the temporal change due to seasonality is based on McGee and Spencer (1997), whereas the difference among populations is based on Neubauer et al. (1997).

Dispersal

We calculated dispersal based on a hydrological model of the Gunpowder River. The hydrological model CH3D is a "particle tracer" developed by the U.S. Army Corps of

Table 16.3. Area, density, and initial abundance of each subpopulation.

Subpopulation	Area (km^2)	Density (individuals/m^2)	Initial abundance (millions)	Relative fecundity	Relative survival
1	6.25	7.50	46.88	1.26940	1.01660
2	0.31	22.67	7.00	0.18976	0.91066
3	0.36	1.00[a]	0.36	0.23636	0.93069
4	1.85	2.50	4.63	1.10248	1.00866
5	0.19	1.00[a]	0.19	0.17581	0.98134
6	1.39	11.29	15.68	1.01794	1.01349
7	0.19	1.00[a]	0.19	0.24789	0.99484
8	2.55	42.37	107.88	0.95954	0.99852
9	4.86	1.67	8.10	0.87208	0.99366
10	3.09	2.50	7.72	0.87977	0.98792
11	6.17	1.67	10.29	0.91569	0.99425
12	0.93	1.00[a]	0.93	1.05209	0.98680

[a]Estimated density was zero; 1.0 was assumed.

Engineers Waterways Experiment Station (CEWES) Environmental Laboratory and Coastal and Hydraulics Laboratory for predicting the movement of discrete items in the water (for historical and technical information about the model, see Bunch et al. 2000). These tools can be used to determine the ultimate destination of a particle or its point of origin. Particle tracers differ from traditional water quality computations in that the particle is a distinct object. Unlike water quality constituents, the particle does not undergo reactions, decay, or diffuse. The particle has no mass. Its movement is purely a function of the hydrodynamic conditions surrounding it.

A recent development in the area of particle tracking is the inclusion of rules in the movement of the particle. Incorporation of rules allows the modeler to mimic the behavior of living organisms. If adequate information is available, rules can be developed that describe the fundamental behavior of a living organism. Then it is possible to simulate the response of that organism to changes in its environment.

The spatial extent of the hydrodynamic model, CH3D, was the northern portion of Chesapeake Bay, including the waters of the Gunpowder River. Eight of the 12 populations shown in figure 16.1 are in the river, and four are in waters and creeks inland of the main channel. The hydrodynamic model grid resolution is relatively fine throughout the upper Chesapeake Bay. However, model resolution still does not include all peninsulas, islands, embayments, and creeks that exist in the real world. Consequently, only the eight populations that are located in the Gunpowder River are in the domain of the hydrodynamic and water quality model. Each of the eight populations was mapped to several CH3D model cells and modeled as a set of 144 particles. Initially, the particles were uniformly distributed in the cells that were determined to have populations. Different conditions and behavior were imparted upon particles, and their movement over time was monitored. Eight of the 12 populations shown in figure 16.1 are located in the Gunpowder River proper. The other four, in creeks and waters off of the main channel, are outside of the model domain and could not be included in these runs with the model as configured.

Horizontal movement in all simulations was passive. The particles did not provide any of their own propulsion; they were simply carried by the flow. Vertical movement was controlled to simulate particle activity/inactivity during the day. Particles were initially given an upward velocity of 0.1 cm/s (3.6 m/hr), which would be adequate to move a particle from the bottom to the surface in most of the waters of the Gunpowder River in approximately 30 minutes. During this time, the particle is also being transported in the horizontal direction. After a period of time, the vertical velocity was reversed and the particles are forced to settle.

A settling velocity of 0.1 cm/s forced particles to the bottom. Once the particles reached bottom, they stayed in place until the next time a resuspension velocity was imposed. The most important feature in particle movement appeared to be the duration that the particle is in the water column. Due to the shallow nature of the system, a combination of resuspension/settling rates that allowed the particles to remain in the water column for the longest duration possible allowed for the greatest possible exchange.

Consequentially, later simulations were made with a suspension rate of 0.1 cm/s applied for 1 hr, 0.0 cm/s for 4 hr, and –0.1 cm/s for 1 hr. This arrangement was repeated daily and allowed particles to move for approximately 5.5 hr/day.

The output of the hydrological model was the number of particles moving from one population to each other population during 52 one-week time steps. Based on this output, the proportion dispersing from each population to each other population was calculated (table 16.4).

Because the hydrological model included only the main river populations, and excluded the four tributary populations, the dispersal rates to and from these four populations are given as zero in table 16.4. Because these four populations have the lowest relative fecundity and relative survival values (see table 16.3), assuming zero dispersal to these populations is likely to underestimate risks of decline. In reality, dispersal into these populations may cause them to become sink populations, increasing the risk of decline of the metapopulation. To explore this possibility, we ran additional simulations with dispersal into these four populations. In these simulations, we assumed a dispersal rate of 0.076 (average, over the other eight populations, of the maximum dispersal rate) to and from these four populations and the nearest main-river population (population 4 or 6; see table 16.4).

Initial abundances and density dependence

The initial abundance of each subpopulation was estimated based on density estimates in Neubauer et al. (1997, appendix F). The density was multiplied by an estimate of the total area of each subpopulation, based on the map of the Gunpowder River. For some of the populations, the estimated density was zero, due to the limitations of sampling. Because these populations are unlikely to have zero individuals, we assumed that the density was below the detection limit. For these populations, we assumed a density of one individual per square meter (table 16.3).

The laboratory population on which our model is based showed sensitivity of the growth rate to a doubling in the amount of food, which might indicate density dependence. However, we could not find any other studies relating to density dependence in this species that would inform the type and strength of density dependence. Therefore, we used a very simple form of density dependence ("ceiling"), by truncating the population abundance at 10 times the initial abundance, because the maximum

Table 16.4. Dispersal rates from each population (column) to other populations (rows).

Target populations	Source populations											
	Pop 1	Pop 2	Pop 3	Pop 4	Pop 5	Pop 6	Pop 7	Pop 8	Pop 9	Pop 10	Pop 11	Pop 12
Pop 1	0	0	0	0.0073	0	0	0	0.0207	0	0	0	0
Pop 2	0	0	0	0.076[a]	0	0	0	0	0	0	0	0
Pop 3	0	0	0	0.076[a]	0	0	0	0	0	0	0	0
Pop 4	0.0215	0.076[a]	0.076[a]	0	0	0.0079	0	0.0083	0	0	0	0
Pop 5	0	0	0	0	0	0.076[a]	0	0	0	0	0	0
Pop 6	0	0	0	0.0908	0.076[a]	0	0.076[a]	0.02	0.0116	0	0	0
Pop 7	0	0	0	0	0	0.076[a]	0	0	0	0	0	0
Pop 8	0.0031	0	0	0.0099	0	0.0047	0	0	0.0057	0.0154	0	0
Pop 9	0	0	0	0	0	0.0549	0	0.0138	0	0.0036	0.0343	0
Pop 10	0	0	0	0	0	0	0	0.059	0.0524	0	0.1186	0.0354
Pop 11	0	0	0	0	0	0	0	0	0.1822	0.0397	0	0.0003
Pop 12	0	0	0	0	0	0	0	0	0	0.0443	0.0023	0

[a] Simulations were run with no dispersal to or from the four tributary populations (2, 3, 5, 7), and also with the assumed dispersal rate of 0.076 to and from tributary populations and the nearest in-river populations (4 or 6). The assumed rate was based on the average maximum dispersal. See text for details.

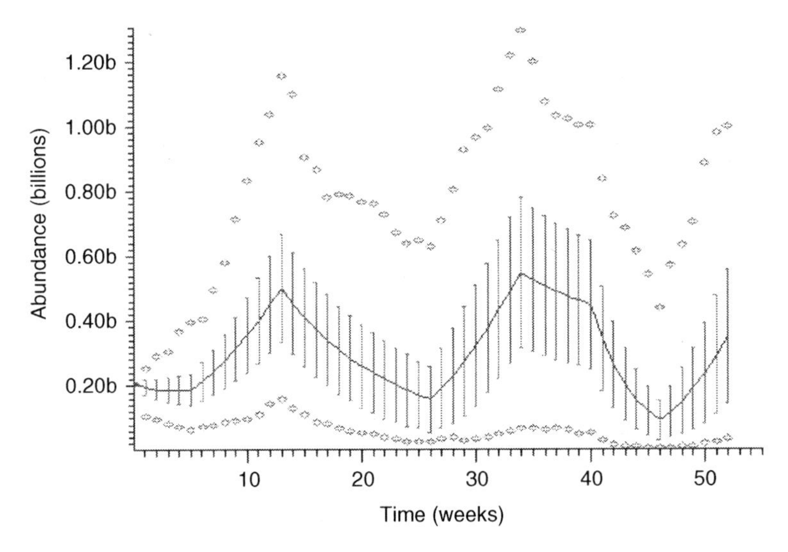

Figure 16.2. Summary of model predictions. The curve shows the total metapopulation size (total number of individuals in all populations, averaged over 1,000 replications). The vertical bars indicate plus and minus one standard deviation of the metapopulation size, and diamonds give minimum and maximum metapopulation sizes over the 1,000 replications.

number of individuals observed in the field data is about 10 times the initial number of individuals.

Results

In the field data, most of the variation is clearly in the second half of the year, which is also the pattern in the simulated population abundance (figure 16.2). The risk of extinction was zero, but there were local extinctions. With the "control" model, the number of occupied populations at the end of 52 weeks ranged from 8 to 12, with an average of 9.4 (out of 12 populations). Four of the populations (2, 3, 5, and 7) remained unoccupied for an average of 4–9 weeks and a maximum of 30–44 weeks (out of 52 weeks). With the "impact" model, the number of occupied populations at the end of 52 weeks ranged from 8 to 11, with an average of 8.5. Four of the populations (2, 3, 5, and 7) remained unoccupied for an average of 8–13 weeks and a maximum of 33–45 weeks (out of 52 weeks).

Although the "impact" model predicted lower population sizes than the "control" model in general, the large amount of variability in the model resulted in substantial overlap in the abundances predicted by the two models. The difference between the two models was more pronounced in terms of the risk of decline (figure 16.3, curves a and c). Each point on a risk curve in figure 16.3 gives the probability that the metapopulation abundance will decline by the percentage (in the x-axis) from the initial abundance anytime during the 52-week simulation. The upper curve is the result of simulation with the "impact" model. The vertical bar (at x = 85%) indicates the maximum difference between the two risk curves. The risk of a substantial (85%)

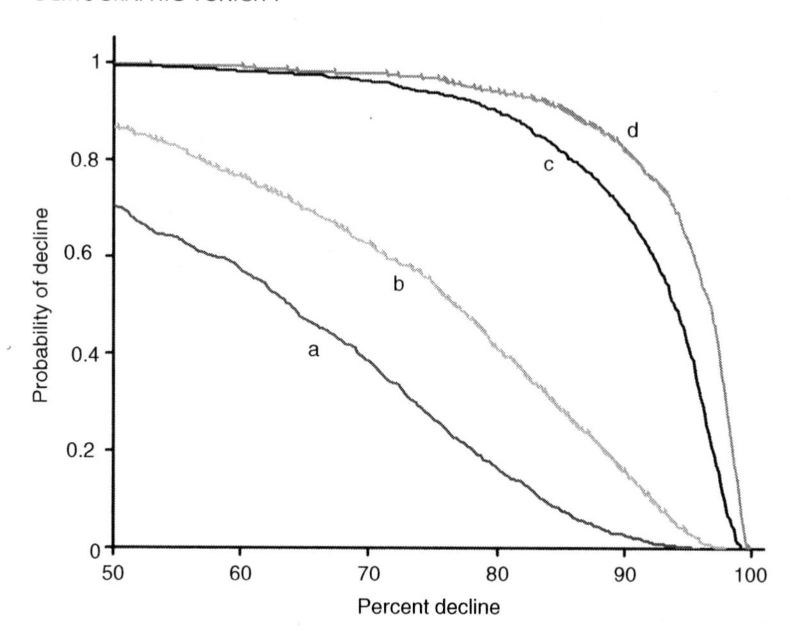

Figure 16.3. Risk of decline curves for simulations with control model, no dispersal to/from tributaries (a); control model with dispersal to/from tributaries (b); impact model, no dispersal to/from tributaries (c); and impact model with dispersal to/from tributaries (d). Each point on a curve gives the probability that the metapopulation abundance will decline by the percentage (on the x-axis) from the initial abundance anytime during the simulated time (52 weeks). The differences between curves were statistically significant (Kolmogorov-Smirnov test).

decline from the initial abundance was only about 9% with the "control" model, but about 83% with the "impact" model. The difference between these two probabilities was highly significant (Kolmogorov-Smirnov test). Simulations with dispersal to and from tributary populations gave higher risks than did simulations with these populations isolated (figure 16.3, curve a vs. b and curve c vs. d). With dispersal to and from tributary populations, there was a difference between control and impact models (figure 16.3, curves b and d) that was comparable to simulations without dispersal to and from these populations.

Discussion

In this study, we demonstrated the use of a stochastic metapopulation model for ecological risk assessment for sediment toxicity at the population level. We used several sources of data, including two sets of laboratory bioassays, a field study, and results of a hydrological model. Both sets of laboratory data were needed because they provided different types of information. One was a long-term (30-week) bioassay, which is necessary to estimate the age-specific parameters (survival and fecundity) of the model. The other bioassay (Neubauer et al. 1997) was short term but included sediments from multiple locations, which was necessary to reflect the local conditions for each population in the metapopulation model. The field data (Spencer and McGee 2001) were necessary to model the temporal changes in the matrix during the course of a year. Finally, the results of the hydrological model were necessary to estimate dispersal parameters of the model.

Although the variety of data that were used may appear to be large, it is important to remember that none of the studies used in developing this model was designed with the purpose of building a stochastic metapopulation model, or with the goal of metapopulation-level risk analysis in general. Thus, even in cases where a single set of experiments does not allow probabilistic risk assessment at the population or metapopulation levels, our study demonstrates that results from multiple studies can be used to make these types of assessments by using realistic, spatially structured metapopulation models.

There are several advantages of the type of model we used in this study. Unlike standard bioassays, the results of these models are at the level of population, or metapopulation, rather than the ecologically less relevant individual level. And, unlike assessments (e.g., life table response experiments) that are based on deterministic measures such as the population growth rate, these stochastic metapopulation models do not make unrealistic assumptions about the lack of temporal and spatial variability. Their temporal variability means that the results are best expressed in probabilistic terms, such as the risk of decline, and their spatial structure allows incorporating life history traits such as dispersal and spatially varying survival and fecundity rates.

One of the unique features of our study is the link between a physical model of hydrology and an ecological model of metapopulation dynamics. We believe that as the necessity of making spatially explicit ecological risk assessments increases, these types of links between models in different domains will become more common.

Acknowledgments We thank Beth McGee and Matthew Spencer for allowing us to use their data.

References

Akçakaya, H.R. 1991. A method for simulating demographic stochasticity. Ecological Modelling 54:133–136.

Akçakaya, H.R. 2005. RAMAS Metapop: Viability Analysis for Stage-Structured Metapopulations (version 5.0). Applied Biomathematics, Setauket, New York.

Akçakaya, H.R, Mills, G., and Doncaster, C.P. 2007. The role of metapopulations in conservation. Pages 64–84 in Key Topics in Conservation Biology. D.W. Macdonald and K. Service, editors. Blackwell, Oxford, UK.

Ares, J. 2003. Time and space issues in ecotoxicology: Population models, landscape pattern analysis, and long-range environmental chemistry. Environmental Toxicology and Chemistry 22:945–957.

Bunch, B.W., Cerco, C.F., Dortch, M.S., Johnson, B.H., and Kim, K.W. 2000. Hydrodynamic and Water Quality Model Study of San Juan Bay Estuary. ERDC TR-00-1, U.S. Army Engineer Research and Development Center, Vicksburg, MS. Available: el.erdc.usace.army.mil/elpubs/pdf/tr00-1.pdf.

Chaumot, A., Charles, S., Flammarion, P., and Auger, P. 2003. Do migratory or demographic disruptions rule the population impact of pollution in spatial networks? Theoretical Population Biology 64:473–480.

Ferraro, S.P., and Cole, F.A. 2002. A field validation of two sediment-amphipod toxicity tests. Environmental Toxicology and Chemistry 21:1423–1437.

Hallam, T.G., and Lika, K. 1997. Modeling the effects of toxicants on a fish population in a spatially heterogeneous environment: I. Behavior of the unstressed, spatial model. Nonlinear Analysis: Theory, Methods, and Applications 30:1699–1707.

Johnson, A.R. 2002. Landscape ecotoxicology and assessment of risk at multiple scales. Human and Ecological Risk Assessment 8:127–146.

Landis, W.G. 2002. Uncertainty in the extrapolation from individual effects to impacts upon landscapes. Human and Ecological Risk Assessment 8:193–204.

Lotufo, G.R., Farrar, J.D., Duke, B.M., and Bridges, T.S. 2001a. DDT toxicity and critical body residue in the amphipod Leptocheirus plumulosus in exposures to spiked sediment. Archives of Environmental Contamination and Toxicology 41:142–150.

Lotufo, G.R., Farrar, J.D., Inouye, L.S., Bridges, T.S., and Ringelberg, D.B. 2001b. Toxicity of sediment-associated nitroaromatic and cyclonitramine compounds to benthic invertebrates. Environmental Toxicology and Chemistry 20:1762–1771.

McGee, B., and Spencer, M. 1997. A Field-Based Demographic Matrix Model for the Estuarine Amphipod Leptocheirus plumulosus. Report prepared for U.S. Army Corp of Engineers, Waterways Experiment Station.

McGee, B.L., and Spencer, M. 2001. A field-based population model for the sediment toxicity test organism Leptocheirus plumulosus: II. Model application. Marine Environmental Research 51:347–363.

Neubauer, R.J., Emery, V., Moore, D., Bridges, T., Thebeau, L., and McKown, G. 1997. Investigation of the Sediments in the Gunpowder River Study Area. DAAA15-91-D-0014, Vicksburg, MS.

Spencer, M., and McGee, B.L. 2001. A field-based population model for the sediment toxicity test organism Leptocheirus plumulosus: I. Model development. Marine Environmental Research 51:327–345.

Spromberg, J.A., John, B.M., and Landis, W.G. 1998. Metapopulation dynamics: Indirect effects and multiple distinct outcomes in ecological risk assessment. Environmental Toxicology and Chemistry 17:1640–1649.

U.S. Environmental Protection Agency. 2001. Method for Assessing the Chronic Toxicity of Marine and Estuarine Sediment-Associated Contaminants with the Amphipod Leptocheirus plumulosus. EPA 600/R-01/020, Office of Research and Development, Washington, DC.

17

Applications of Life Table Response Experiments to the Evaluation of Toxicant Effects at the Population Level with the Polychaete *Dinophilus gyrociliatus*

ROBERTO SIMONINI
DANIELA PREVEDELLI
MARINA MAURI

Heavy metals are among the most widespread pollutants in harbor environments and coastal habitats subject to anthropogenic impact, and their harmful effects have been demonstrated on various species of polychaetes, depending on the metal and its role in physiological processes (Reish and Gerlinger, 1997; Reish, 1998). An evaluation of the impact of contaminants at both individual and population levels can be obtained through life table response experiments, a method for estimating the effect of pollutants on biological fitness (Caswell, 2000). This approach has been recently adopted to study the biological effects of pollutants on marine environments utilizing the polychaetes *Capitella capitata*, *Streblospio benedicti*, and *Dinophilus gyrociliatus* as bioindicator species (Levin et al., 1996; Hansen et al., 1999; Mauri et al., 2002, 2003). In this chapter, we compare the sensitivity of some population parameters such as the population growth rate, λ, the life expectancy, e^0, and the net growth rate, R_0, estimated for cohorts of *D. gyrociliatus* exposed to zinc and chromium.

Methods

Study species

D. gyrociliatus is a small (0.8–1.2 mm), widely distributed interstitial polychaete with a short life cycle that inhabits harbors and brackish habitats and can be cultured easily in laboratory. It appears to be sensitive to various classes of pollutants, such as metals, detergents, xenobiotics, and ordnance compounds. In particular, both short- and long-term ecotoxicological tests demonstrate that it is sensitive to cadmium, mercury, copper, lead, and zinc (Carr et al., 1986, 1989, 2001; Reish and Gerlinger, 1997; Reish, 1998; Depledge and Billinghurst, 1999; Nipper et al., 2001, 2002; Mauri et al., 2002, 2003). The specimens used in the experiment came from laboratory cultures established in 1995 with organisms taken from the port of Genoa, Italy. Cultures are kept under conditions of constant temperature, photoperiod, and salinity (24 ± 1°C, 12/12-hr light/dark cycle, 30 psu) and fed on Tetra Min (Tetra Werke, Melle, Germany). Additional details regarding laboratory breeding are reported in Prevedelli and Zunarelli Vandini (1999). In *D. gyrociliatus*, the sex determination is progamic, and the sex can be easily recognized from the zygote stage. The adult females lay capsules containing large "female-type" eggs (80–100 μm diameter) and small "male-type" eggs (40 μm diameter) that develop, respectively, into females and males. Normally, only females hatch from the capsule and give rise to the free-living population.

Experimental methods

Four experimental cohorts of newborn *D. gyrociliatus*, two exposed to 1.00 μg/ml nominal zinc and chromium, respectively (added as $ZnCl_2$ and $CrCl_3$), and two as control groups (one for each treated group), were analyzed. These concentrations of zinc and chromium were considerably lower than those found in previous experiments to be capable of having lethal effects (Mauri et al., 2002, 2003). Artificial seawater (pH = 8.1, 0.35 mg/l total organic carbon, 0.04 mg/l total Zn, 0.40 mg/l total Cr) was obtained by dissolving commercial sea salt (Instant Ocean's Reef Crystals; Aquarium Systems, Wickliffe, OH, USA) in highly purified water. The water obtained was then 0.45-μm Millipore filtered. Experimental cohorts were monitored every other day from the zygote stage until all individuals in the cohort died; at every check, the survivors and the female-type eggs laid were counted.

Experiments were performed at constant temperature, photoperiod, and salinity (24°C, 12/12-hr light/dark cycle, 30 psu), feeding animals on dry fish food (Tetra Min; Tetra Werke). All experimental solutions were replaced every other day after a feeding of 0.1 mg food/ml for 6 hr, in an attempt to minimize the effects of time and addition of food on the metal species present. Description of materials and methods is also given in Mauri et al. (2002, 2003).

Population model

The survival (l_x) and maternity (m_x) data were used to build a complete age-classified population model and the relative matrices (Leslie matrices) using a projection interval of two days. Following Caswell (2000), in order to obtain the age-specific survival (P_i) and fecundity (F_i), we approximated the survival and the maternity within each age class by their averages over the interval $i - 2 \le x \le i$, so the survival probabilities (P_i) on the subdiagonal were calculated as

$$P_i = \frac{l_{(i+2)} + l_{(i)}}{l_{(i)} + l_{(i-2)}},$$

and the age-specific fecundity (F_i) in the first row was calculated as

$$F_i = (l_{(0)} l_{(2)})^{1/2} \frac{(m_i + P_i m_{i+2})}{2}.$$

The values of l_x, m_x, F_i, and P_i for each experimental cohorts are given in appendix 17.1.

The experiments were performed under optimal laboratory conditions and with ample resources so that density dependence was not a factor. For each experimental group, the net growth rate, R_0, for female-type eggs was calculated as

$$R_0 = \sum_{x=0} l_x m_x.$$

Moreover, the life expectancy, e^0 (the average life span of a newborn, in days), was obtained by trapeze method integration and was calculated as:

$$e^0 = 2 \cdot \left(0.5 + \frac{l_2 + l_4 + l_6 + \cdots l_n}{l_0} \right).$$

This formula is derived from the standard expression of e^x (i.e., Hummon, 1974) but takes into account the length of the projection interval (two days).

The λ values, the sensitivities, and the elasticities of λ to changes in the matrix entries were obtained with RAMAS Metapop (Model → Populations → Display). The differences in the population growth rate between treated and control groups (Δλ) were broken down into the contributions of each term of age-specific survival and fecundity using the decomposition analysis technique (Caswell, 2000). Let λ^c and λ^d denote the values of λ for population c and d, respectively. Then,

$$\Delta\lambda = \lambda^{(c)} - \lambda^{(d)} = \sum_i \sum_j \Delta a_{ij} \frac{\partial \lambda}{\partial a_{ij}},$$

where a_{ij} denotes F_i or P_i. Each term in the summation is the contribution of the difference in the matrix entry a_{ij} of population c with respect to d on Δλ. This method allows identification of the life-history traits that had the greatest influence on λ for each group and comparison of the relative contribution of each trait to the Δλ between treated and control groups.

Statistical analysis of the demographic data was performed using resampling methods. The 95% confidence intervals of the parameters under consideration were estimated by the percentile method, based on 2,000 bootstrap-generated estimates. Permutation tests were used to highlight significant differences ($\alpha < 0.05$) in the demographic parameters between treated and control groups (Levin et al., 1996).

Results

Survival and maternity patterns of *D. gyrociliatus* were strongly affected by exposure to metals, whereas no differences were observed in the no-treatment groups. Survival

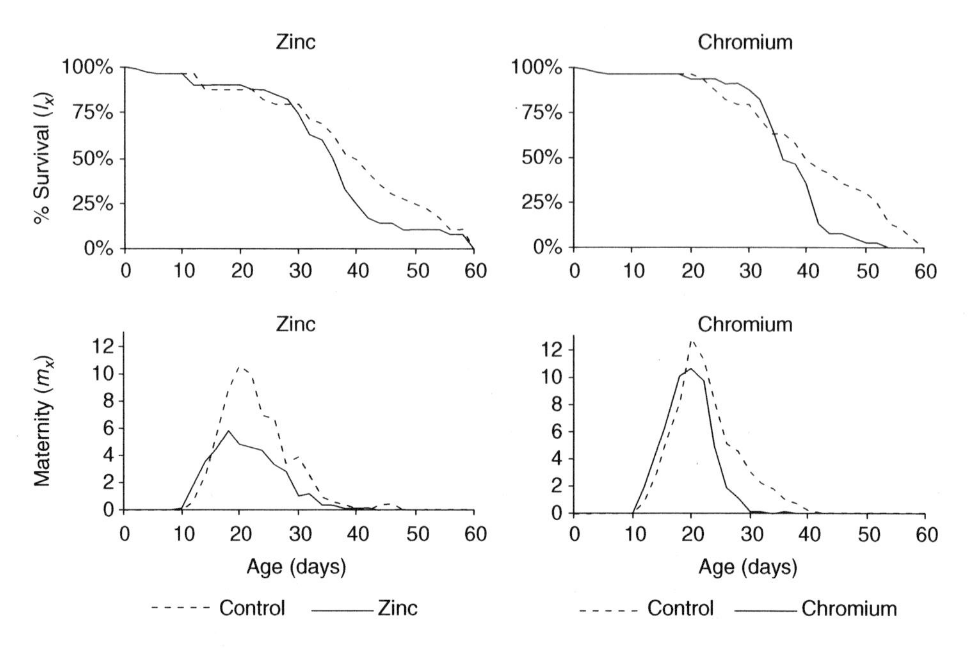

Figure 17.1. Survivorship (l_x) and individual maternity (m_x) of control and treated groups of *Dinophilus gyrociliatus* exposed to zinc and chromium.

during the first 11 days (embryo and juvenile phases) exceeded 90% in all experimental groups and then declined gradually. The survival of cohorts exposed to zinc and chromium enrichment decreased only after the fifth week of life compared to the control groups (figure 17.1). The effects of metal enrichment on maternity are more evident than those observed for survival. In all experimental groups, *D. gyrociliatus* reached sexual maturity 10–12 days after the zygote stage. The number of eggs laid during early reproductive events was greater for groups exposed to zinc and, especially, to chromium with respect to the control groups. Later, metals caused a considerable reduction of maternity in both exposed groups compared to control groups (figure 17.1). The net growth rate, R_0, appeared to be the demographic parameter most sensitive to metal enrichment; in fact, both the zinc and chromium groups exhibited a sharp decline in R_0 with respect to controls (–38% and –19%, respectively). To a lesser extent, life expectancy, e^0, was also reduced by zinc and chromium (–10% for both metals), while metal enrichment did not affect the population growth rate, λ (table 17.1). The reductions in R_0 were determined by the sharp decrease in fecundity observed for individuals exposed to zinc and chromium from day 15 and 20, respectively. The reductions in e^0 were caused mainly by the marked decline in survival observed from day 30 of life in the groups exposed to zinc and chromium with respect to controls. The slight variations of λ induced by metals were due to life table variation that occurred from day 10 to day 30 of life. Decomposition analysis indicated that the contribution of the differences in survival to $\Delta\lambda$ were negligible for both metals

Table 17.1. Demographic parameters obtained for laboratory cohorts of *Dinophilus gyrociliatus* exposed to zinc and chromium and percent variation (Δ%) compared to control (Mauri et al., 2002, 2003).

	Zinc			Chromium		
	Control	Treatment	Δ%	Control	Treatment	Δ%
Net reproductive rate (R_0, offspring)	52.9	32.9	−38%	59.8	48.8	−18.4%
Lower–upper confidence interval	47.2–58.6	28.6–37.1		57.6–61.9	44.2–53.8	
p-values[a]	<0.01			<0.001		
Life expectancy (e^0, days)	38.5	34.9	−9%	39.7	35.9	−10%
Lower–upper confidence interval	36.0–40.5	31.6–38.2		38.7–40.7	34.1–37.2	
p-values[a]	>0.14			<0.001		
Population growth rate (λ, 2 days^{-1})	1.47	1.43	−2%	1.48	1.51	+2%
Lower–upper confidence interval	1.44–1.48	1.41–1.45		1.47–1.49	1.48–1.53	
p-values[a]	>0.13			>0.06		

[a] p-Values refer to permutation test, treatment versus control for Zn and Cr.

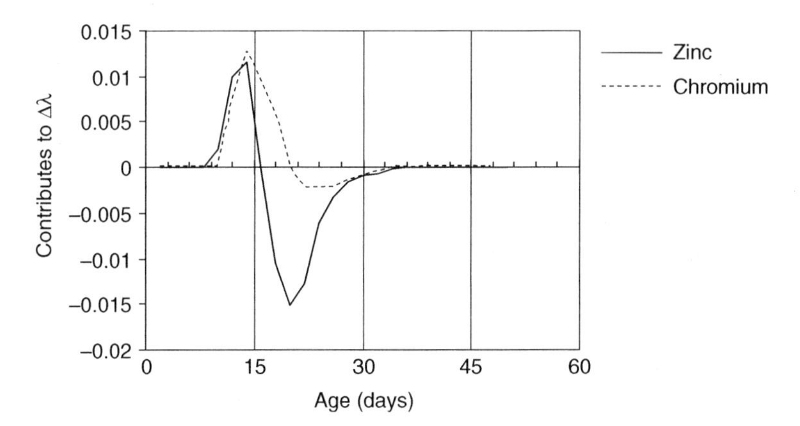

Figure 17.2. Decomposition analysis of the effects of zinc and chromium exposure to *D. gyrociliatus*. The lines represent the sum of the contributions of F_i and P_i to Δλ (the P_i contributions are negligible).

and the differences in population growth rate between treated and control animals were due to the differences in fecundity observed during the first three weeks of life. In particular, the contributions of the greater fecundity in the early depositions compensated (Zn) or overcompensated (Cr) for those due to the subsequent greater decline in reproduction of the treated cohorts with respect of the control groups (figure 17.2). The life tables of the four experimental cohorts are reported in appendix 17.1.

Appendix 17.1. Life tables of the four experimental cohorts of *D. gyrociliatus* exposed to zinc and chromium.

	Zn Control				Zn 1.00				Cr Control				Cr 1.00			
x	l_x	m_x	P_i	F_i	l_x	m_x	P_i	F_i	l_x	m_x	P_i	F_i	l_x	m_x	P_i	F_i
0	1.000	0	0.987	0	1.000	0.987	0	0	1.000	0	0.987	0	1.000	0	0.987	0
2	0.987	0	0.987	0	0.987	0.987	0	0	0.987	0	0.987	0	0.987	0	0.987	0
4	0.974	0	0.993	0	0.974	0.993	0	0	0.974	0	0.993	0	0.974	0	0.993	0
6	0.962	0	1.000	0	0.962	1.000	0	0	0.962	0	1.000	0	0.962	0	1.000	0
8	0.962	0	1.000	0	0.962	0.971	0	0.083	0.962	0	1.000	0	0.962	0	1.000	0
10	0.962	0	0.957	0.258	0.962	0.971	0.171	0.860	0.962	0	1.000	0.397	0.962	0	1.000	0.908
12	0.962	0.543	0.955	1.486	0.907	1.000	1.606	2.544	0.962	0.800	1.000	1.874	0.962	1.829	1.000	3.194
14	0.879	2.563	1.000	3.990	0.907	1.000	3.515	3.929	0.962	2.971	1.000	4.230	0.962	4.600	1.000	5.706
16	0.879	5.469	1.000	7.095	0.907	1.000	4.394	5.028	0.962	5.543	1.000	6.841	0.962	6.886	0.986	8.374
18	0.879	8.813	1.000	9.594	0.907	0.985	5.727	5.217	0.962	8.229	0.986	10.356	0.962	10.114	0.986	10.180
20	0.879	10.500	0.969	9.999	0.907	0.985	4.848	4.640	0.962	12.800	0.957	11.712	0.934	10.529	1.000	10.067
22	0.879	9.938	0.952	8.230	0.879	0.984	4.563	4.391	0.934	11.265	0.939	9.417	0.934	9.735	0.985	7.226
24	0.824	6.967	0.983	6.711	0.879	0.968	4.344	3.710	0.879	8.188	0.952	6.431	0.934	4.882	0.985	3.345
26	0.797	6.655	1.000	5.019	0.852	0.934	3.226	2.933	0.824	5.000	0.983	4.707	0.907	1.879	0.985	1.497
28	0.797	3.448	0.948	3.549	0.824	0.877	2.867	1.892	0.797	4.552	0.948	3.674	0.907	1.152	0.954	0.646
30	0.797	3.897	0.927	3.052	0.742	0.900	1.074	1.078	0.797	3.000	0.891	2.478	0.879	0.156	0.871	0.135
32	0.714	2.423	0.941	1.653	0.632	0.889	1.217	0.745	0.714	2.231	0.939	1.940	0.824	0.133	0.778	0.066
34	0.687	0.960	0.875	0.742	0.604	0.750	0.318	0.282	0.632	1.783	0.957	1.402	0.659	0.000	0.833	0.046
36	0.632	0.609	0.881	0.464	0.495	0.700	0.333	0.224	0.632	1.087	0.886	0.855	0.495	0.111	0.857	0.055
38	0.522	0.368	0.892	0.232	0.330	0.714	0.167	0.122	0.577	0.714	0.872	0.451	0.467	0	0.600	0
40	0.495	0.111	0.848	0.055	0.247	0.733	0.111	0.116	0.495	0.222	0.912	0.110	0.357	0	0.444	0
42	0.412	0.000	0.857	0.164	0.165	0.909	0.167	0.083	0.440	0	0.903	0	0.137	0	0.750	0
44	0.357	0.385	0.875	0.389	0.137	0.900	0	0	0.412	0	0.893	0	0.082	0	0.833	0
46	0.302	0.455	0.905	0.226	0.137	0.889			0.357	0	0.920	0	0.082	0	0.600	0
48	0.275	0	0.895	0	0.110	1.000			0.330	0	0.870	0	0.055	0	0.667	0
50	0.247	0	0.824	0	0.110	1.000			0.302	0	0.700	0	0.027	0	0.500	0
52	0.220	0	0.714	0	0.110	0.875			0.247	0	0.643	0	0.027	0	0	0
54	0.165	0	0.800	0	0.110	0.857			0.137	0	0.667	0	0	0		
56	0.110	0	0.500	0	0.082	0.500			0.110	0	0.333	0				
58	0.110	0	0	0	0.082	0.000			0.055	0	0	0				
60	0	0		0	0				0	0						

Discussion

The population growth rates observed in the laboratory strains of test organisms such as the opportunistic polychaetes *Capitella capitata*, *Streblospio benedicti*, and *Dinophilus gyrociliatus* are higher than those recorded in the field, probably owing to the absence of predators and parasites, the reduction in density-dependent competition, the abundance of resources, and the constancy of conditions during the laboratory experiments (Levin et al., 1987, 1996; Hansen et al., 1999; Mauri et al., 2002, 2003). When the population growth rates are so high, the comparison of the λ values is not very informative in extinction risk assessment.

On the other hand, biotic and abiotic factors, and their interactions, could complicate assessment of effects of a toxic substance in the field, particularly in harbors, lagoons, and brackish habitats characterized by marked and unpredictable oscillations in the main chemical and physical factors. In this context, the application of demographic analysis to evaluate the toxic effects on organisms that live in polluted marine environments may serve to integrate current practices in ecotoxicology, which focus largely on short-term individual responses.

The increase of reproductive effort observed in the earlier stages of *D. gyrociliatus* exposed to sublethal concentrations of Zn and Cr could be interpreted in terms of hormesis (a stimulation of organism performance occurring at low levels of exposure to agents that are harmful or toxic at high levels of exposure; Calabrese and Baldwin, 1998; Forbes, 2000; Cook and Calabrese, 2006). This stimulation does not appear to translate into stimulation at the population level because there is a trade-off between reproduction in earlier stages and survivorship/reproduction in the adult stages. Life table response experiments on polychaetes have shown that λ is determined mainly by age at maturity, fecundity of the young adults, and juvenile survival (Levin et al., 1987; Caswell, 2000). Sublethal concentrations of contaminants frequently affect mainly the growth, fecundity, and survival of adults: they often have no significant effects on λ, whereas they determine a marked reduction in R_0 and e^0 (Levin et al., 1996; Hansen et al., 1999; Mauri et al., 2002).

From a population ecology perspective, the marked reduction of R_0 should not be underestimated because it indicates a decline in the reproductive potential of the individual, eventually leading to negative consequences at higher ecological levels in complex systems (Forbes, 2000). Considering all demographic effects of a pollutant on target organisms would appear to be a precautionary means for the decision maker to ensure the maintenance of the viability of wild populations in polluted environments.

References

Calabrese, E.J., and Baldwin, L.A. 1998. Hormesis as a biological hypothesis. *Environmental Health Perspectives* 106: 357–362.

Carr, R.S., Curran, M.D., Mazurkiewicz, M. 1986. Evaluation of the archiannelid *Dinophilus gyrociliatus* for use in short-term life-cycle toxicity tests. *Environmental Toxicology and Chemistry* 5: 703–712.

Carr, R.S., Nipper, M., Biedenbach, J.M., Hooten, R.L., Miller, K., and Saepoff, S. 2001. Sediment toxicity identification evaluation (TIE) studies at marine sites suspected of ordnance contamination. *Archives of Environmental Contamination and Toxicology* 41: 298–307.

Carr, R.S., Williams, J.W., Fragata, C.T.B. 1989. Development and evaluation of a novel marine sediment pore water toxicity test with the polychaete *Dinophilus gyrociliatus*. *Environmental Toxicology and Chemistry* 8: 533–543.

Caswell, H. 2000. *Matrix population models*. Sinauer, Sunderland, Massachusetts.

Cook, R., and Calabrese, E.J. 2006. The importance of hormesis to public health. *Environmental Health Perspectives* 114: 1631–1635.

Depledge, M.H., Billinghurst, Z. 1999. Ecological significance of endocrine disruption in marine invertebrates. *Marine Pollution Bulletin* 39: 32–38.

Forbes, V.E. 2000. Is hormesis an evolutionary expectation? *Functional Ecology* 14: 12–24.

Hansen, F.T., Forbes, V.E., and Forbes, T.L. 1999. Effects of 4-n-nonylphenol on life-history traits and population dynamic of a polychaete. *Ecological Application* 9: 482–495.

Hummon, W.D. 1974. Effects of DDT on longevity and reproductive rate in *Lepidodermella squamata* (Gastrotricha, Chaetonotida). *American Midland Naturalist* 92(2): 327–339.

Levin, L.A., Caswell, H., De Patra, K.D., and Creed E.L. 1987. Demographic consequences of larval development mode: planktotrophy vs. lecitotrophy in *Streblospio benedicti*. *Ecology* 68: 1877–1886.

Levin, L.A., Caswell, H., Bridges, T., Di Bacco, C., and Plaia, G. 1996. Demographic responses of estuarine polychaetes to pollutants: life-table response experiments. *Ecological Application* 6: 1295–1313.

Mauri, M., Simonini, R., and Baraldi, E. 2002. Demographic responses of the polychaete *Dinophilus gyrociliatus* to chromium exposure. *Environmental Toxicology and Chemistry* 21: 1903–1907.

Mauri, M., Baraldi, E., and Simonini, R. 2003. Effects of zinc exposure on the polychaete *Dinophilus gyrociliatus*: a life-table response experiment. *Aquatic Toxicology* 65: 93–100.

Nipper, M., Carr, R.S., Biedenbach, J.M., Hooten, R.L., Miller, K., Saepoff, S. 2001. Development of marine toxicity data for ordnance compounds. *Archives Environmental Contamination and Toxicology* 41: 308–318.

Nipper, M., Carr, R.S., Biedenbach, J.M., Hooten, R.L., Miller, K. 2002. Toxicological and chemical assessment of ordnance compounds in marine sediments and porewaters. *Marine Pollution Bulletin* 44: 789–806.

Prevedelli, D., and Zunarelli Vandini, R. 1999. Survival, fecundity and sex ratio of *Dinophilus gyrociliatus* (Polychaeta: Dinophilidae) under different dietary conditions. *Marine Biology* 132: 163–170.

Reish, D.J. 1998. The use of larvae and small species of polychaetes in marine toxicological test. Pages 383–393 in P.G. Wells, K. Lee, and C. Blaise (eds.), *Microscale testing in aquatic toxicology. Advances techniques and practice*. CRC Press, Boca Raton, Florida.

Reish, D.J., and Gerlinger, T.V. 1997. A review of the toxicological studies with polychaetous annelids. *Bulletin of Marine Science* 60: 584–687.

Appendix

Using RAMAS GIS

This book comes with a CD-ROM that includes a demonstration version of RAMAS GIS version 5.0 and the input files for all the models included in this book. The demonstration version is intended only for viewing models developed with the regular version and not for developing, creating, or modifying models. It differs from the regular version in three ways:

1. It does not include a manual. However, a complete set of Help files is provided and contains most of the material in the manual.
2. It does not allow users to create a new model or to modify a model by changing its parameters. Users can view and type in parameters (e.g., to see the list of options available for a parameter) but must click the "Cancel" button to exit a dialog box or window.
3. Users can run simulations and review the results but cannot save the results.

The rest of this appendix lists hardware and software requirements for this demonstration version of RAMAS GIS and briefly describes how to use it. For more detailed information, see the Help files.

Required Hardware and Software

The program requires an IBM-compatible personal computer with a CD drive, running Microsoft Windows 95 or later.

Starting the Program

The demonstration version of RAMAS GIS will *not* be installed on your computer's fixed drive; it must be run from the CD-ROM.

Put the CD-ROM in a drive. If the program does not start automatically, double-click on the file RAMASGIS.EXE on the CD-ROM. This will start a shell program that provides access to all programs of RAMAS GIS.

RAMAS GIS consists of five programs, represented by icons in this shell program: the metapopulation program RAMAS Metapop (the main program for RAMAS GIS), as well as programs for Spatial Data, Habitat Dynamics, Sensitivity Analysis, and Comparison of Results. Below, RAMAS Metapop is described in detail, followed brief descriptions of the other four programs.

Exploring and Running Metapopulation Models

The main program is named Metapopulation Model or RAMAS Metapop; it is used to build stage-structured, spatially explicit metapopulation models, to run simulations with these models, and to predict the risk of species extinction, time to extinction, expected metapopulation abundance, its variation, and spatial distribution. The main features and parameters of this program are summarized in chapter 1 and in the Help files. This program requires both demographic data (e.g., survival rates, fecundities, density dependence) and spatial data (e.g., location and size of subpopulations). Metapopulation model files used by this program have the extension .MP. All models described in this book are represented by metapopulation model (.MP) files.

In the RAMAS GIS Shell program (see figure A.1), click on the icon for the Metapopulation model (also called RAMAS Metapop) to start this program. The main window of RAMAS Metapop consists of a Title bar, Menu bar, Tool bar, Model summary, and Status bar. The Menu bar includes File, View, Model, Simulation, Results, and Help.

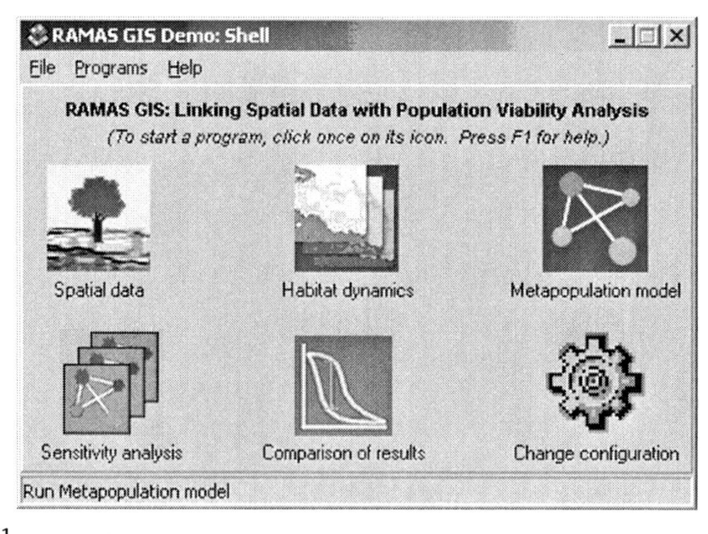

Figure A.1

Click on the File menu, and select Open. If the "Open" dialog box does not point to the CD drive in which you put the CD-ROM, then navigate to this drive by using "Look in" on the toolbar.

The sample files are arranged in folders. To examine the model files, navigate to one of these folders after you select "Open" from the File menu, select a subfolder for a species, and double click on a metapopulation model (.MP) file. In many cases, there is a single model in the species folder, with more in an "Additional Models" folder.

Once the file is opened, select each item under the Model menu, starting with "General Information." This will open a dialog box with the input parameters. The parameters are explained in the Help file (click the "Help" button, or press F1).

In the General Information dialog, click on the "Additional Information" button if it is present. This will start your Web browser and open an associated .HTM file that includes background information, pictures, and links about the species and the model.

Return to RAMAS Metapop (using Alt-Tab, or selecting from the taskbar) and examine the input parameters in the other dialog boxes under the Model menu. Click the "Cancel" button to exit each dialog box.

You can run a model by selecting "Run" under the Simulation menu. While the simulation is running, click on the buttons in the toolbar at the row to display the metapopulation trajectory or the metapopulation map. See the Help file for details.

To see the model results, click on the Results menu and select one of the results. See the Help file for a discussion of each result type and its interpretation.

Note that the sample files or the data they contain may not be used in any publication or research without prior written permission of Applied Biomathematics (www.ramas.com).

Using the Other Programs

In some models described in this book, the spatial structure of the metapopulation model is based on habitat data. This link between habitat data and the metapopulation model is made possible by the Spatial Data program. This program uses spatial data on habitat requirements of a species, such as GIS-generated maps of vegetation cover, microclimate, and land use. It combines these data into a map of habitat suitability (HS) with a user-defined function. This map is then used to find habitat patches by identifying areas of high suitability where a population might survive. The program determines the spatial structure of the metapopulation (locations of patches and distances among them) and calculates demographic parameters (e.g., carrying capacities, vital rates, initial abundances) of populations in each patch, with user-defined functions of the HS in that patch. Both the spatial structure and the demographic parameters are used as input for the Metapopulation model.

To use this program, click on the icon for Spatial data in the RAMAS GIS Shell program. The main window consists of Title bar, Menu bar, Tool bar, Model summary, and Status bar. The Menu bar includes File, View, Model, Simulation, Results, and Help. Click on the File menu, and select "Open." The file type (extension) for this program is .PTC. Note that Spatial data files are available only for some of the species. As in RAMAS Metapop, the input data are organized in several dialog boxes under the Model menu. Click on the "Help" button in each of these dialog boxes for more information about the program (also see chapter 1).

The Habitat Dynamics program allows the Metapopulation model to incorporate temporal dynamics in the habitat, with maps input as time series. The model files for the chapters in this book do not include such time series. In the regular version of RAMAS GIS, this program is accessed by clicking on the icon for Habitat Dynamics in the RAMAS GIS Shell program. Click on the "Help" button in the dialog boxes for more information about the program.

Two additional programs can be used to support population viability analysis, risk assessment, and sensitivity analysis: Sensitivity Analysis and Comparison of Results. The Sensitivity Analysis program is not functional in the demonstration version of RAMAS GIS. In the regular release version, it is used to run several simulations of a metapopulation model to analyze the sensitivity of results to parameters.

The Comparison of Results program is used to compare different metapopulation models by superimposing their results (e.g., graphs of metapopulation abundance and occupancy, risk curves, and time-to-extinction distributions). It also allows statistical comparison of different risk curves. It can be used to view the results of a sensitivity analysis, to compare management options, or to assess anthropogenic impact. Start this program by clicking on its icon in the RAMAS GIS shell program.

From the main program window, press Ctrl-O (i.e., hold the Ctrl key down, then press "O") to display the "Load files" dialog box. If files are listed in this window, highlight the first file and then click "Remove" until the list is empty. Then click "Add." In the "Open" dialog box, navigate into a species folder that has at least two metapopulation model (.MP) files. Click on the first file and click "Open." Click "Add" again; this time select the other file and click "Open" (you can repeat this to add up to a total of five files). Finally, click "OK" in the "Load files" dialog box.

Select the Results menu (or press Alt-R). Select each type of result, especially risk-related results, and examine the graphs. While displaying a result, you can press F1 to learn about it.

Index